Principles of
JIG AND TOOL DESIGN

Principles of
JIG AND TOOL DESIGN

M. H. A. KEMPSTER
C.Eng., M.I.Mech.E., A.F.R.Ae.S., M.I.Prod.E.
*Lecturer in Production Technology,
The Bristol Aeroplane Technical College*

Asheville-Buncombe Technical Institute
LIBRARY
340 Victoria Road
Asheville, North Carolina 28801

HART PUBLISHING COMPANY, INC.
NEW YORK CITY

© 1968 M. H. A. KEMPSTER
FIRST AMERICAN PUBLICATION 1969
PRINTED IN THE UNITED STATES OF AMERICA
NO PART OF THIS BOOK MAY BE REPRODUCED WITH-
OUT THE WRITTEN PERMISSION OF THE PUBLISHER.

PREFACE

The design of jigs and tools is important because the rate of production and the quality of the work produced by machine tools and presses can only be as good as the cutting tools, press tools and holding equipment will allow.

This book assumes that the reader has a sound knowledge of basic machine tools and processes, and aims at the application of this knowledge to produce the sound design of production equipment with due regard to economics.

Although intended mainly for students preparing for the City and Guilds of London Institute Full Technological Certificate in Jig and Tool Design, this book will be suitable for other students of Production Engineering. The subject is presented in chapters that cover each topic, rather than in chapters that are of equal length; but each chapter is divided into short sections so that the reader can select portions of a topic for study as required.

Careful thought has been given to the introduction of metric units. At the time of writing, the introduction of the metric system as the British System is in its early stages, and it appears that the complete change-over will take some time. Most of the data is therefore presented in both metric and inch units, and some of the components that are used as exercises are dimensioned in metric units; but the details of metric British Standards have not been anticipated in this edition, and units of force and the strength of materials are in tons-force units.

The reader is reminded that a subject can only be learned by active participation, and that this is particularly true of any subject involving design; he should therefore examine production equipment designs and consider both the good and the bad features of each, and then follow this up with the preparation of as many original designs as possible, which should be critically examined by a competent designer.

The writer wishes to acknowledge the assistance given by the many manufacturers who are mentioned in the text, and also the work done by his wife, who checked the original manuscript and generally assisted in the preparation of this book.

M. H. A. KEMPSTER

Bristol

CONTENTS

1. **Introduction**   1
    Review of the work of the Jig Design and Planning Offices. Cutting Tools. Planning and Tool Layouts. Design of Jigs and Fixtures: principles of jig and fixture design. Broaching. Design Study. Gauges. Presswork. Organisation and Economics. The Study of Jig and Tool Design.

2. **Cutting Tools**   6
    The Simple Wedge-Shaped Cutting Tool: rake and clearance angles. Chip Formation: the formation of built-up edge. The Connection between Cutting Conditions and Tool Shape: single-point cutting-tool angles. Chip Breakers. Cutting-Tool Materials: carbon tool-steels; high-speed steels; cobalt-base alloys; cemented carbides; diamond; sintered oxides. Hole-Producing Cutting Tools: spade drills; twist drill; 'D' bit; trepanning. Hole-Finishing Cutting Tools: boring; floating cutter; counterboring, spotfacing and countersinking cutters; reamers. Milling Cutters: up-cut and down-cut milling; facing cutters; interlocking cutters. Thread-Producing Tools. Gear Cutters: gear planing; hobbing; gear shaping.

3. **Form Tools**   40
    Introduction. Flat Form Tool: form 'correction'. Dovetail Form Tool. Circular Form Tool; form 'correction'. Notes on Form Tool Calculations. The Skiving Tool.

4. **Process Planning**   49
    Introduction. The Choice of Method and Equipment. Planning Method. Specimen Operation Layouts.

5. **Tooling and Cam Layouts**   56
    Introduction. Capstan and Turret Lathes. The Automatic Screw-Type Machine: tool and cam layout for screw-type automatic machines. The Swiss-Type Automatic Machine: tool and cam layout. Multi-Spindle Automatic Machine.

6. **Location**   77
    Introduction. The Six Degrees of Freedom. The Extent of Location and the Choice of Location System. Redundant Location. Foolproofing. The Six-Point Location Principle. Location Devices: location from plane surfaces; location from a profile; location from a cylinder; vee location. Workpiece Ejectors.

7. **Clamping**   94
    Introduction: requirements of the clamping system; lever systems. Plate Clamps. Pivoted and Latch-Type Clamps. Direct Clamping: floating

CONTENTS vii

pads; direct clamping using a post; quick-action nuts. Hook Bolts. Clamping Plates. Clamping More Than One Workpiece: equalising clamps. Differential Clamping. Cam-Actuated Clamping. Toggle Clamps. Pneumatic Clamping. Hydraulic Clamping.

8. **Drilling Jigs** 113

Introduction. Location. Clamping. Handling the Jig. Feet. Controlling the Cutters. Control of Depth. Burr Grooves. Drill Over-Run. Handling Clearances. Swarf and Cutting-Fluid Clearances. Typical Drilling Jigs: plate jigs and channel jigs; solid jigs; post jigs; pot jigs; sandwich jigs; nutcracker jigs; latch jigs; box jigs; trunnion jigs.

9. **Milling Fixtures** 136

Introduction. Milling Methods: straddle milling; gang milling; string or line milling; pendulum milling; rotary-table milling; profile milling. The Location and Clamping of the Fixture to the Machine Table. Workpiece Location. Workpiece Clamping. Tool Setting. General Features of Milling Fixtures. Special Vice Jaws. Milling Fixtures.

10. **Miscellaneous Workholding Fixtures** 147

Workholding Devices for Turning: collet chucks; jaw chucks; expanding posts; turning fixtures. Workholding Devices for Grinding: surface grinding; cylindrical grinding. Workholding Devices for Assembly. Welding Fixtures.

11. **Indexing Jigs and Fixtures** 156

Introduction: applications of indexing. The Basic Features of an Indexing Jig or Fixture. Indexing Devices. Typical Indexing Jigs and Fixtures.

12. **Broaching** 163

Introduction: advantages claimed for broaching; the limitations of the broaching process. Types of Broaching. Broaching Machines. Internal Broach Design: broach material; face angle; tooth width; back off; side clearance; tooth space; chip breakers; cut per tooth; burnishers; number of teeth; calculation to obtain a constant load per tooth; push broaching; pull broaching; broach pullers; broach pull-end; spiral broaching. Surface Broaching: surface broach design; surface broach holders. Broaching Fixtures: fixtures for internal broaching; fixtures for surface broaching.

13. **Producing the Design and the Working Drawings of Production Equipment** 187

Design Study. Consideration of the Method of Construction; casting; welding; fabricating from several parts; use of plastics materials. Working Drawings of Production Equipment.

14. **Limit Gauges** 197

Introduction. The Taylor Principle. Design of Limit Gauges. Limit Gauge Tolerances. Allowance for Wear. Materials for Limit Gauges. Typical Limit Gauges: plain plug gauges; gauges for shafts; gauging of screw threads; thickness and length gauges; recess gauges; step gauges; position gauges; receiver gauge.

15. Presswork   212

Introduction. Presswork Operations: piercing; blanking; piercing-and-blanking; bending and forming; drawing; embossing; coining; crimping. Presses: hand- and foot-operated; power presses; single-acting presses; double-acting presses; triple-acting press; cut-and-cupping press; press capacity; feeding; ejection of finished workpieces. Standard Die-Sets. Piercing and Blanking: simple blanking; piercing-and-blanking; material for punches and dies; punch retention; piercing and blanking dies; clearance between punch and die; angular die clearance; stripper and guides; pressure pads; pilots; feed methods; stock stops; piercing and blanking pressures; application of shear; blanking layout; blank layout for economy. Bending and Forming: ejectors; punch and former material; material size; springback allowance; bending forces. Drawing: punch material; die material; die shapes; combination tools; blank size; pressure required for drawing; cupping on a double-acting press.

16. Organisation and Management   248

The Organisation of an Engineering Concern. The Function of the Planning and Jig Design Departments. The Organisation of the Planning and Jig Design Departments. The Planning of Product Manufacture. The Planning of Tooling, and Tool Records: title block and parts list; equipment numbering; tool record cards; component record cards; the ordering and manufacture of equipment; the issue of drawings; progress. Modifications: modification notes; modifications to processes and equipment. Information Library and Records: capacity charts; manufacturers' catalogues and handbooks; national standards; company standards; stores holding. The Contract Tool Drawing Office: the placing of contracts; company standards and the contract tool drawing office; modification action. The Tool Room.

17. The Economics of Tooling   266

Manufacturing Systems: job production; batch production; mass production; flow production. Tooling Systems: basic tooling; skeleton tooling; complete tooling. The Calculation of 'Break-Even' Quantity. Other Factors that Influence the Tooling Expenditure. Factors that Influence the Purchase of Special Equipment. Consideration of Bought-Out and Sub-Contract Parts. The Economical Design of Tooling Equipment. The Economical Use of the Equipment Designer's Time. Batch Size.

Exercises   275

Index   286

Classified Index   292

Chapter 1

INTRODUCTION

The work of the Jig Design Office and the Planning Office includes the selection of standard cutting tools, the design of cutting tools, the preparation of machining instructions, the design of tooling equipment and gauges, and the design of press tools.

In this chapter some of the basic principles of design are stated, and contents of the chapters that follow are outlined.

1.1. Cutting Tools (Chapters 2 and 3)

The geometry of all cutting tools can be related to the simple wedge tool; any complication of shape is introduced to suit the direction of cutting or the form to be produced. Shapes can be produced by forming, copying, generating or by a combination of these three basic methods. **Forming** implies that the tool is fed in radially, so that the shape produced is a 'negative' of the tool shape; **copying** implies that the tool produces a special shape by being caused to move in a way that is directed by a template, profile plate or cam; **generating** implies that the shape is produced by a combination of tool shape and tool movement caused by using the basic machine drives. The shape of the cutting tool is therefore related to the method to be used to produce the shape.

1.1.1. A range of cutting-tool materials is available to suit the cutting conditions; many of these materials are brittle, and the tool and workpiece rigidity must be considered when designing the tool and the workpiece holding device.

1.1.2. Chapter 2 deals with the tool geometry and materials, and then considers the tool shapes and mounting methods to suit the principle cutting methods.

Form tools must be dimensioned to suit their manufacture; Chapter 3 explains the method by which these dimensions are calculated or obtained by drawing.

1.2. Planning and Tool Layouts (Chapters 4 and 5)

The planning of the operation sequence to produce a component demands a thorough understanding of the capabilities and the limitations of machine tools. Chapter 4 assumes that the student has this knowledge, and explains how the operation planning can be done systematically.

Turning, using capstan, turret and automatic machines, demands a tool layout so that the tool clearances can be checked, and this information is also used as the basis of the detailed tool layout that is passed to the tool setter.

Cams are required for automatic machines; the procedure followed when preparing the tool and cam layout is usually outlined in a manufacturer's handbook, and Chapter 5 describes the layout and cam design for some typical machines.

1.3. The Design of Jigs and Fixtures (Chapters 6–11)

The workpiece is held in a jig or a fixture during the machining operation; a **jig** incorporates means of locating and clamping the workpiece and also means of guiding the cutting tool during the actual cutting; a **fixture** incorporates means of locating and clamping the workpiece but does not incorporate means of guiding the tool, but may incorporate tool-setting arrangements. The derivation of the terms **jig** and **fixture** is described on page 113, but jigs are usually used for drilling and associated operations, whilst fixtures are used for other machining operations; the term **jig** is often used in assembly work, to describe the equipment used to hold parts that are to be joined by welding or fastening devices.

1.3.1. The principles of jig and fixture design

Location

    (*a*) Ensure that the workpiece is given the desired constraint.
    (*b*) Position the locators so that swarf will not cause mal-alignment.
    (*c*) Make the location points adjustable if a rough casting or forging is being machined.
    (*d*) Introduce foolproofing devices, such as fouling pins and projections, to prevent incorrect positioning of the workpiece.
    (*e*) Make all location points visible to the operator from his working position.
    (*f*) Make the location progressive (i.e. locate on one locator and then on to the other).
    (*g*) Ensure that redundant location is not present (i.e. two location points to control one constraint).

Clamping

    (*a*) Position the clamps to give best resistance to the cutting forces.
    (*b*) Position the clamps so that they do not cause deformation of the workpiece.
    (*c*) Design the clamps so that they are not deformed by the clamping forces.

INTRODUCTION

(d) If possible, make the clamps integral with the fixture body.
(e) Make all clamping and location motions easy and natural to perform.

Clearance
(a) Allow ample clearance to allow for variation of workpiece size.
(b) Allow ample clearance for the operator's hands.
(c) Ensure that there is ample swarf clearance and cutting-fluid clearance.
(d) Allow clearance so that the workpiece can be removed after machining, when burrs may be present.

Stability and rigidity
(a) Provide four feet so that uneven seating will be obvious, and ensure that the forces caused by the mass of the workpiece and the cutting action act within an area enclosed by a line joining the seating points.
(b) Make the equipment as rigid as is necessary for the operation to be performed.
(c) Provide means of positioning and bolting the equipment to the machine table or spindle if required.

Handling
(a) Make the equipment as light as possible, particularly if it is to be moved about by the operator for loading, etc.
(b) Consider the shape of the equipment so that it can be handled easily; ensure that there are no sharp corners or awkward projections.
(c) If possible, provide lifting handles or lifting hooks.

General
(a) Keep the design simple in order to minimise cost and to avoid breakdown caused by over-complication.
(b) Utilise standard and 'bought out' parts as much as possible.
(c) Ensure that the workpiece can be loaded into and removed from the equipment. When designing jigs and fixtures, it is usual to draw the component, and to design the equipment around it; it is a common error to produce a design that does not allow for the passage of the workpiece.

1.3.2. The basis of jig and fixture design is the location and the clamping of the workpiece, and Chapters 6 and 7 deal with the basic principles involved, and the established methods of location and clamping are described.

In the chapters that follow, the essential features of the more common equipment are stated, and the various types of jigs and fixtures are illustrated. The examples given are only representative, and endless variations are possible.

1.4. Broaching (Chapter 12)

Broaching is a specialised operation, but is commonly used in large-scale production. This chapter deals with the design of broaches and broaching fixtures, but it must be emphasised that the design of this equipment depends very largely upon the recording of past experience because the variables are so numerous that design formulae cannot be used with success.

1.5. Design Study (Chapter 13)

In the early stages of the study of this subject the student can learn much by designing jigs and tools on the lines of the examples shown in the text, but as soon as the basic principles are understood, design should be developed by a careful analysis of the problem, allowing the final design to be gradually developed; the folly of commencing a design with preconceived ideas cannot be over-emphasised. Chapter 13 describes the design technique, and also compares the various methods of construction that are available.

1.6. Gauges (Chapter 14)

Limit gauges are used so that certain dimensions can be checked to ensure that they are within the limits of size stipulated by the designer; the disadvantage associated with limit gauging is that no indication of the actual size is given, but the advantage is that the workpieces can be checked by semi-skilled personnel. The correctness of work produced by automatics and similar machines is often checked by actual measurement in conjunction with statistic methods to avoid having to measure every piece produced; the machine can be given 'running adjustments' to prevent scrap. Chapter 14 illustrates the basic types of limit gauges, many of which are the subject of British Standards. Receiver gauges as used to check castings for completeness, and to ensure that mating components produced in different places will assemble together correctly. Position gauges are special fixture-like gauges that enable the relative positions of many features to be checked.

1.7. Presswork (Chapter 15)

The design of press tools is usually the work of a specialist, but the tool designer employed by a small, multi-activity company, or by a contract tool office (see page 263), may be called upon to design press tools. Chapter 15 describes the principal presswork operations and presses, and then considers the design of punches and dies, and finally the press-tool set as a whole.

1.8. Organisation and Economics (Chapters 16 and 17)

During the early stages of his studies the student may feel that the study of organisation lies outside of his field of interest; it must be realised that an

INTRODUCTION 5

understanding of the functions of the various departments will be invaluable, and also that the jig and planning offices are usually the 'training ground' of the production management. Chapter 16 describes the organisation of a typical engineering concern and shows how the jig and planning offices fit into the organisation. The organisation of the jig and planning offices is then considered in some detail, but it must be realised that the precise organisation structure varies from company to company.

Economy must be studied by all departments if a concern is to be a financial success, and the final chapter of this book links up the economic considerations with the purely technical considerations.

1.9. The Study of Jig and Tool Design

The examples shown in this book illustrate designs that are typical; although the student may do well to base his initial work upon these examples, it is considered that he would be unwise to slavishly follow them at all times, and that he would be wise to produce original work, based, if necessary, upon existing practice. It is much better to try out new ideas, even if they are rejected, than not to attempt original work.

1.9.1. Finally, the student is again reminded that this subject cannot be learned by only reading a book, and that the only way to learn to be a designer is to examine actual equipment in the workshop and to note their good and bad features, and also to submit as many designs as possible to one's lecturer for his comments and criticism.

Chapter 2
CUTTING TOOLS

2.1. Simple Wedge-Shaped Cutting Tool

A cutting tool can be considered to be a simple wedge as shown in fig. 2.1; in this figure the cutting edge is straight, perpendicular to the tool movement and is wider than the chip produced. This type of cutting is known as **orthogonal cutting**, and is the operation used in most experimental work. The applications of orthogonal cutting are limited; fig. 2.2 shows a more

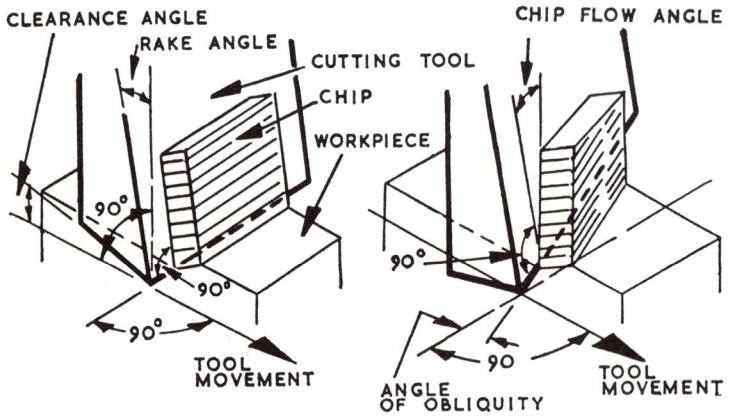

Fig. 2.1 Orthogonal Cutting   Fig. 2.2 Oblique Cutting

practical consideration of the cutting operation known as **oblique cutting**; in this operation the cutting edge is straight, wider than the chip produced, but is not perpendicular to the tool movement; the chip flows at an angle to the cutting edge as shown.

2.1.1. Rake and clearance angles. Fig. 2.3 illustrates the rake and clearance angles in orthogonal cutting. The rake angle controls the chip formation and is, in turn, governed by the mechanical properties of the material being cut. Fig. 2.4 shows that the tangential force is reduced as the rake angle is increased; this is because the length of the shear plane (see fig. 2.8 on page 8) is reduced as the rake angle is increased. The cutting tool is weakened by increasing the rake angle, and so increasing the rake angle ceases to be effective when the weakening effect upon the tool exceeds the beneficial effect of reducing the tangential force acting upon it. In general, cutting

CUTTING TOOLS

tools for use on brittle materials require a small rake angle, and those for use on ductile materials require a large rake angle. The clearance angle is introduced to prevent the cutting tool from rubbing the workpiece, and is associated with the tool and workpiece geometry; it is usually kept as

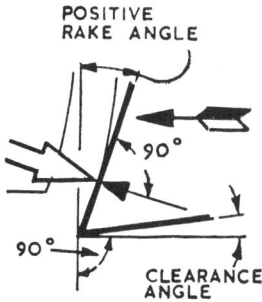

Fig. 2.3 Rake and Clearance Angles

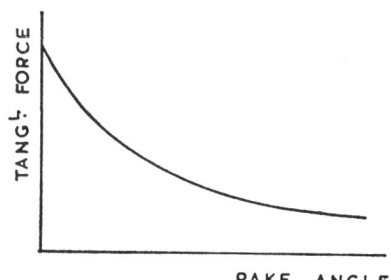

Fig. 2.4 The Connection between Tangential Force and Rake Angle

small as possible to avoid weakening the cutting tool and may be in stages (see fig. 2.17 on page 12).

Fig. 2.3 shows a cutter with a positive rake angle, but in certain cases the rake may be **negative** as shown in fig. 2.5. The tool may be shaped as

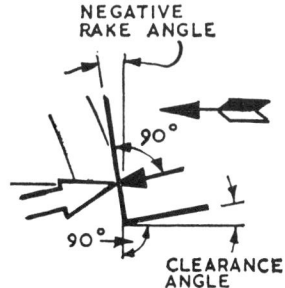

Fig. 2.5 The Application of Negative Rake Angle

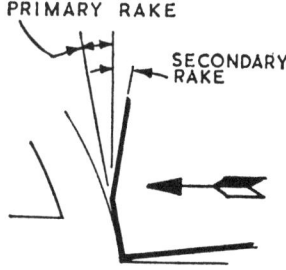

Fig. 2.6 Tool Shaped to Reduce Chip Friction

shown in fig. 2.6 to reduce the friction of the chip on the tool face, and still have a negative rake angle.

2.2. Chip Formation

Research workers have classified chip formation as **tear and shear**. Fig. 2.7 illustrates the formation of a chip by tear; in this formation the workpiece metal adjacent to the face of the tool is compressed until a crack is produced that runs ahead of the cutting tool and towards the body of the workpiece. This

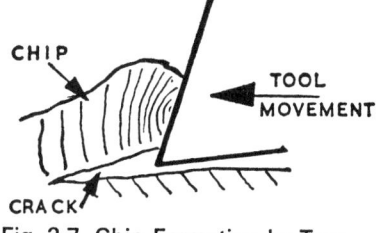

Fig. 2.7 Chip Formation by Tear

crack separates the highly deformed chip from the relatively undeformed workpiece material. Cutting takes place intermittently, and there is no movement of workpiece material over the tool face.

2.2.1. In the formation of the chip by shear there is a general movement of the chip over the tool face; chip formation by shear may be continuous or discontinuous. In continuous chip formation the pressure on the workpiece builds up until the material fails by slip along the line S–S (see fig. 2.8); in practice, the inside of the chip displays the 'steps' produced by the intermittent slip, but the outside of the chip is burnished smooth by the action of the chip rubbing upon the tool face. Continuous chip formation is associated with ductile workpiece materials and cutting tools with a large rake angle.

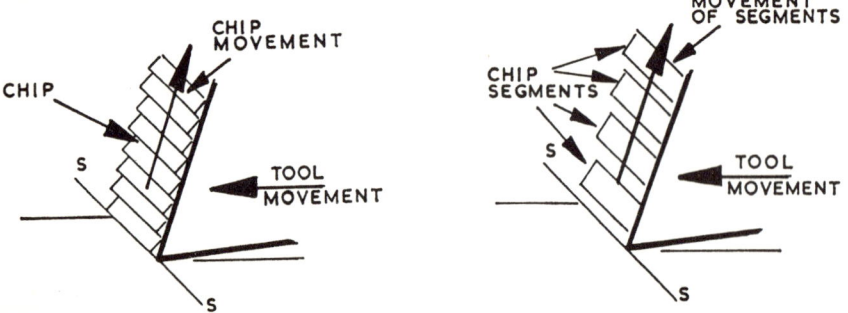

Fig. 2.8 Continuous Chip Formation   Fig. 2.9 Discontinuous Chip Formation

2.2.2. In the discontinuous chip formation the pressure builds up, but produces complete shear, so that the chip is in the form of segments (see fig. 2.9). Discontinuous chip formation is associated with brittle workpiece material and cutting tools with a small rake angle.

2.2.3. The formation of built-up edge (see fig. 2.10). Built-up edge is associated with continuous chip formation; it is produced by the underside of the chip becoming elongated, and as a result of the combination of high

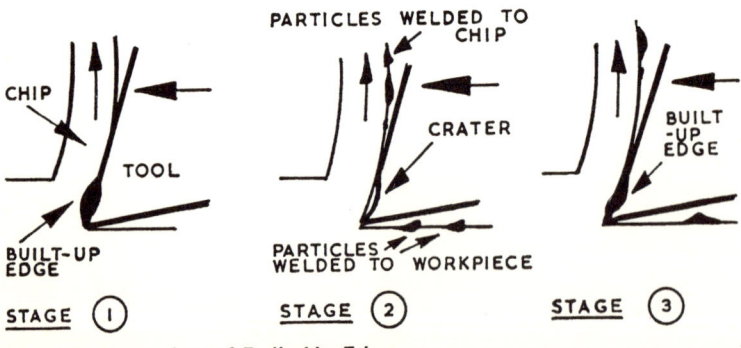

Fig. 2.10 Formation of Built-Up Edge

CUTTING TOOLS

temperature and pressure, some of the chip material becoming welded to the tool face. As cutting continues, the material piles up on the tool face, forming a dome, over which the chip flows as it leaves the parent material. The chip movement and pressure causes the dome to become work-hardened so that particles of it break away and become welded to the chip and to the body of the workpiece. As the dome breaks down, some of the tool material breaks away as well, causing a crater to be produced on the tool face. After the breakdown of the dome, the built-up edge forms again; the formation of the built-up edge, and its breakdown, takes place very rapidly. The built-up edge is caused by the combination of heat and pressure—its formation can therefore be delayed or prevented by making the tool face smooth, using a tool material with a low coefficient of friction with the workpiece material, and by using an efficient cutting fluid; the pressure can be reduced by having a large rake angle, and the tendency of welding to occur can be eliminated by making the cutting tool from a material that is non-metallic. The actual solution used in a specific case will depend upon the prevailing conditions.

2.3. The Connection between Cutting Conditions and Tool Shape

It has already been stated that the magnitude of the rake angle depends mainly upon the mechanical properties of the workpiece material, and that of the clearance angle depends upon the tool and workpiece geometry. The location of the rake and clearance angles depends upon the tool and workpiece geometry and the direction of the tool feed. The 'development' of tool shape from a simple wedge will be considered by taking turning tools as examples.

2.3.1. Fig. 2.11 illustrates orthogonal cutting applied to turning the periphery of a flange; this illustration indicates that a simple wedge tool with a rake angle and clearance angle is adequate. If it is required to form a

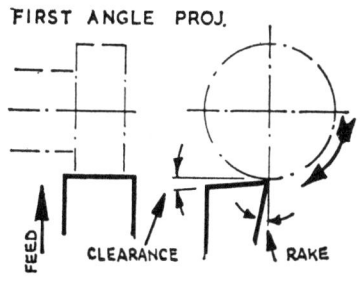

Fig. 2.11 Simple Tool for Plunge Cutting a Narrow Flange

Fig. 2.12 Lathe Tool for Turning a Groove by Plunge Cut

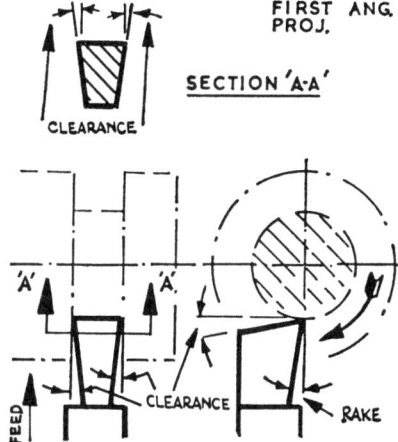

groove (or to part-off) it is necessary to introduce additional body clearance angles to prevent the tool from rubbing the sides of the groove (see fig. 2.12).

2.3.2. In the previous examples the feed was in a radial direction. Fig. 2.13 shows the tool shape required when the feed is parallel to the axis of workpiece rotation. It will be seen that a clearance (called the plan trail angle) is

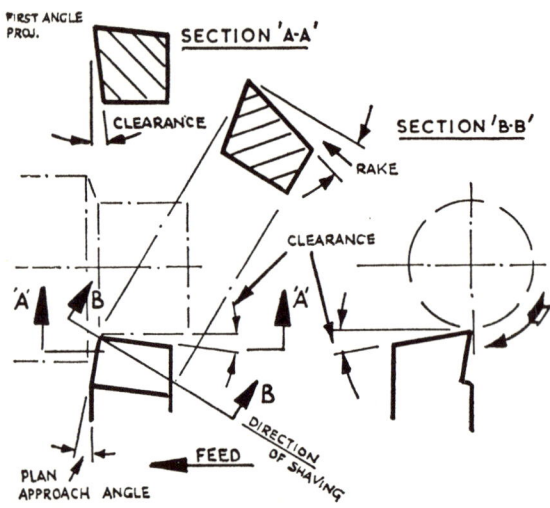

Fig. 2.13 Lathe Tool for Turning a Cylinder by Feed Parallel to Axis

introduced to prevent the tool from rubbing on the workpiece; the cutting-edge clearance must be large enough to prevent the tool from rubbing on the face of the 'step' produced on the workpiece during cutting (this clearance angle must be larger than the helix angle of the face of the step, and is therefore associated with the feed per revolution of the work). The direction of maximum rake determines the direction in which the chip

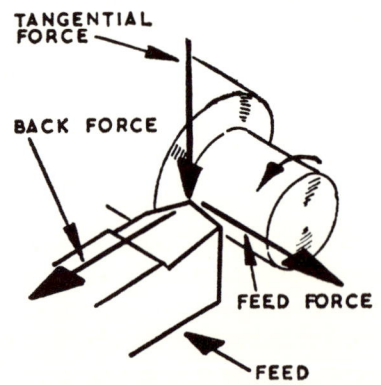

Fig. 2.14 Forces Acting upon a Lathe Tool

CUTTING TOOLS

leaves the workpiece (called the 'direction of shaving'). It will also be seen that the leading edge of the tool is not at right angles to the direction of feed (see plan approach angle on fig. 2.13). The effect of approach angle upon the forces acting upon the tool can be determined experimentally. Fig. 2.14 illustrates the three forces acting upon the tool, and fig. 2.15 shows the connection between these three forces and the plan approach angle.

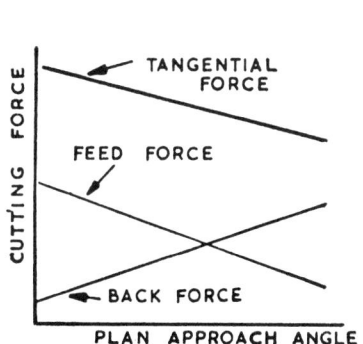

Fig. 2.15 Connection between the Plan Approach Angle and Forces Acting upon a Lathe Tool

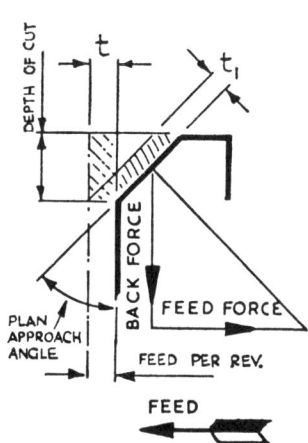

Fig. 2.16 Effect of Increasing the Approach Angle

Reference to fig. 2.16 will show that as the plan approach angle is increased, the feed force will be reduced, but the back force will increase. It will also be seen that if the depth of cut and the feed per revolution of the workpiece are kept constant, the thickness t of the chip will be reduced as the plan approach angle is increased; but the width of the chip will increase. The tangential force acting upon the tool is reduced as the chip thickness is reduced, and so the horsepower required to remove the chip will be reduced as the plan approach angle is increased; the limit to this is usually the chatter produced when the chip width is too large.

2.3.3. Single-point cutting-tool angles. The nomenclature for single-point cutting-tool angles is laid down in B.S. 1886; this specification is based upon the direction and magnitude of the maximum rake angle. Fig. 2.17 shows the main features of B.S. 1886.

2.4. Chip Breakers

It has already been stated that a continuous chip is formed when a ductile material is being machined; if the chip is allowed to remain in a long length it will be difficult to dispose of, and may even cause injury to the machine operator. By introducing a chip breaker the chip can be made to curl up or, if required, become severed so that it is cut up into short lengths. To do this a step is introduced on the tool surface to cause the chip to change its

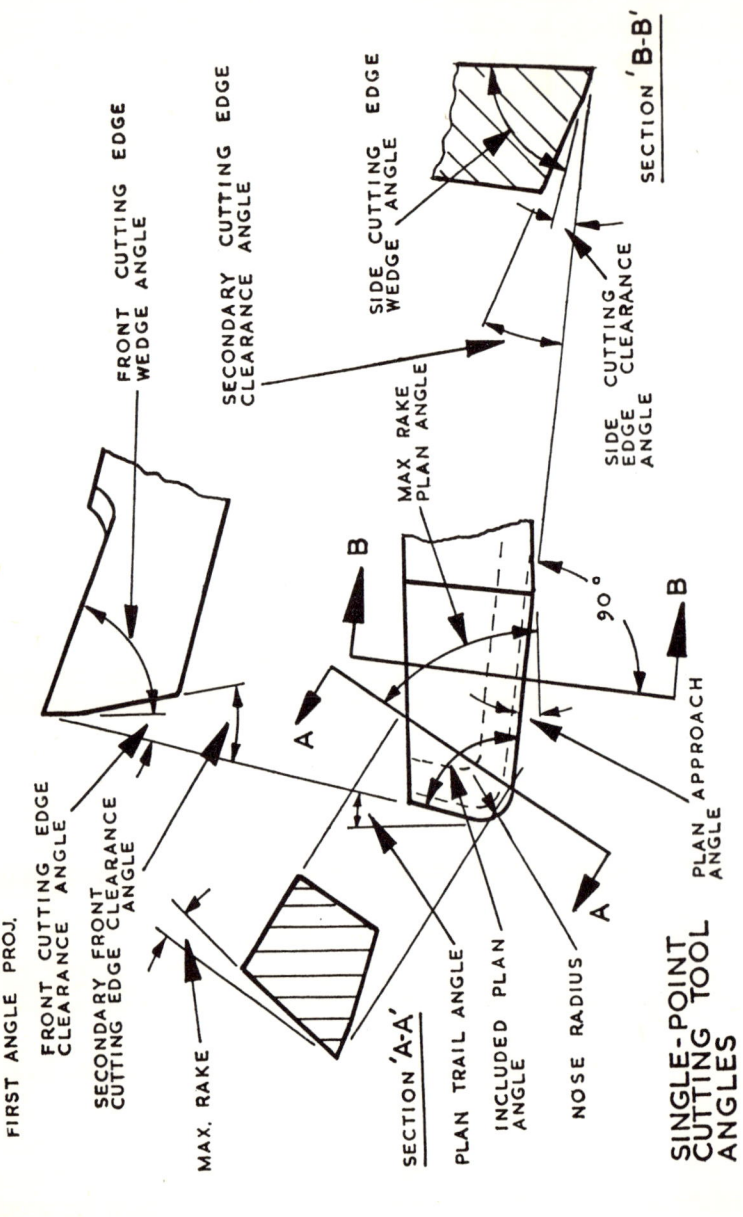

Fig. 2.17 SINGLE-POINT CUTTING TOOL ANGLES

CUTTING TOOLS

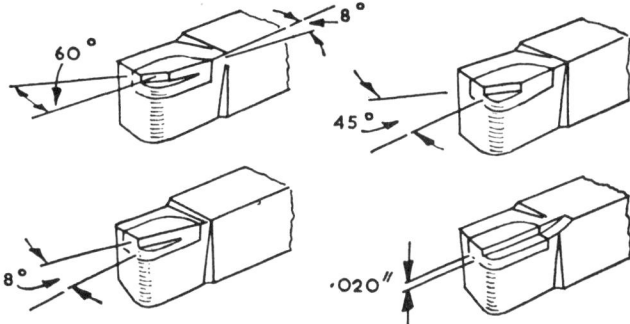

Fig. 2.18 Chip Breakers

flow direction abruptly. Fig. 2.18 shows simple steps machined on the tool face, and fig. 2.19 shows a step produced by clamping a block of metal on to the tool face. Fig. 2.20 shows an alternative type of chip breaker that is produced by grinding a small groove on the tool face.

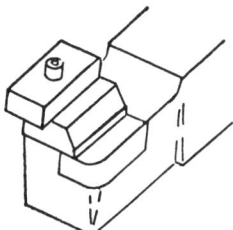

Fig. 2.19 Clamped-on Chip Breaker

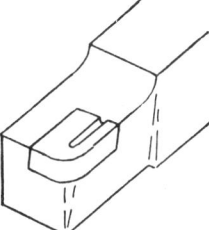

Fig. 2.20 Chip Breaker

2.4.1. Chip breakers are also used to reduce the width of the chip to make it easier to handle, or to reduce the chatter associated with a wide chip. Fig. 12.11 on page 170 shows chip breakers of this type applied to a broach.

2.5. Cutting-Tool Materials

The rate of metal removal can be increased by (*a*) increasing the depth of the cut, (*b*) increasing the feed rate and (*c*) increasing the cutting speed. Some benefit can be obtained by increasing the depth of cut, but this will only increase the force acting upon the tool, and also leave the workpiece in a state of some considerable stress if excessive; it may be the cause of service failure of the product. Similarly, some benefit can be obtained by increasing the feed rate, but this will increase the force acting upon the tool, and is limited mainly by the quality of the surface to be produced. Fig. 2.21 shows that the tangential force acting upon the tool is reduced by increasing the cutting speed from a very low speed to a slightly higher one, and remains constant when the speed is further increased; this illustration also indicates that the tangential force is reduced as a result of reducing the depth of cut from d_1 to d_4. It therefore appears that the most effective

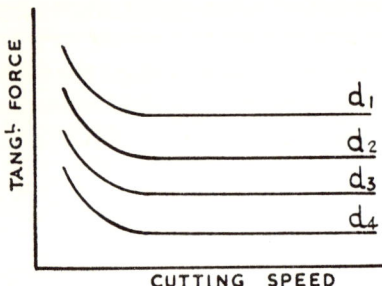

Fig. 2.21 Connection between Tangential Force and Cutting Speed

way of increasing the rate of metal removal is to increase the cutting speed.

The problem is, however, complicated because of the effect of pressure and chip movement (see again fig. 2.8 on page 8); this will produce heat and cause the tool to loose its hardness and strength. Developments in tool materials have therefore been directed towards producing cutting materials that retain their hardness and strength at higher temperatures, so that they are suitable for high-speed cutting; the property of retaining hardness at high temperature is known as **red-hardness**. Other properties required of a cutting-tool material are low coefficient of friction so that the overheating is minimised, resistance to the formation of built-up edge, toughness and good compressive strength. Not all these most desirable properties are present in one cutting-tool material, and the choice of material depends to a large extent upon the conditions expected during cutting and the machine tool to be used.

2.5.1. The principal cutting-tool materials can be classified as follows:

1. **Ferrous metals**
   (*a*) Carbon tool-steels
   (*b*) High-speed steels
2. **Non-ferrous metals**
   (*a*) Cobalt-base alloys
   (*b*) Cemented carbides
3. **Non-metallic materials**
   (*a*) Diamond
   (*b*) Sintered oxide

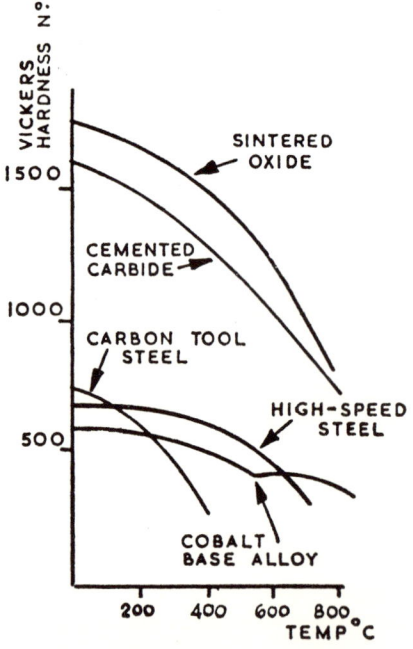

Fig. 2.22 Effect of Heating upon the Hardness of Cutting-Tool Materials

Fig. 2.22 shows the effect of heat upon the hardness of these materials; this diagram indicates that carbon tool-steels loose their hardness at a comparatively low temperature, and that high-speed steel and cobalt-base alloys, whilst not particularly hard, retain their hardness up to a higher temperature. If a greater red-hardness is required it is necessary to use a material that is much harder so that some hardness can be lost as a result of heating, and yet leave sufficient hardness to allow the material to cut efficiently. These very hard materials

CUTTING TOOLS 15

are very brittle, and so great care must be taken when using them; for example, the cutting tool and the workpiece must both be held securely to prevent vibration, and the machine tool must be able to run at the high speeds that are required when these materials are used. Fig. 2.23 indicates how the toughness of cutting-tool materials is reduced as the red-hardness is increased.

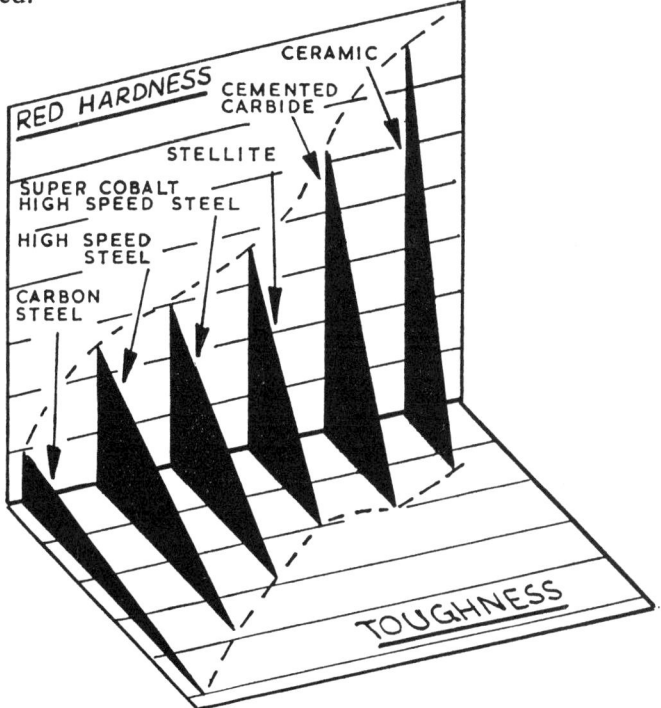

Fig. 2.23 The Connection between the Red-Hardness and Toughness of Cutting-Tool Materials

2.5.2. Carbon tool-steels. Carbon tool-steels contain more than about 0.7% carbon and are hardened by quenching them in water; if a small quantity of chromium is included they can be oil-quenched. They are unsuitable for cutting at high speeds because they soften at temperatures over about 250°C; but are used for large taps and dies, and are useful for prototype form tools.

2.5.3. High-speed steels. These steels contain between about 14 and 22% tungsten and about 4% chromium. They retain their hardness at temperatures of up to about 660°C; about 5% cobalt is included if the material to be cut has a high abrasive characteristic. They must be forged at a high temperature in order to break down the complex carbides that are present, and when hardened must be slowly heated to about 870°C, rapidly heated to about 1 250°C and then quenched. The rapid heating at the high temperature is necessary to prevent the surface hardness from being reduced following decarburisation. High-speed steels can be used for complete tools and cutters, can be applied as tips or butt-welded to carbon-steel shanks (see fig. 2.24); application as tips and butt-welded ends is done to reduce the tool cost.

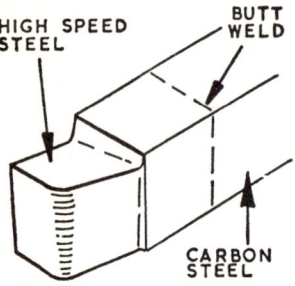

Fig. 2.24 Tool Produced by Butt-Welding

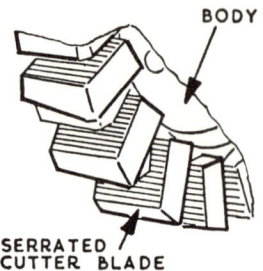

Fig. 2.25 Application of Inserted Blades

2.5.4. Cobalt-base alloys. A typical cobalt-base alloy contains about 33% chromium, 20% tungsten, 3% carbon and the remainder cobalt. It is very hard as cast, and cannot be forged; no heat treatment is necessary to develop its hardness. It is extremely brittle, and can only be used as tips, applied to carbon-steel shanks or as blades inserted into a cast-iron milling

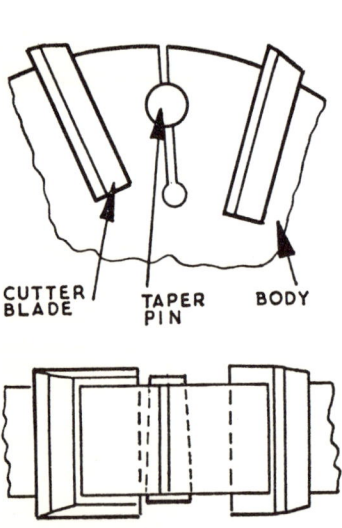

Fig. 2.26 Retention of Inserted Blades

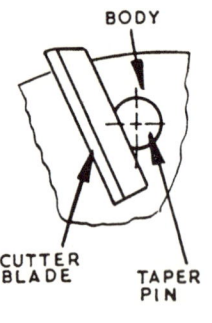

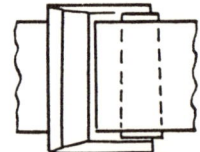

Fig. 2.27 Retention of Inserted Blades

cutter body. Fig. 2.25 shows part of a milling cutter with serrated cutter blades that are held in serrated slots in the cutter body. Fig. 2.26 shows how inserted blades can be secured in position by the springing action produced by thin slots between the blade-location slots and taper pins. Fig. 2.27 shows a similar locking device using a flatted taper pin.

2.5.5. Cemented carbides.
Cemented carbides are produced by powder metallurgy, and consist of particles of tungsten carbide in a matrix of metal with a lower melting point; this matrix metal is usually cobalt. The carbide and cobalt, in particle form, is mixed wet, compressed in dies and, after being cut up into the required shape, sintered by heating at about

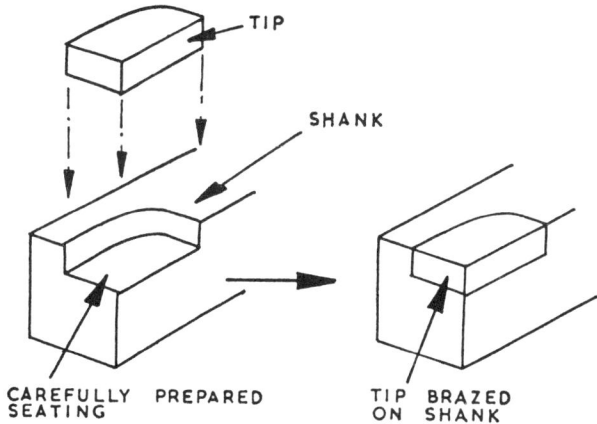

Fig. 2.28 Application of Tool Tip

1 500°C. Tungsten carbides are used for cutting irons and bronzes, and similar hard but weak materials. Steels, and similar materials that tend to form built-up edge, are cut with titanium tungsten carbides; titanium tungsten carbide is produced in a similar way to tungsten carbide, but titanium carbide in particle form is introduced into the wet mixture before

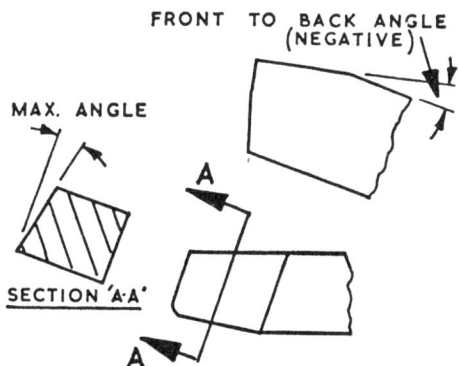

Fig. 2.29 Tool Shape Showing Front-to-Back Angle

compressing. Cemented carbides are available in a range of grades to suit workpiece materials and cutting conditions. Cemented carbides are very brittle and are applied as tips that are clamped or brazed to a carbon-steel shank, or as inserted blades. Fig. 2.28 shows the application of a brazed-on tip; the seating must be very carefully prepared because it is necessary that the tip be completely supported by the shank to prevent its

18   JIG AND TOOL DESIGN

fracture during its use. Fig. 2.29 shows the shape of a typical cemented carbide cutting tool; it will be seen that a negative front-to-back angle is introduced to direct the cutting force towards a larger section of the tool and to cause the point of workpiece impact to be some distance behind the cutting edge; the effect of this is to put the tip material in a condition of compression instead of tension.

2.5.6. Diamond. Diamond is used to finish-turn and bore non-ferrous metals and non-metallic materials, and produces an excellent surface finish. Very high cutting speeds and fine feeds are used; the machine tool is designed especially for diamond cutting tools, and the bearings and drive are such that vibration is minimised; if a fixture is used it must be balanced

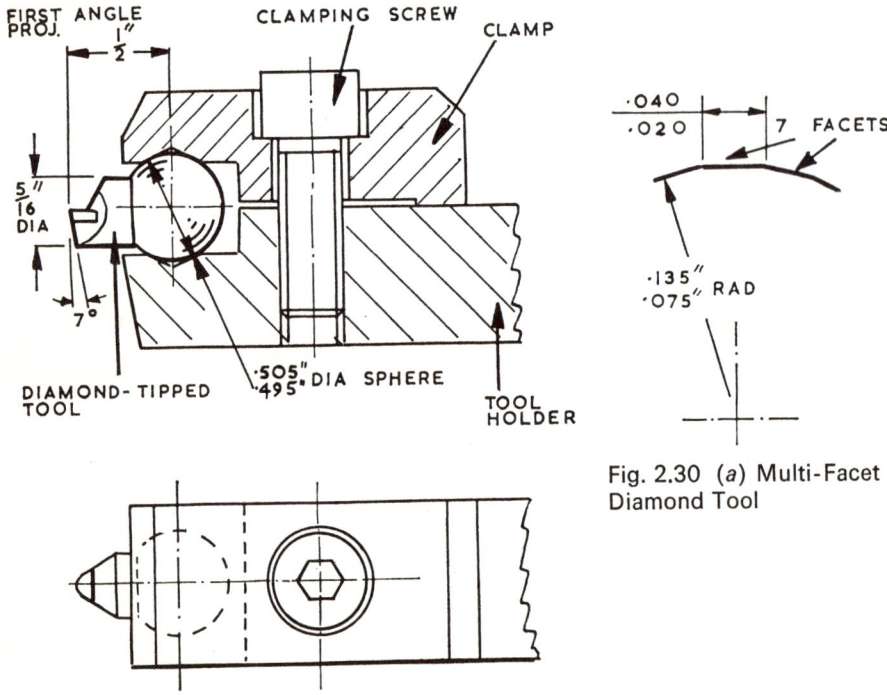

Fig. 2.30 (a) Multi-Facet Diamond Tool

Fig. 2.30 Application of Diamond Tool

carefully. The diamond is usually brazed to a holder that is, in turn, clamped to a tool holder. Fig. 2.30 shows the design recommended in B.S. 1128:1943; this design incorporates a holder with a spherical-end to allow the tool to be positioned easily. The tip can be of single facet or may be multi-facet (as shown in fig. 2.30 (a)); a multi-facet tool has the advantage that each facet can be used in turn, before the tool is sent back to the manufacturer for reconditioning.

2.5.7. Sintered oxides. These ceramic cutting-tool materials consist of at least 85% aluminium oxide with other oxides, carbides or nitrides, to give

CUTTING TOOLS

improved strength and to allow control of the sintering process. These materials have a high red-hardness, and are chemically inert so that the formation of built-up edge is minimised. Sintered oxides can cut at very high speeds, but demand even more rigidity and freedom from vibration than required for cemented carbides.

Because of their extreme brittleness, sintered oxides are used in the form of tips that are attached to the shank by metallising the tip and then brazing

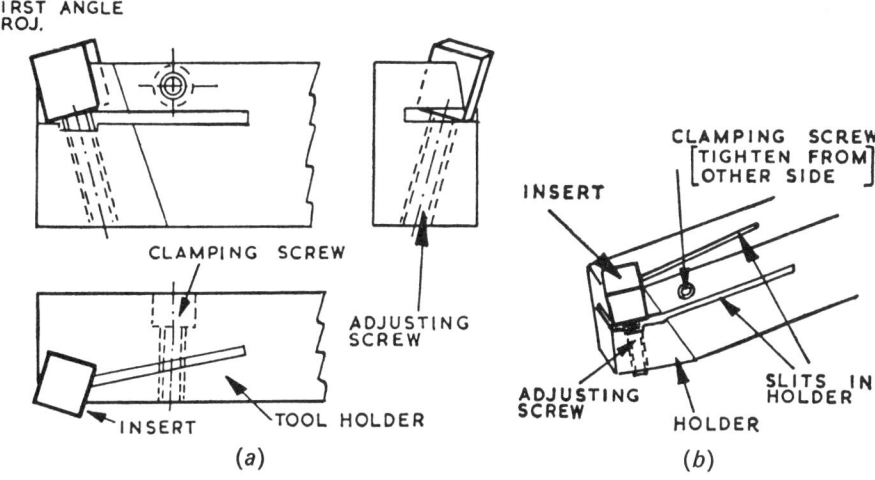

Fig. 2.31 Application of Throw-Away Tool Tip

it to the shank, by bonding with epoxy resin or by mechanical clamp; the latter is considered to be the best method. Sintered oxide tips are particularly suited to the throw-away principle. In this method the multi-edged tip is clamped to the holder that produces the required cutting angle; each edge is used in turn, and when the tip is worn out it is thrown away instead of being reground. Fig. 2.31 (a) and (b) illustrates the application of a prism-type tip; it will be seen that the tool holder is designed to produce a negative rake, and to permit fine adjustment of the cutting-edge height before the tip is clamped in position. In the illustration the tip has four 'lives' at each end; other forms include a diamond-shaped tip for use as a knife tool (with two 'lives' at each end), and a cylindrical tip that is used as a round-nosed tool (with about five 'lives' at each end). Fig. 2.32 illustrates the application of the 'flat' type of tip.

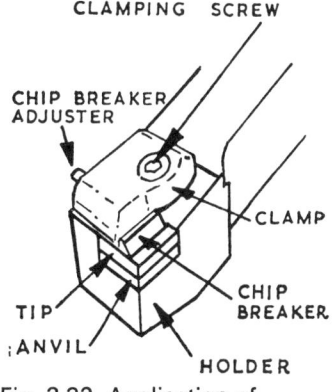

Fig. 2.32 Application of Throw-Away Tool Tip

2.6. Hole-Producing Cutting Tools

2.6.1. Spade drills. Fig. 2.33 shows a spade drill with a rake angle (some spade drills are flat and so have a zero rake angle). If this diagram is compared with fig. 2.13 on page 10 it will be seen that the spade drill is similar to a single-point tool used to turn a cylinder, except that the spade drill has two cutting edges and is used to produce an internal cylinder. The spade drill shown in fig. 2.33 is ground to a point, so that each cutting edge has an approach angle, and if required the spade drill can be made with a back taper to produce body clearance (similar to the 'plain trail angle' of a single-point tool). Fig. 2.34 (*a*) shows a spade drill for producing a flat-bottomed hole, and fig. 2.34 (*b*) shows a general-purpose spade drill (the approach angle of this tool is about 17°). These spade drills are of plate form, located in a bar, but a small spade drill can be forged in one piece from a piece of bar. Fig. 2.34 (*c*) shows a spade drill used for forming; usually sets of about three spade drills are used to produce a form in stages. The illustration shows the spade drill used for the roughing stage, and incorporates chip breakers to reduce the width of the

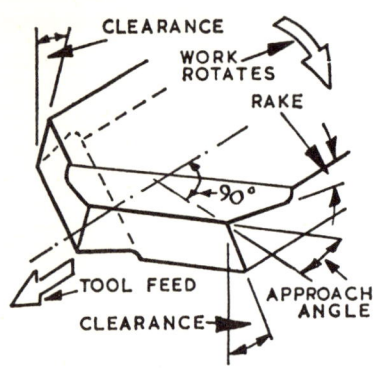

Fig. 2.33 Spade Drill Angles

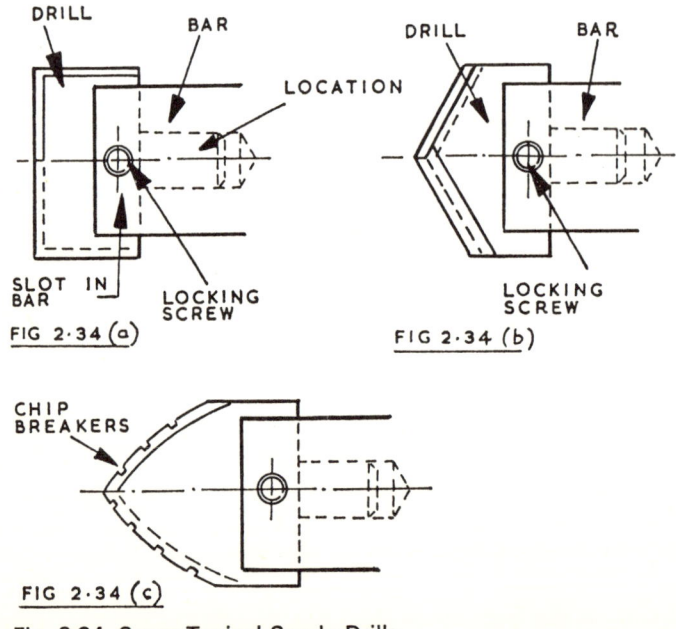

Fig. 2.34 Some Typical Spade Drills

CUTTING TOOLS

chip and so reduce chatter and the power required. The spade drill, in the form of a blade (as shown in fig. 2.34), can be of carbon tool-steel, high-speed steel or be of carbon steel tipped with cemented carbide. The blade must be located relative to the axis of the holder.

2.6.2. Twist drill. The geometry of a twist drill is easier to understand if it is regarded basically as a flat spade drill that is twisted about its axis to form a helix; the helix angle produces the rake angle. The twist drill is

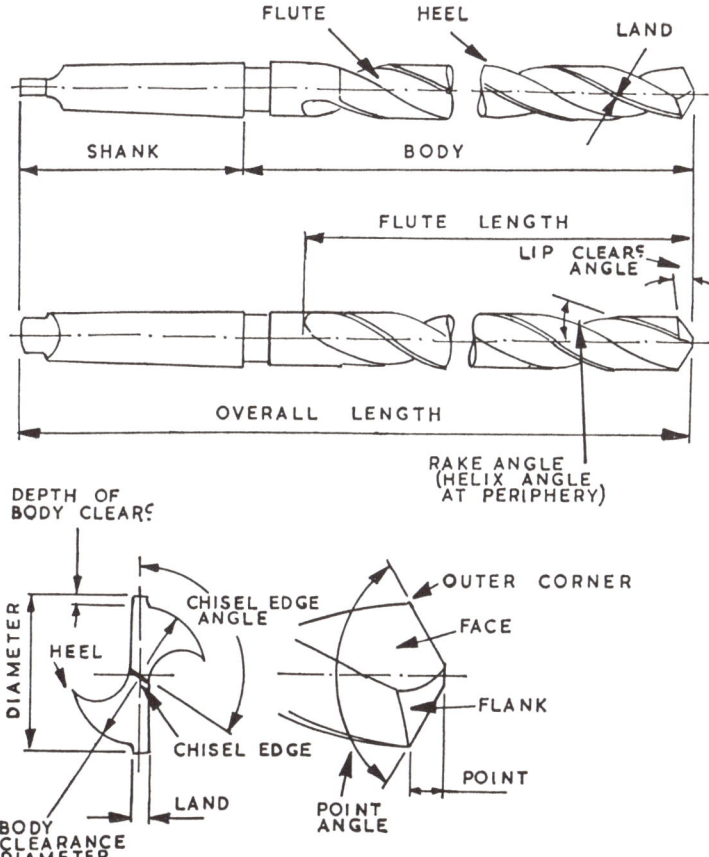

Fig. 2.35 Twist Drill Nomenclature

more complicated because it is machined from a cylindrical bar (refer to fig. 2.35). The reader will understand the geometry of the twist drill much quicker if he examines a large twist drill at this stage. The end of a twist drill is pointed to produce an approach angle; the point angle is about 118°, but it may be designed for a specific operation, and may be between 90° and 140°. The point of the drill cannot be part of a cone because the flanks will rub on the bottom of the hole; the flanks are therefore ground back to an angle of about 10° to produce cutting clearance. The body of the drill must be shaped to prevent it from rubbing on the side of the hole; its diameter is therefore reduced, leaving a land, and also given a small back taper towards the shank (between 0·000 5 and 0·001 in per inch). The helix angle is usually between about 22° and 27°, but for special work may be between 10° and 45°. The drill nomenclature shown in fig. 2.35 is based upon B.S. 328: Part 1:1959, Twist Drills and Combined Drills and Countersinks. Core drills, with three flutes, are used to open out cored holes; the additional cutting point ensures that the hole is concentric whatever the condition of the cored hole. Twist drills may be made from carbon tool-steel, but are usually made completely from high-speed steel, from high-speed steel butt-welded to a carbon-steel shank or from a cemented carbide tip brazed to a carbon-steel shank.

2.6.3. 'D' bit (or 'D' drill). The cutting tool is used to produce long and very accurate holes; it is so-named because of the shape of its section (see fig. 2.36). A 'D' bit will accurately follow an existing hole, and so a start is

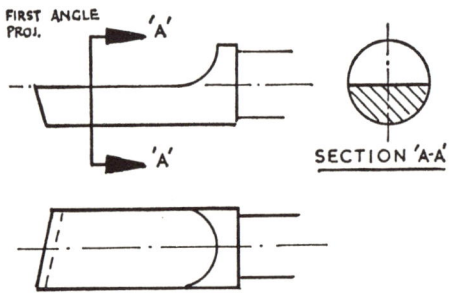

Fig. 2.36 Simple 'D' Bit

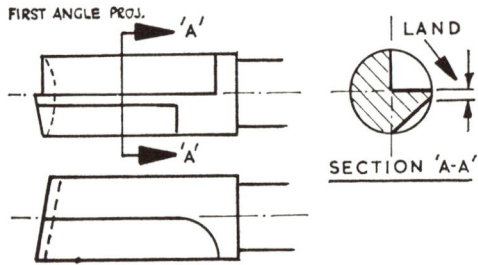

Fig 2.37 'D' Bit with Clearance Angles

often given by drilling and boring. The illustration fig. 2.36 shows a simple 'D' bit, fig. 2.37 shows a 'D' bit that incorporates a body clearance and that shown in fig. 2.38 includes a coolant hole to allow the coolant to be directed down the hole being produced, so that it is effective at the point of cutting. The 'D' bits illustrated are all designed to fit into the bore of a tubular boring bar.

CUTTING TOOLS

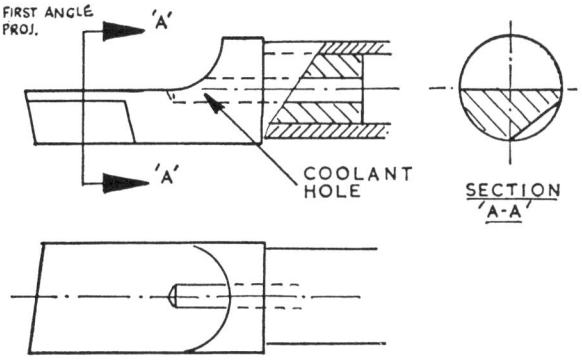

Fig. 2.38 'D' Bit with Coolant System

2.6.4. Trepanning. In the trepanning operation a cylindrical groove is cut in the face of the workpiece to remove a cylindrical 'core' and so produce a cylindrical hole in such a manner that most of the metal is removed in one piece and can be used to make another part instead of finding its way into

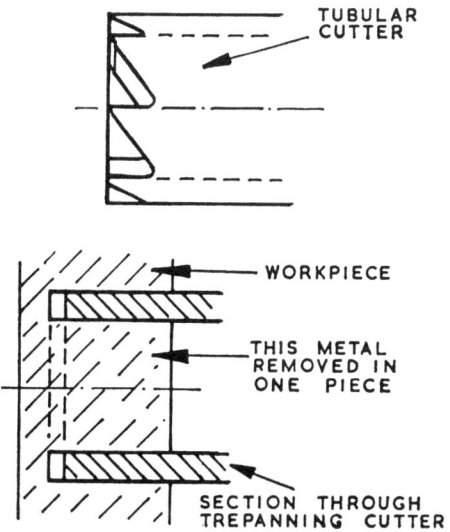

Fig. 2.39 Trepanning Tool

the swarf bin. Holes can be trepanned in reasonably small-section workpieces by using a boring tool, but a multi-toothed cutter of tubular form is used when the hole is long. Trepanning using this type of cutter is illustrated in fig. 2.39; if the hole is exceptionally long it may be necessary to introduce a bearing bush so that the cutter is supported by the bore already produced by the cutter.

2.7. Hole-Finishing Cutting Tools

A hole cut may be opened out by using a boring tool, and may be finished if required by reaming; drilling may be followed directly by reaming if the difference between the drill and reamer sizes is small.

2.7.1. Boring. Boring can be regarded as internal turning; boring presents greater problems than turning does because there is less room for the

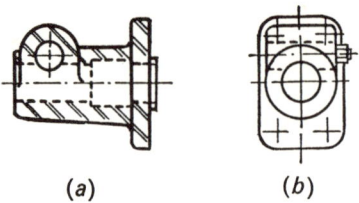

Fig. 2.40 Boring Bar Holder

cutting tool, and because it is more difficult to convey the cutting fluid to the point of cutting. The boring tool can be forged in one piece from a bar, but it is more usual to attach small boring tools to a boring bar; when used in a turret lathe the boring bar is held in a holder that is, in turn, bolted to

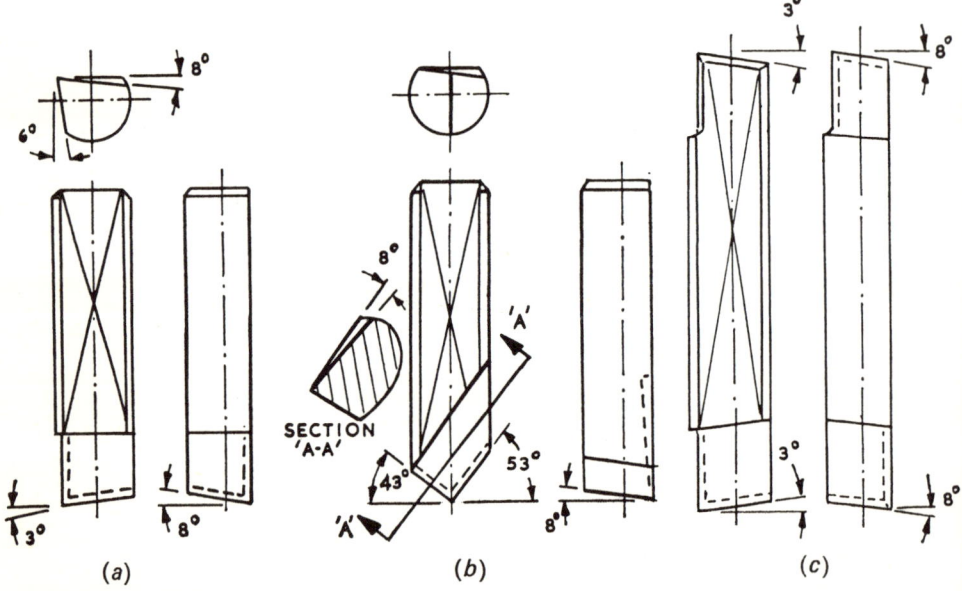

Fig. 2.41 Typical Boring Tools

a face of the turret; fig. 2.40 shows a typical holder of this type. Unless the boring bar is short and of heavy section, it should be located and supported by a bush that forms part of the fixture. Fig. 2.41 shows three typical boring tools made from bar; those shown in fig. 2.41 (a) and (b) are single-point, but that shown in fig. 2.41 (c) is double-point. Fig. 2.42 shows two

CUTTING TOOLS

methods of securing the boring tool to the boring bar; both methods incorporate an adjusting screw as well as a locking screw. The arrangement shown in fig. 2.42 (a) is best if it is necessary to control both the diameter of the bored hole and its depth relative to another tool; but if possible, the method shown in fig. 2.42 (b) should be used because this allows the feed force to be directed across a larger section of the boring tool.

Fig. 2.43 shows several boring tools held in one boring bar, and fig. 2.44

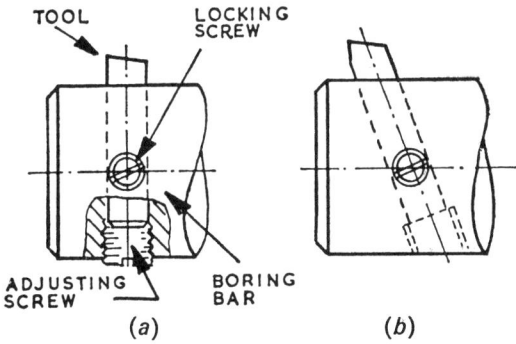

Fig. 2.42 Retention of Boring Tools

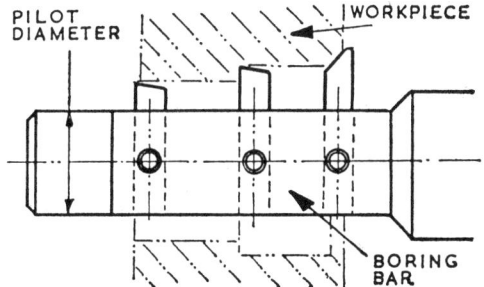

Fig. 2.43 Several Boring Tools in One Bar

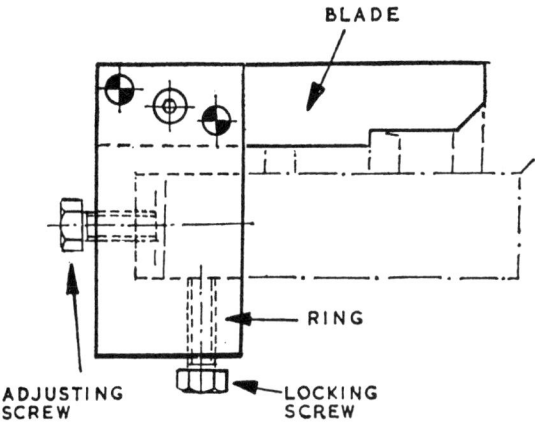

Fig. 2.44 Boring-Tool Setting Piece

B

shows a simple setting gauge that can be used to save setting time. When deep holes are bored, the boring bar usually has a copper tube fitted to it so that the cutting fluid is conveyed to the tools effectively.

A flat cutter can be used for boring as shown in figs. 2.45 and 2.46. The cutters shown have two cutting edges and so must be of the same size as the bore to be machined, and be located relative to the boring-bar axis. The arrangement shown in fig. 2.45 is for a flat cutter that is located from location flats machined on the outside of the boring bar, and secured in the

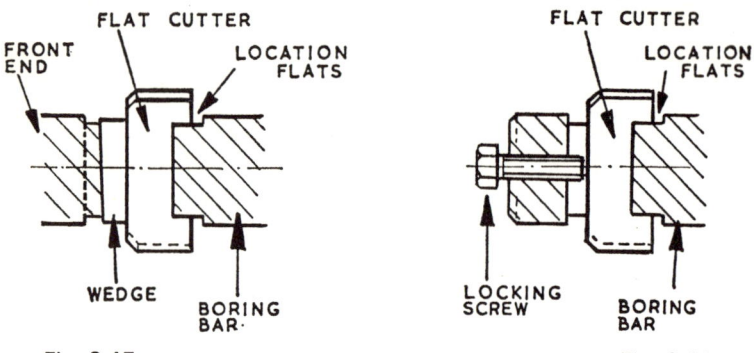

Figs. 2.45 and 2.46 Retention of Flat Boring Cutter

location slot in the bar by a wedge; the arrangement shown in fig. 2.46 is similar, but locked by a locking screw. In both these arrangements the feed force is taken by the boring bar and not by the locking device. The locking device is only placed behind the cutter if a blind hole, or similar shape, is produced that will not permit the locking device to be in front of the cutter.

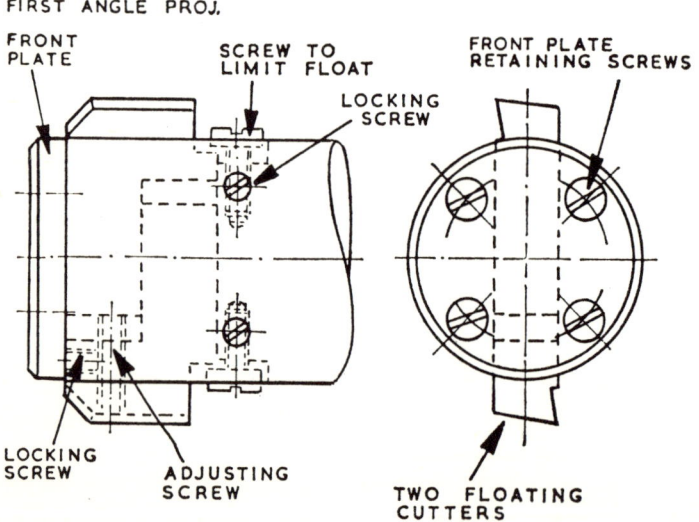

Fig. 2.47 Floating Boring Cutter

CUTTING TOOLS

2.7.2. A *floating cutter* (or floating reamer) can be used to finish holes accurately. The arrangement shown in fig. 2.47 used two cutters that can be set for maximum and minimum position, and yet 'float' at right angles to the axis.

2.7.3. Counterboring, spotfacing and countersinking cutters. A counterboring cutter is similar to an end mill; it can be in one piece including a taper shank, or be attached to a holder as shown in fig. 2.48. The cutter

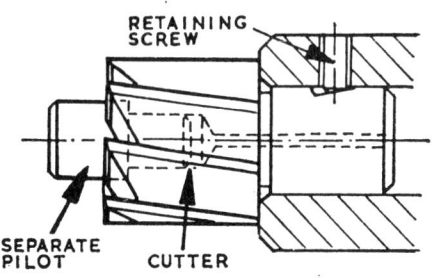

Fig. 2.48 Counterboring Cutter

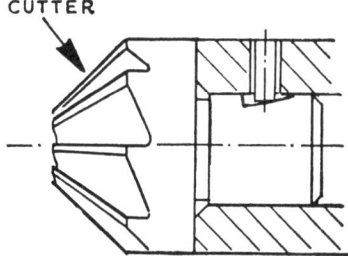

Fig. 2.49 Countersinking Cutter

shown has a pilot that is separate so that the end of the cutter can be ground easily. The counterboring cutter can be hollow (like a 'shell' milling cutter) and assembled with a drill, so that drilling and counterboring can be performed at one setting. Helical flutes are introduced to prevent chatter. A spotface is really a counterbore of small depth, and so a spotfacing cutter is similar to a counterboring cutter, but when used to spotface a surface where there is no hole, one tooth must cut to the centre. Fig. 2.49 shows a countersinking cutter; if necessary a pilot can be incorporated.

2.7.4. Reamers (see B.S. 122, Milling Cutters and Reamers). A reamer is used to finish a drilled or bored hole; the similarity between a reamer and milling cutters is the reason why they are covered by the same British Standard. The reamer rake angle is small, zero (as shown in fig. 2.50) or of small negative angle; when a reamer is required to finish a hole to depth the teeth are made end-cutting. The flutes are usually of left-hand helix to prevent the reamer from feeding into the bore. Reamers are classified as hand-type, machine-type and taper-type; they may be 'solid' or 'shell' and held in a cutter bar.

2.8. Milling Cutters

Milling cutters are the subject of British Standard B.S. 122, Milling Cutters and Reamers; in this standard, milling cutters are classified into three groups, these are: non-form-relieved cutters, end mills and form-relieved cutters. Fig. 2.51 illustrates the teeth of a peripheral-cutting milling cutter of the non-form-relieved type. The rake angle is determined by the

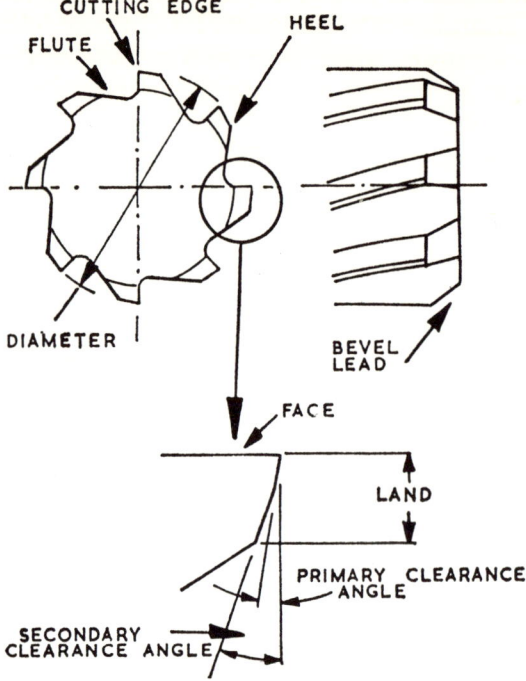

Fig. 2.50 Reamer Teeth

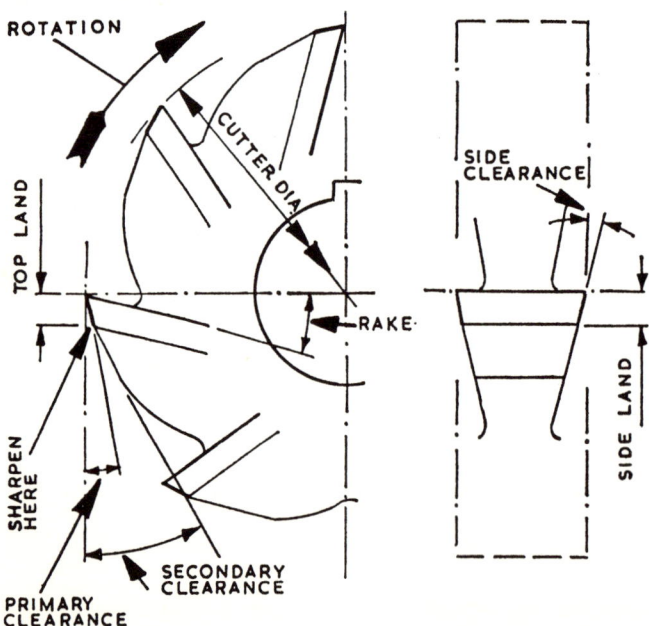

Fig. 2.51 Teeth of a Non-form-Relieved Peripheral-Cutting Milling Cutter

CUTTING TOOLS

mechanical properties of the workpiece material, and the cutting-tool material. The clearance is in two parts (the primary clearance angle and the secondary clearance angle); this enables the actual cutting clearance to be small, thus producing a strong tooth, but allows adequate body clearance so that the cutter does not touch the workpiece. Reference to fig. 2.56 will show that the curved path taken by the cutter relative to the workpiece is not circular, and so the 'heel' of the tooth is likely to foul the workpiece. The two-stage clearance produces a land of about 0·030 in in the case of a new cutter; this land increases during the life of a cutter because it is sharpened by regrinding the land. Behind the secondary clearance the cutter is shaped to reduce its weight and to produce a chip space or 'gullet'. When a milling cutter is to cut on the sides, a side clearance is also introduced (see the R.H. view on fig. 2.51). The rake and primary clearance angles are usually kept as small as possible to produce maximum tooth strength and also to give good heat dissipation.

2.8.1. There is no fixed rule regarding the number of teeth, but the tendency is to use as small a number of teeth as possible, to allow them to be very large and strong, and also to allow a large chip clearance between them.

2.8.2. The teeth of peripheral-cutting teeth are often cut on a helix so that they perform 'oblique cutting'; this reduces the vibration because each tooth engages the workpiece before the previous one has completed its cut; fig. 2.52 (*a*) and (*b*) shows straight-fluted and helical-fluted cutters.

The teeth of a form cutter are 'form relieved' (see fig. 2.53) so that clearance is introduced, but allowing the form to be preserved after the cutter has been reground. The radial form at, say, OA, OB, OC and OD is the

(*a*) (*b*)

Fig. 2.52 Comparison between Straight and Helical Fluted Cutters

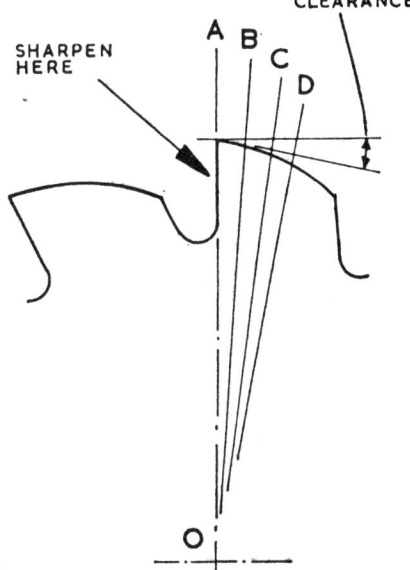

Fig. 2.53 Form-Relieved Cutter Teeth

same, and so when the face of each tooth is reground, the cutter will continue to cut correctly provided the same amount of metal is removed from the face of each tooth.

2.8.3. Up-cut and down-cut milling. Figs. 2.54 and 2.55 show the difference between these two methods of milling. The path of the tooth relative to the

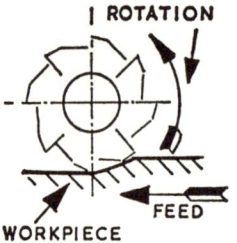
Fig. 2.54 Up-Cut Milling

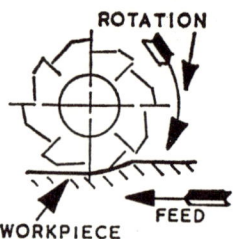
Fig. 2.55 Down-Cut Milling

workpiece and the shape of the chip produced by each of these methods is shown in figs. 2.56 and 2.57; these illustrations also show why peripheral milling produces a characteristic wavy surface. Figs. 2.58 and 2.59 show that the rake and clearance angles alter during the cut; fig. 2.58 illustrates up-cut milling, and shows that the **actual** rake increases during the passage

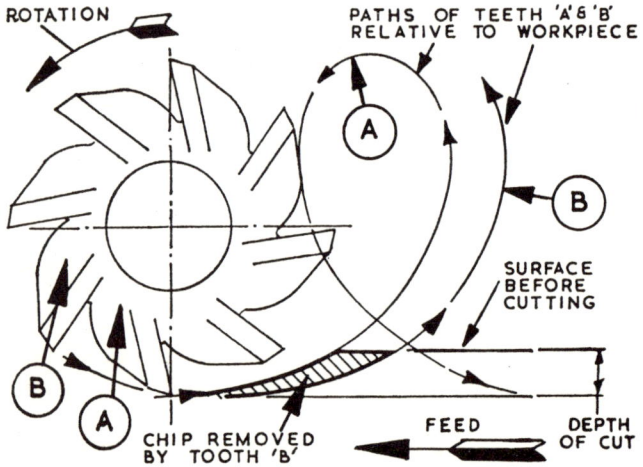
Fig. 2.56 Path Taken by Cutter Teeth when Up-Cut Milling

of each tooth, and that the **actual** clearance decreases during the same time (it must be emphasised that the actual rake and clearance must be considered relative to a tangent to the workpiece surface at the point during the cutting that is under consideration). Fig. 2.59 illustrates down-cut milling, and it will be seen that in this case the **actual** rake decreases during the cut and that the **actual** clearance increases during the cut. If the two methods are compared it will also be seen that the up-cut method produces a chip that starts small and increases in thickness, but that down-cut milling

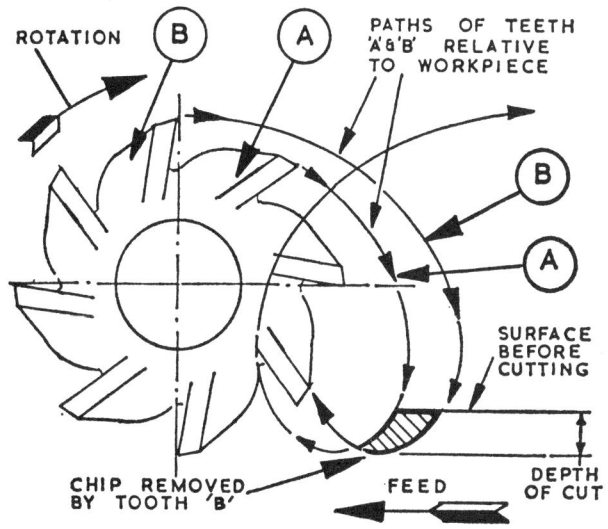

Fig. 2.57 Path Taken by Cutter Teeth when Down-Cut Milling

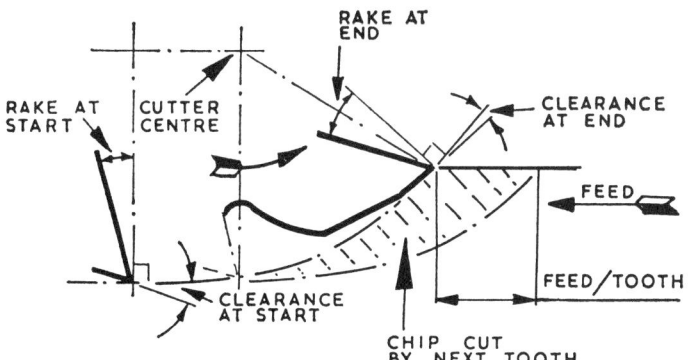

Fig. 2.58 Cutting Angles when Up-Cut Milling

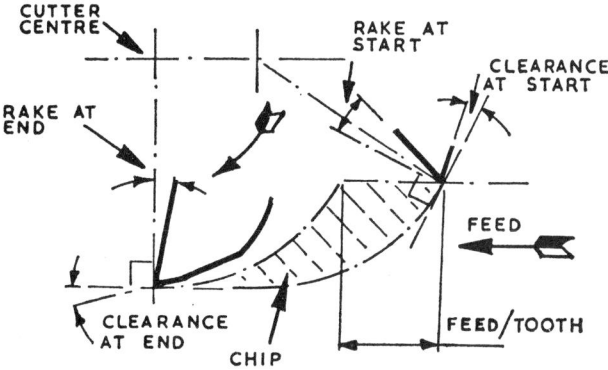

Fig. 2.59 Cutting Angles when Down-Cut Milling

produces a chip that starts thick and ends thin. The difference between these two methods is very marked when a ductile material is being machined because built-up edge produces a bad surface in the case of down-cut milling; down-cut milling is considered to be the better method when thin cutters are used, but is less efficient when a casting is machined because the greatest thickness of chip is produced by 'attacking' the characteristic hard case associated with a cast product. Down-cut milling, unlike up-cut milling, forces the workpiece down on to the machine table, and is better in this respect, but it demands a machine whose leadscrew is in first-class condition.

2.8.4. Facing cutters. Fig. 2.60 illustrates the workpiece–cutter relationship during a facing operation, and fig. 2.61 illustrates the main features of a facing cutter.

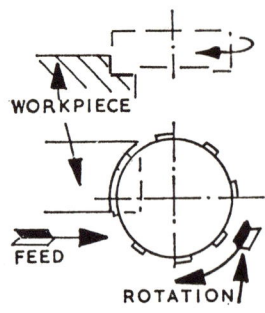

Fig. 2.60 Facing

The helix angle is equivalent to the front-to-back angle of a single-point tool, and is negative when a cemented carbide, or similar hard cutting-tool material, is used. The bevel angle is equivalent to the approach angle of a single-point tool and is usually about 15°. Provided that the cutter is correctly ground, a facing cutter produces a flatter surface than a peripheral-cutting cutter because the path of the teeth relative to the workpiece does not produce the characteristic wavy surface associated with peripheral cutting.

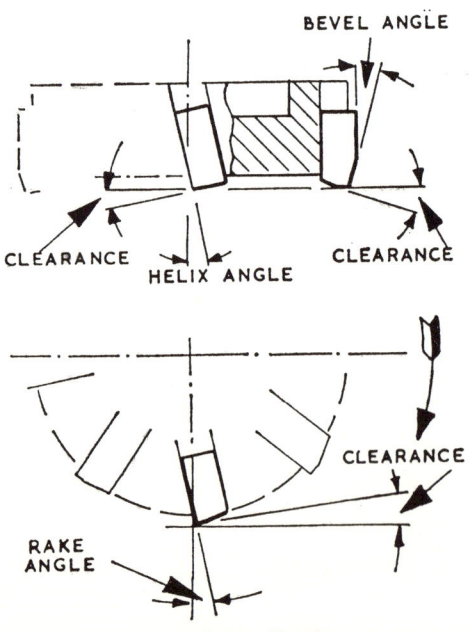

Fig. 2.61 Facing Cutter Teeth

CUTTING TOOLS

2.8.5. Interlocking cutters. Side-and-face milling cutters are often used in pairs and made interlocking (see fig. 2.62) so that the width of the slot produced will not alter following regrinding of the sides of the cutters; a suitable spacer is used to produce the required width. It will be seen from fig. 2.62 that when a pair of helical cutters is used, opposite hand cutters are selected; this is to nullify the side thrust that would be produced if a single helical cutter is used.

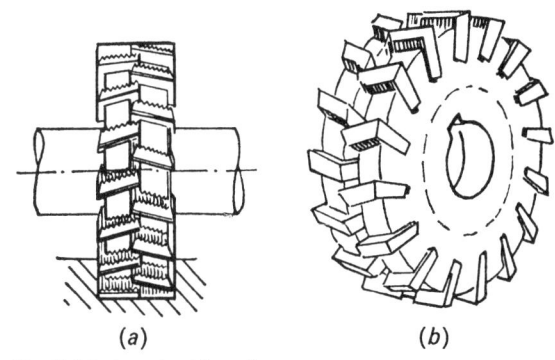

Fig 2.62 Interlocking Cutters

2.9. Thread-Producing Tools

Fig. 2.63 shows the nomenclature used in conjunction with taps. In theory the rake angle should be made to suit the material to be tapped, but in practice a suitable compromise is made so that one tap can be used for

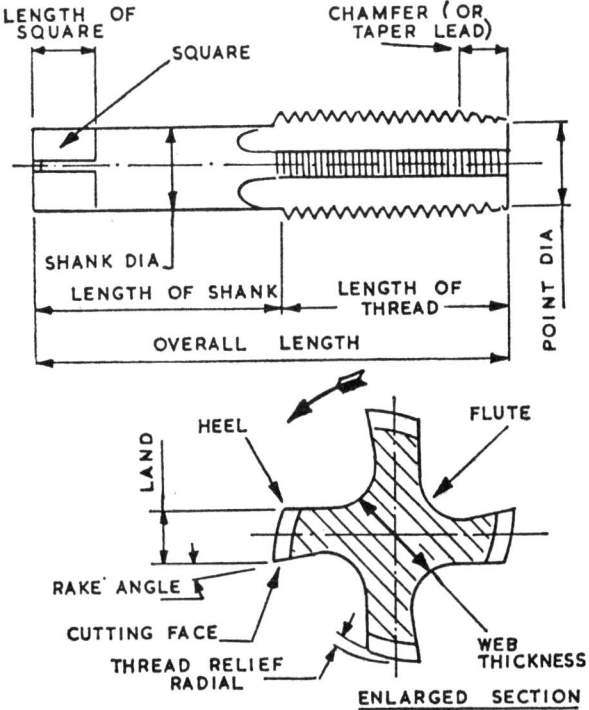

Fig. 2.63 Tap Nomenclature

most materials. The thread relief is usually just sufficient to prevent the tap from rubbing the hole.

Fig. 2.64 illustrates a simple button die used to produce external threads; the disadvantage of this method of thread production is that the die must

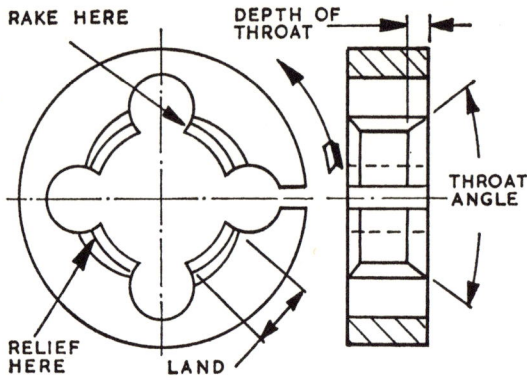

Fig. 2.64 Button Die

be 'unthreaded' from the workpiece. For production work on capstan lathes a self-opening die head is used that opens at the end of the thread and can be withdrawn rapidly from the workpiece. The 'dies' can be arranged radially (fig. 2.65) or tangentially (fig. 2.66), and are specially shaped so that all but the leading teeth of each die engage the thread already cut (by

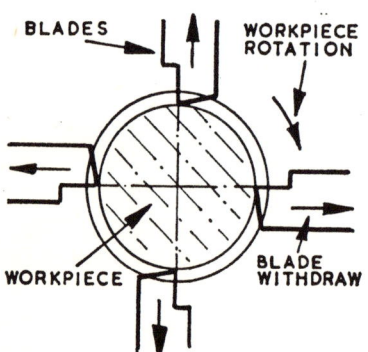

Fig. 2.65 Diehead Arrangement with Radial Dies

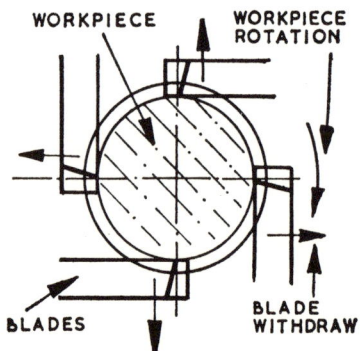

Fig. 2.66 Diehead Arrangement with Tangential Dies

the leading teeth) above the centre-line, so that instead of cutting, they act as a 'nut' causing the die head to generate the rest of the thread; this is termed the 'self-guiding action' (see fig. 2.67).

CUTTING TOOLS   35

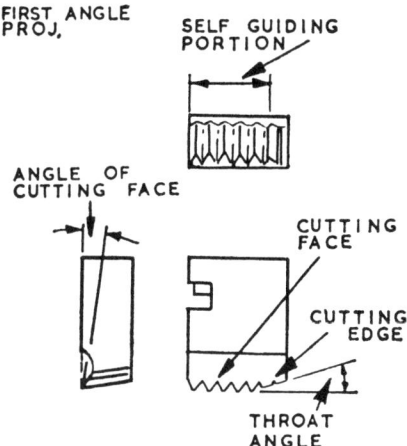

Fig. 2.67 Die (Showing Self-Guiding Portion)

2.10. Gear Cutters

Gear production is specialised, but the main principles of the more common methods of involute gear production are outlined in this section. The reader should consult a textbook on gear production and one on drawing if he requires a more detailed account. The main features of an involute gear are: the **diametral pitch** (defined as the number of teeth per inch of pitch circle diameter), which indicates the size of the teeth; and the **pressure angle**, which controls the shape of the sides of the teeth. The teeth are regarded as being spaced around the **pitch circle**; the number of teeth in the gear, the pitch circle diameter and the diametral pitch are related to each other. Gears of the same diametral pitch and pressure angle will mesh successfully together, but they do not have the same shape teeth unless they have the same pitch circle diameter. A rack is a gear of infinite pitch circle diameter; an involute rack has straight sides.

2.10.1. Gear cutting can be classified as (*a*) forming and (*b*) generating. In the first method, the cutter is shaped like the gap between adjacent teeth, and a series of 'gashes' made in the blank. In the latter method, the cutter is not shaped like the required teeth gap, but by combining its shape with a generating action, the required shape is produced. It has been stated that although gears of same diametral pitch and pressure angle will mesh together, they are only of the same shape if they also have the same pitch circle diameter. If the gear produced by forming is to be accurate, a specially-designed cutter must be used for it. In practice, this method is used only for gears that do not need to move smoothly and rapidly, and so a compromise of shape is possible. Gears for tooling equipment are often made by form milling, using a form-relieved cutter; a basic set of seven cutters is used (additional cutters can be used), so that most of the gears so cut are of approximate shape, but quite accurate enough to run together slowly (note: the shape compromise does not produce variation of velocity ratio). Most gears are therefore produced by generating; the principle generating methods are planing, hobbing and shaping (or cutting with a pinion cutter).

2.10.2. Gear planing. Since involute gears of the same diametral pitch (D.P.) and pressure angle will run together, an involute rack will run with any involute gear of the same D.P. and pressure angle. This is an attractive machining proposition because, as already stated, an involute rack has straight sides, and is therefore easy to produce. Fig. 2.68 illustrates the principle of gear planing; the cutter is shaped like a rack, and has about six teeth. It is reciprocated, and gradually fed into the blank until its pitch line coincides with that to be produced on the gear. At this point the generating action takes place. Before each stroke the blank and cutter are given a small generating movement (see fig. 2.68). The cutter produces

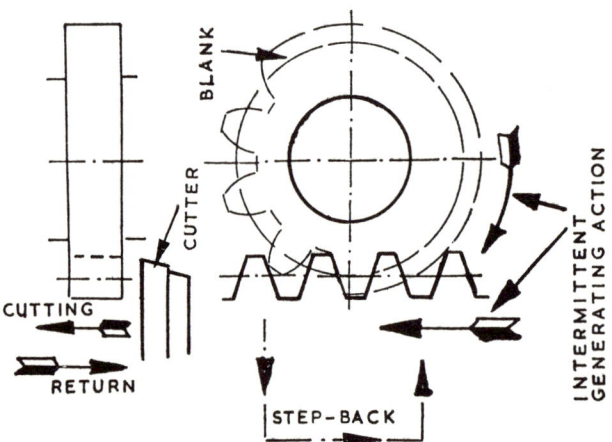

Fig. 2.68 The Principle of Gear Planing

a large number of small 'flats' on the blank surface, so that an involute is produced. Gear cutting is followed by grinding if the gear is to be hardened and when greater accuracy is required, and so these 'flats' will be smoothed out. The generating action cannot produce a complete gear without step-back action because the rack is short; after two or three teeth have been cut (depending upon size), the generating action is stopped so that the cutter can be withdrawn, stepped back the required distance and re-engaged so that the generating action can continue. The cutter rake and clearance angles are about 6°, and a clapper box prevents damage to the workpiece during the return stroke of the cutter. This method can be used for helical gears as well as straight gears, but cannot be used for internal gears.

CUTTING TOOLS

2.10.3. Hobbing. This process uses a continuously-rotating cutter, called a **hob**, the teeth of which are shaped like a rack, and cut on a helix. If the hob is rotated, the 'imaginary rack' will appear to pass continuously along a line parallel to the axis of the hob. The relative hob and blank rotation is

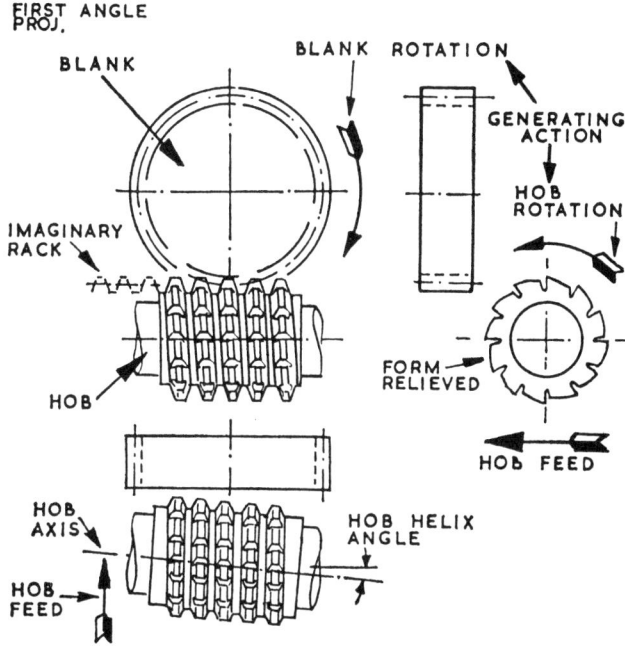

Fig. 2.69 The Principle of Hobbing

set to produce the movement that would be obtained if the gear to be produced is run with a worm with the same characteristics as the hob. The actual hob speed depends upon the mechanical properties of the workpiece material, and its feed across the blank face is very small. When a straight

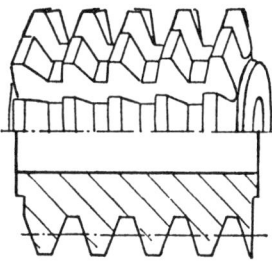

Fig. 2.70 Gear Hob

spur gear is being cut, the hob is inclined by an amount equal to the helix angle of the hob (see fig. 2.69); when a helical gear is being cut, the hob is inclined so that the path of the hob teeth at top dead centre coincides with the teeth of the gear to be cut. Hobbing cannot be used for internal gears.

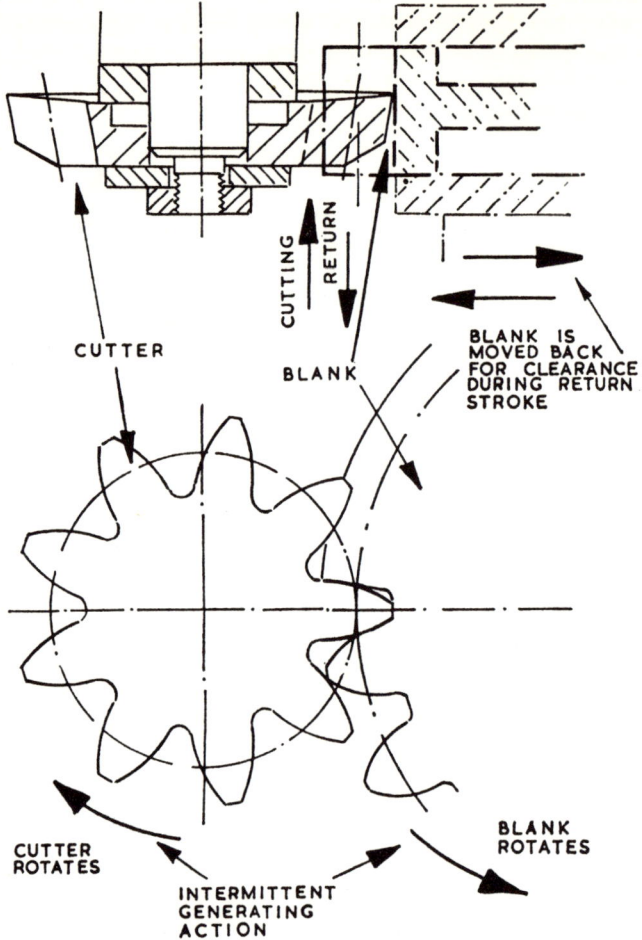

Fig. 2.71 The Principle of Gear Shaping

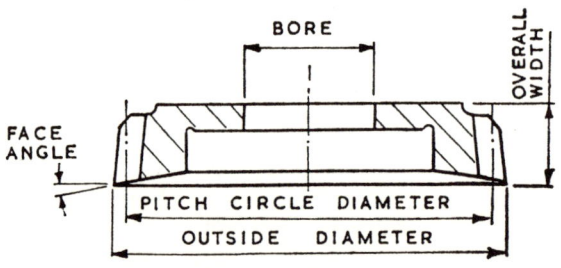

Fig. 2.72 Cutter for Gear Shaping

2.10.4. Gear shaping (or gear cutting with a pinion cutter). In this process the cutter is shaped like a pinion, and reciprocates past the gear blank. The generating action is intermittent, but a step back is not required; the blank must be moved away from the cutter by a small amount before each return stroke to produce clearance. The cutters are standardised at 3 in P.C.D. and 4 in P.C.D., the rake is usually 5° and the clearance 6°. This process can be used for internal gears, but when helical gears are to be cut, the cutter must have a helix of the same angle but opposite hand from that of the gear to be cut, and is operated by special helical guides; the construction of the guide assembly makes this method unsuitable for cutting gears with a helix angle that is more than 35°.

References

British Standards.
'Turret Lathe Work', published by Alfred Herbert Ltd.
Publications by: A. C. Wickman Ltd.; Richard Lloyd & Co. Ltd.; The English Steel Corporation.

Chapter 3

FORM TOOLS

Form tools are used to turn short profiles, usually on a capstan lathe or a turret lathe. The form tool may be designed to be fed radially into the workpiece, or it may be designed to be passed tangentially across the workpiece.

Form tools can be classified as follows:
1. Radial feed (a) Flat form tool
              (b) Dovetail form tool
              (c) Circular form tool
2. Tangential feed using a skiving tool

3.1. Flat Form Tool

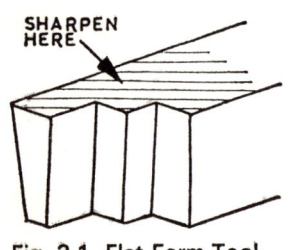

Fig. 3.1 Flat Form Tool

The flat form tool is usually used in the horizontal position. Fig. 3.1 shows the end of a flat form tool, and it will be seen that the top of the tool is reground when the tool is sharpened; the tool must therefore be of adequate section to maintain an acceptable strength during its working life, and the tool holder must incorporate means of adjustment so that the cutting edge of the tool can be set to the workpiece centre-line after re-grinding.

3.1.1. Form 'correction'. For manufacturing purposes it is usual to dimension the tool in a plane that is normal to the front face of the tool, and so it is necessary to make so-called 'form correction' calculations, or to determine the tool shape in the required plane by graphical methods.

Fig. 3.2 illustrates the principles involved in the 'correction' when the rake is zero, and it indicates that the widths W remain constant and the form depths T are altered by an amount depending upon the clearance angle θ. The 'correction' for a curved profile can be determined by considering co-ordinate dimensions such as W_1 and T_1.

Fig. 3.3 shows the basis of calculations when a positive rake angle ϕ is introduced. It will be seen that the form along the top of the tool (plane A–B) is first determined, and then this is expressed as a dimension in a

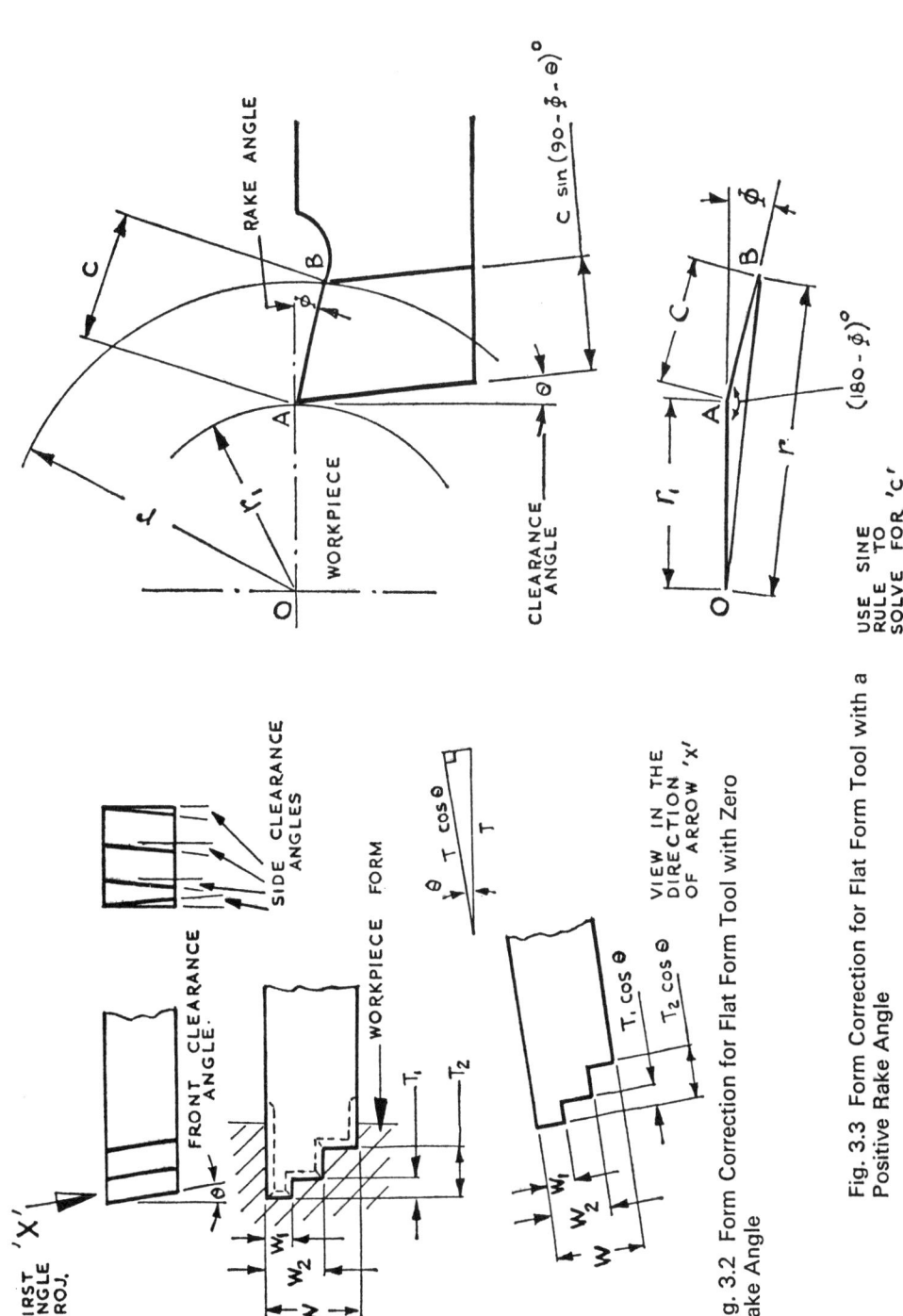

Fig. 3.2 Form Correction for Flat Form Tool with Zero Rake Angle

Fig. 3.3 Form Correction for Flat Form Tool with a Positive Rake Angle

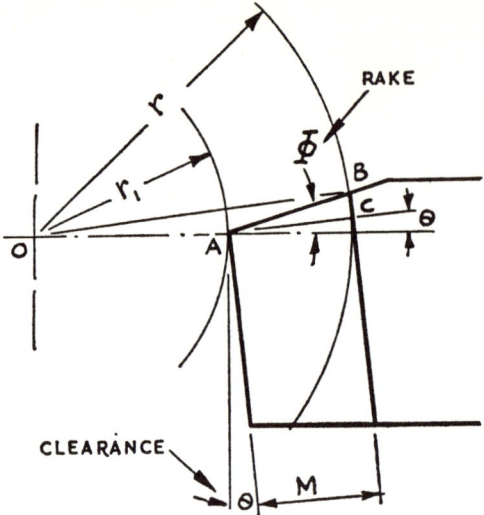

Fig. 3.4 Layout for Form Correction for Flat Form Tool with Negative Rake Angle

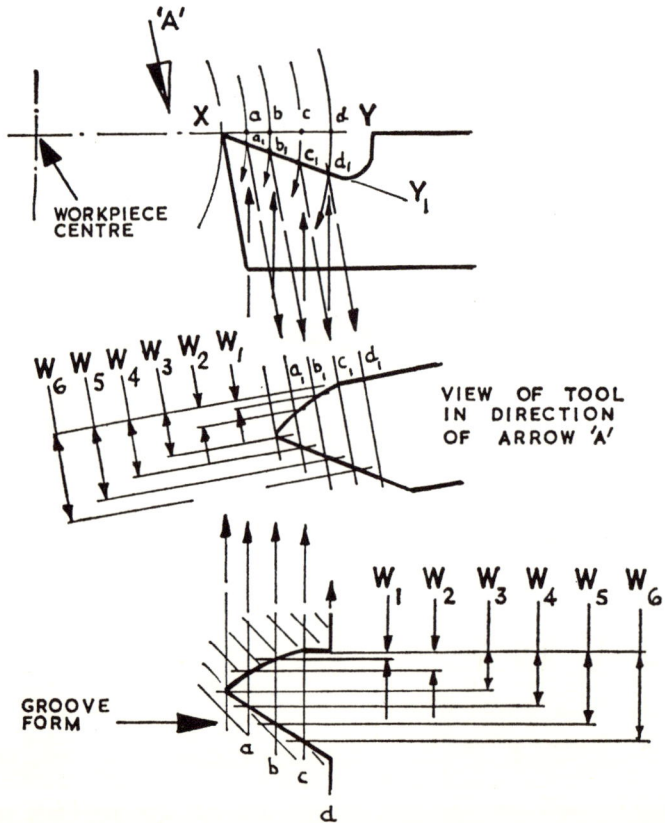

Fig. 3.5 Graphical Method of Form Correction for Flat Form Tool

FORM TOOLS

plane normal to the tool front face. A negative rake is sometimes employed, and fig. 3.4 shows the basis of the calculations involved.

The correction when the tool has a positive rake angle can be determined graphically as shown in fig. 3.5. The points on the form on line $X-Y$ are drawn on the 'tool face line' ($X-Y_1$) and then projected in the direction of arrow A. The correction when the rake angle is negative is obtained in a similar manner.

3.2. Dovetail Form Tool

Fig. 3.6 shows a dovetail form tool mounted in its holder, and fig. 3.7 shows a sketch of the holder. The clearance angle is obtained by the inclination of the tool location face of the holder. The dovetail form tool is reground on the top face, and has a longer life than the flat form tool; much higher cutting forces can be permitted than in the case of the flat

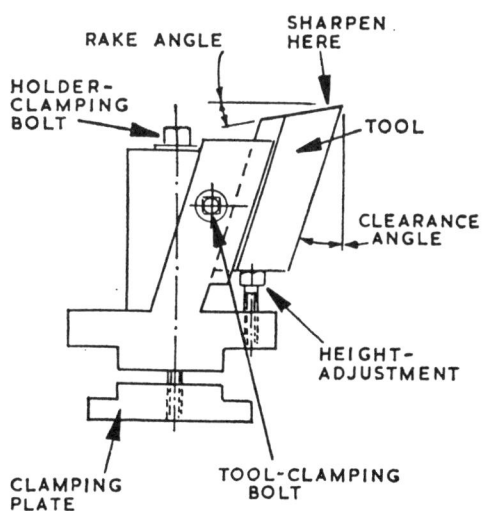

Fig. 3.6 Dovetail Form Tool in Holder

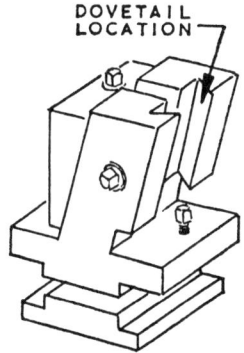

Fig. 3.7 Holder for Dovetail Form Tool

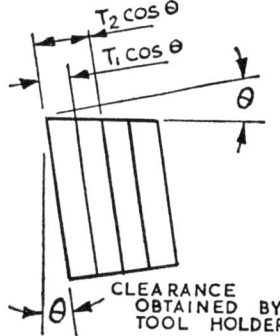

Fig. 3.8 Dovetail Form Tool

form tool because the tool is much stronger, and is held more rigidly. The disadvantage associated with the dovetail form tool is that the front clearance is obtained by the tool holder, and so a change of clearance angle demands another tool holder; in practice it is possible to select one clearance angle that is suitable for a wide range of applications. Fig. 3.8 shows a dovetail form tool for zero rake angle, and if this is compared with fig. 3.2 it will be seen that the flat form tool and the dovetail form tool are very similar. A rake angle can be allowed for in a similar way to the methods used in the case of the flat form tool.

3.3. Circular Form Tool

A circular form tool is a specially-shaped disc that is gashed to produce a cutting edge. Fig. 3.9 (a) shows a circular form tool, and it will be seen that the top face of the 'gash' is below the tool centre-line so that a front clearance is obtained when the cutting edge is set to the workpiece centre-line. The top face of the 'gash' is re-ground to sharpen the tool, which is then re-set by rotating it about its axis to bring the new cutting edge into position. A circular form tool maintains its accuracy and 'life' over about 270° of tool profile (a circular form tool after repeated re-grinding is shown in fig. 3.9 (b)). Circular form tools are between about 2 and 5 in in diameter, and are held in a special holder of the type shown in fig. 3.10.

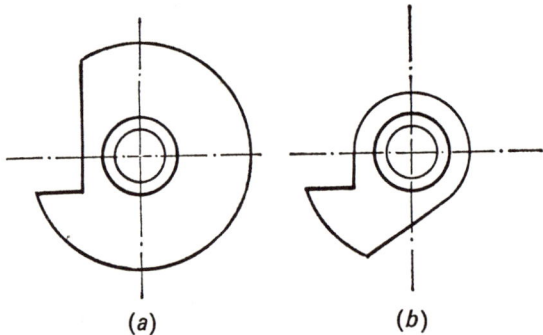

Fig 3.9 (a) Circular Form Tool before Re-grinding
(b) Circular Form Tool after Repeated Re-grinding

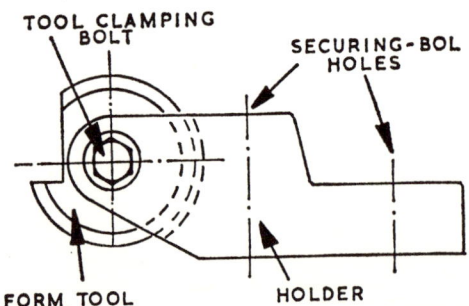

Fig. 3.10 Holder for Circular Form Tool

FORM TOOLS

3.3.1. When a straight shoulder is to be produced it is necessary to provide some form of clearance to prevent the side of the tool from rubbing on the workpiece. This may be obtained by making the form of helical shape, or by inclining the tool and feeding as shown in fig. 3.11.

3.3.2. Form correction. The tool form is produced before the 'gash' is made, and so the form must be expressed radially. Fig. 3.12 shows the layout of form tool and workpiece, and fig. 3.12 (a) shows the triangles that must be solved to obtain the tool centre height H, and one radial dimension R_1. Again, a complex form is dealt with by considering a number of co-ordinates. Fig. 3.12 also shows how the correction can be done graphically.

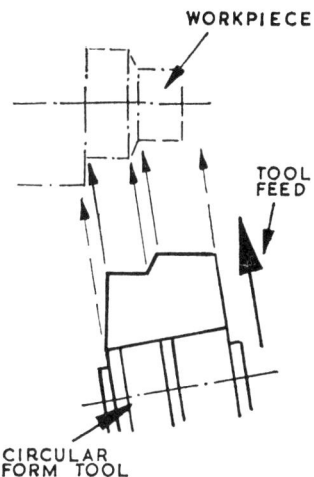

Fig. 3.11 Obtaining a Side Clearance

When the tool is to have a rake angle, the correction is made as shown on fig. 3.13 (for a positive rake angle ϕ). Two triangles must be solved; the

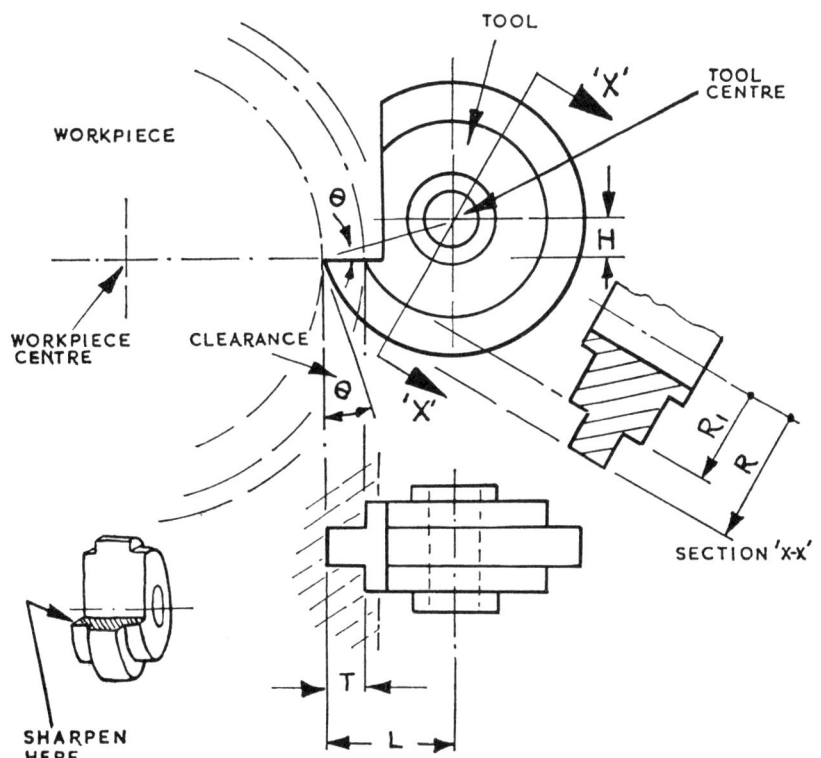

Fig. 3.12 Form Correction for Circular Form Tool with Zero Rake Angle

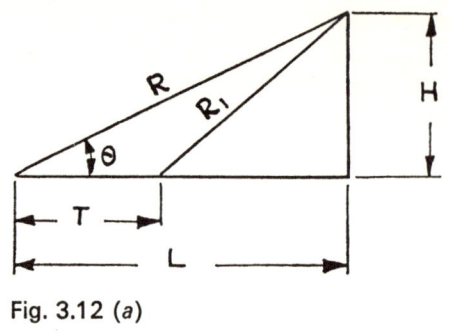

Fig. 3.12 (a)

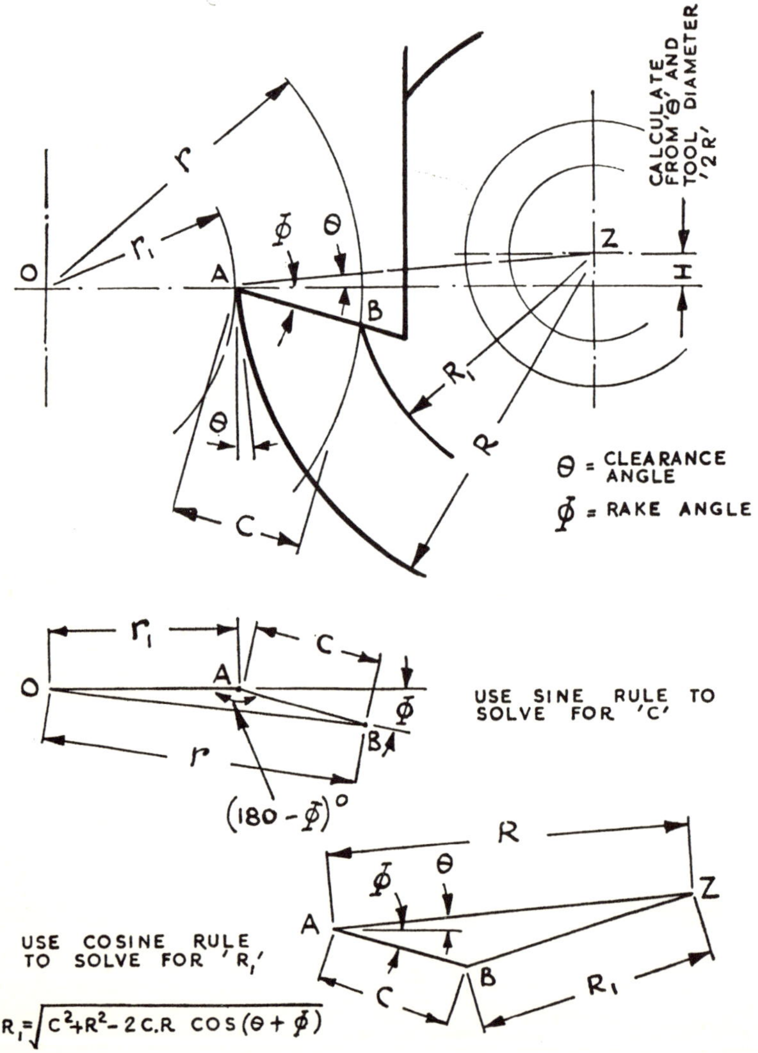

θ = CLEARANCE ANGLE
ϕ = RAKE ANGLE

USE SINE RULE TO SOLVE FOR 'C'

USE COSINE RULE TO SOLVE FOR 'R_1'

$$R_1 = \sqrt{C^2 + R^2 - 2C \cdot R \cos(\theta + \phi)}$$

Fig. 3.13 Form Correction for Circular Form Tool with Positive Rake Angle

FORM TOOLS

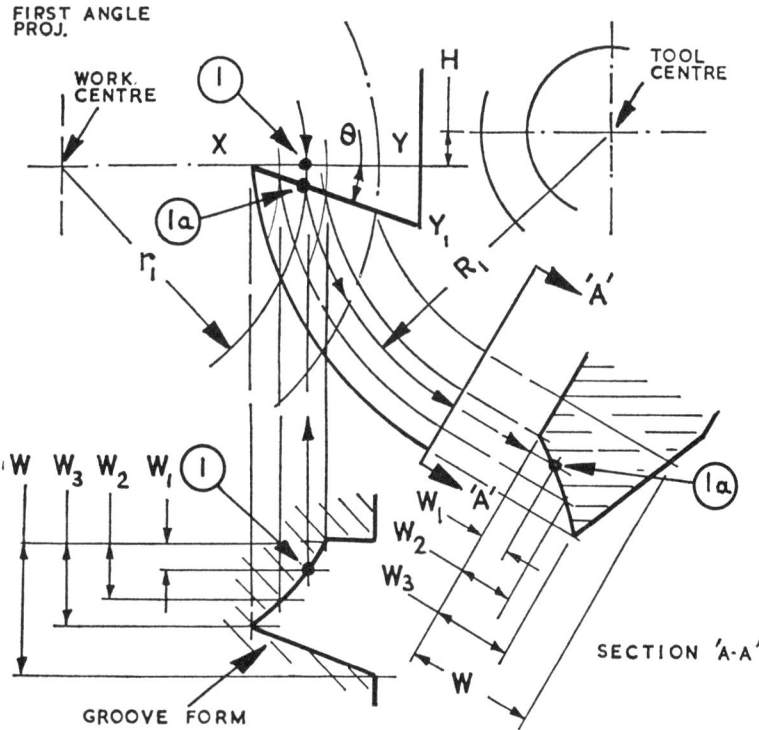

Fig. 3.14 Form Correction for Circular Form Tool by Graphical Method

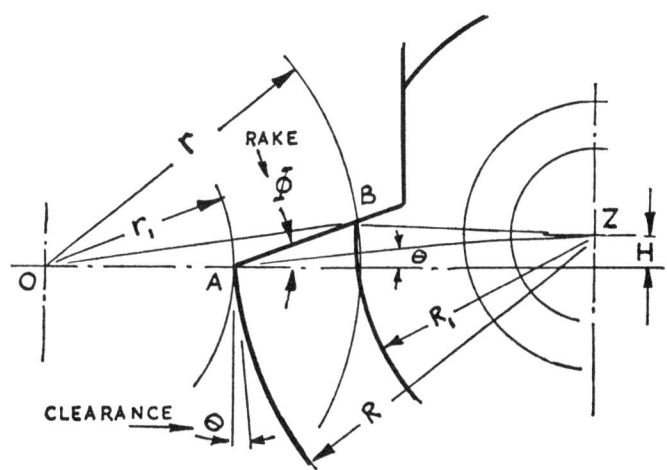

Fig. 3.15 Layout for Circular Form Tool with Negative Rake Angle

first to obtain the form on the tool face *A–B*, and the second to obtain the radial form. Fig. 3.14 shows how this correction can be done graphically.

The layout for a circular form tool with a negative rake angle is shown in fig. 3.15.

3.4. Notes on Form Tool Calculations

The differences between the workpiece shape and the form-tool shape will be very small, and so the mathematical tables used must be at least five-figure tables; errors will be reduced if 'logarithms of trigonometrical ratios' are used directly. Should the difference between the workpiece dimensions and the form-tool dimensions be small compared with the workpiece tolerances, the tool can be made to the workpiece dimensions; this is important where the correct tool profile is more difficult to produce than the 'correct' tool form (for example, where the workpiece profile is made up of circular arcs).

3.5. The Skiving Tool

Fig. 3.16 shows a typical skiving tool, which cuts by passing tangentially across the workpiece. The tool can be set at an angle as shown to produce

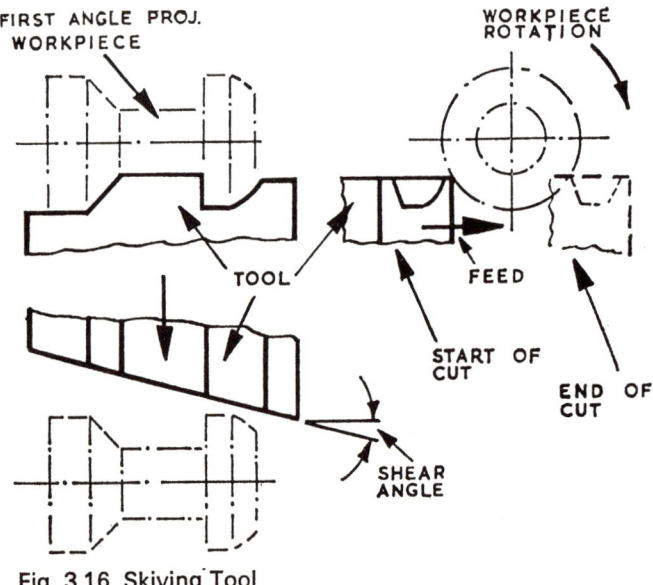

Fig. 3.16 Skiving Tool

oblique cutting, and also to reduce the length of form being cut at any one time, and thereby to reduce chatter. The example shown in fig. 3.16 shows a tool with a zero rake angle and no clearance. The clearance is only introduced if there is a tendency for the workpiece material to 'pick up', but even then is rarely more than 10°. The advantage of this type of form tool is that it is easy to produce, and no 'correction' calculations are required.

Chapter 4

PROCESS PLANNING

4.1. Introduction

The ideal process is that which enables the required accuracy to be obtained in the allowed time and with the minimum cost. This means that when a process is prepared it is necessary to consider the volume of production, the delivery date and the existing equipment; these considerations are dealt with in Chapter 17.

4.1.1. Preliminary process planning is done some time before machining is due to commence so that (i) the raw-material dimensions can be settled, (ii) the machine-tool requirements can be assessed, (iii) the jigs, fixtures, tools and gauges can be designed and manufactured, (iv) the labour requirements can be studied, (v) an accurate estimate of the time taken to manufacture the component can be made and (vi) changes in the component that may assist production can be considered.

4.1.2. The machining sequence and the information given in the form of a process will depend upon the size of the machine shop, the machines that are available and the class of operator to be employed. Planning for a tool room or prototype shop will give the operation sequence, and an indication of the amount of metal to be 'left on' to suit grinding or when some surface has to be 'left soft' when the part is case-hardened. It is necessary to prepare a form of process so that the 'supervision' can allocate the work, and prepare time estimates if required, but detailed instructions are not required because the class of machinist employed in such a shop can work without detailed information, and because, since jigs and fixtures will not be used, there will be no need to specify 'intermediate' sizes for location purposes.

When the process is for the production shop it is necessary to produce detailed instructions, and 'intermediate' dimensions given to suit the jigs and fixtures. It is usual to produce a drawing showing the work done at each stage (see page 255), and the parts are viewed to these dimensions until, at the final inspection stage, the 'master' drawing is used.

4.2. Choice of Method and Equipment

The method of machining and the equipment to be used will depend largely upon the volume of production and the class of labour employed on the work. The following table shows some of the differences between the methods used in the tool room and those used in a production shop.

Tool room	Production shop
Castings and forgings are marked out	Marking out only used for 'trial' material
Work located by 'setting up' using a dial indicator	Work located in jigs and fixtures
Work held in a vice, clamped to machine table, in a chuck or on a stump that is turned by the machinist	Work located and held in special vice jaws, in a fixture or special collet, etc.
Turning done on a centre lathe	Turning done on a capstan lathe or turret lathe
Grinding done when tolerances are fine, or after heat-treatment	Extensive use of grinding because unless tolerances are wide, other methods of machining by semi-skilled machinists will not produce the required accuracy
Circular table used for many profiling operations	Profile plates used extensively
Dividing head used for spacing of holes, etc.	Indexing jigs and fixtures used for spacing
Awkward shapes finished by filing	All shapes machined (if not 'as cast')

Table 4.1

Although the method adopted for the production shop will be different from that for the tool room, the differences will be mainly of detail and the precise equipment. The fundamental methods themselves will not be very different, and the method used for production can be regarded as a variation to suit the particular requirements of production.

4.3. Planning Method

An operation layout must be planned in a methodical way because its object is to produce method. If the following procedure is adopted the work of process planning will be simplified and be more effective. It is important that the work be approached with an open mind, and that the process be developed gradually.

1. Study the component drawing in order to understand the duty of the component so that the relative importance of its features can be deter-

mined; this study will show the feature used to locate the component upon assembly with mating parts (the feature used to locate the component is usually the one best used as a datum feature during machining), and will also show if tolerances on dimensions are applied to produce desired fits, or to ensure clearance between the component and mating parts upon assembly. As a result of this study the planner should also be familiar with the size, shape and weight of the component, and know if it is likely to produce problems of handling and balancing during machining.

2. List or ring the dimensions of features that are to be machined, and indicate if these are to be produced by roughing followed by final machining. Also indicate the features that can be used as the location system.

3. Prepare a rough draft process with due regard to the following basic rules.

(*a*) Establish **at least** one datum at the first opportunity. For example: when turning, face the end, so that distances can be measured from it, and when milling, face a large surface, so that it can serve as a 'seating face' for location purposes. Very often a second datum can be produced at the same time; for example, when turning, a bore can be produced square with the end face.

(*b*) Produce as much as possible from one setting; it must be realised that dimensions can vary, so every time the workpiece is re-set, the possibility of accuracy is reduced. Clamping can cause distortion, and so re-clamping can also cause inaccuracies.

(*c*) In order to ensure very accurate relationship between the features, the original datum features must be used as long as possible. When they can no longer be used, for example, when they themselves are finish-machined, location must be from features that were machined as a result of location from the original location features.

(d) If possible, group similar operations together. This will reduce the time taken in moving the workpiece from section to section, and assist the progress department.

(*e*) Perform accurate operations (for example, grinding) at the end of the machining sequence so that damage to these surfaces is minimised. *Note:* a feature may have material left on for finishing, and still be used for location provided that it is machined sufficiently accurately. In these cases the rough size is stipulated, together with the tolerance.

(*f*) Introduce inspection operations at strategic stages in the sequence to reduce the wasted effort and scrap that will be produced if incorrectly machined parts are allowed to proceed.

(*g*) Introduce a burr-removal operation, if necessary, immediately before a feature is used for location, and again, if necessary, at the end of the sequence.

(*h*) Ensure that all features are position-controlled.

4. Check the draft process to ensure that all machining is covered, and then finalise the process.

4.4. Specimen Operation Layouts

The following two operation layouts will illustrate planning techniques. These examples do not include heat-treatment operations, but when hardening is done, it is usually followed by grinding to remove distortion caused by quenching.

4.4.1. Operation layout for special bolt (see page 53). In this example the shank is made more accurately than demanded by the drawing because it is to be used as a location point. The workpiece is symmetrical about the shank axis until the head is milled or the shank is drilled; it therefore follows that the first of these features to be machined becomes the second locator, and will be used to control the rotational location of the bolt about the axis of the shank. The hole in the shank is unsuitable as a location feature because it is very small and because it would demand a retractable locator; the slot in the head is obviously unsuitable, but **one** of the flats can be used with the shank to produce complete location. *Note:* only **one** of the flats must be used for location because two flats and the shank would produce redundant location (see page 79). The burrs caused by milling must be removed before the head can be used for location.

4.4.2. Operation layout for fulcrum pin (page 54). In this example the stem is at an angle to the flange and the spigot. The spigot is therefore the main location feature and is machined first. The 0·500-in-diameter hole is the second locator, and is machined directly after the spigot to produce a rotational location; it is therefore necessary to split the turning operations (operations 1 and 3). The $\frac{7}{16}$-in-diameter holes and their spotfaces are produced at the same time as the 0·500-in hole. The spigot and the 0·500-in-diameter hole form the location system for all the machining operations that follow, except for the operation at which the spigot itself is finish-machined (operation 8). The spigot and the stem diameters must be finished by grinding in order to obtain the required accuracy. The flange profile is used for location about the spigot axis until the 0·500-in-diameter hole is machined.

PROCESS PLANNING

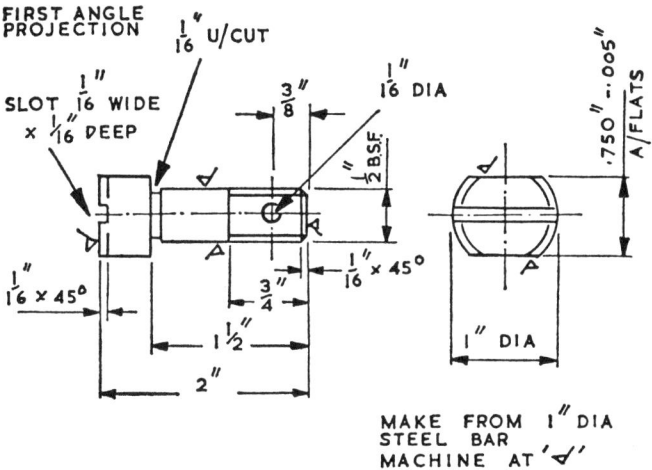

Fig. 4.1 SPECIAL BOLT

Operation number	Description	Machine
1	In collet; face end, turn shank diameter 0·500–0·005 in for location. Form undercut and shoulder. Chamfer end and screw to length. Part off to length $+\frac{1}{32}$ in	Capstan lathe
2	Reverse. In collet; face end and chamfer	Capstan lathe
3	View	View bench
4	In fixture; locating from shank. Gang mill flats and slot	Horizontal milling machine
5	Remove burrs	Burr bench
6	In jig; locating from shank and one flat. Drill $\frac{1}{16}$-in-dia hole in shank	Sensitive drill
7	Remove burrs from hole	Burr bench
8	Final view	View bench

Operation Layout for Special Bolt
Quantity required: continuing batches of 500 off

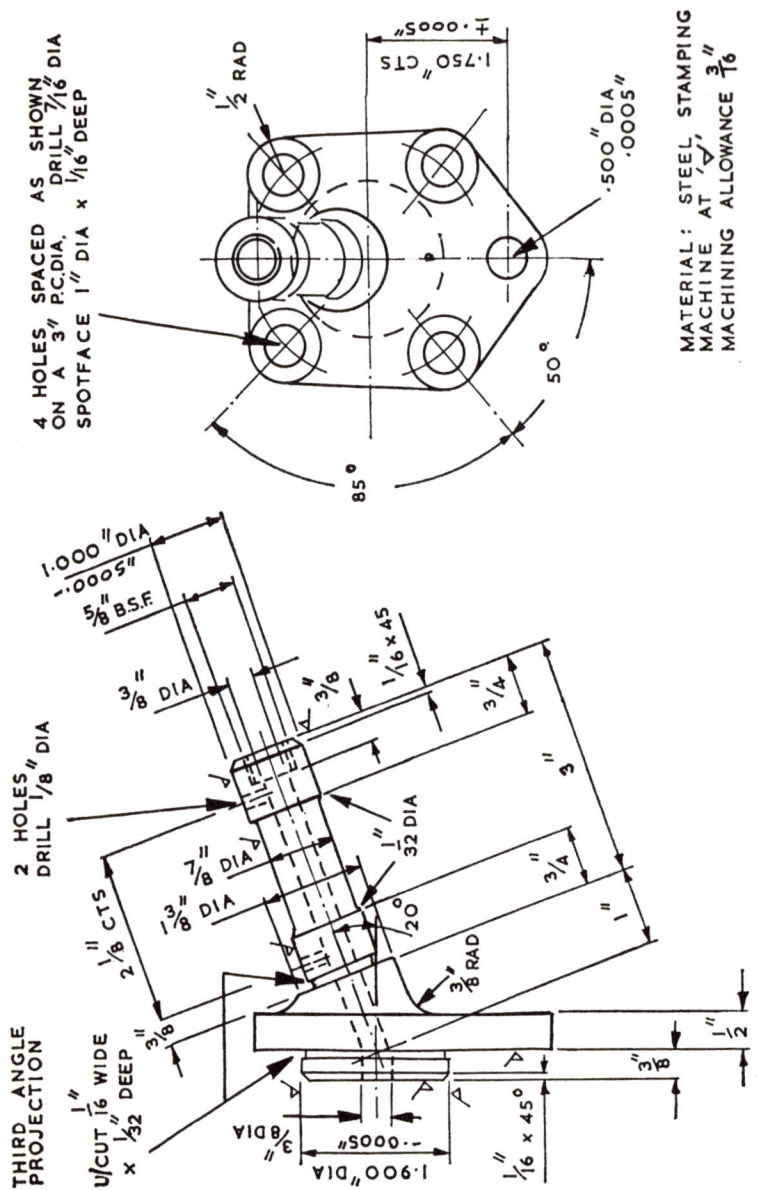

Fig. 4.2 Fulcrum pin

Operation number	Description	Machine
1	In fixture; locate from flange profile. Face spigot end, drill $\frac{3}{8}$ in dia to depth, chamfer and turn spigot diameter 1·915–0·001 in for location	Turret lathe
2	View	View bench
3	Reverse; in jig, locate from spigot and flange profile. Drill and spotface 4 holes $\frac{7}{16}$ in dia. Drill and ream 0·500-in-dia hole	Radial drill
4	Remove burrs	Burr bench
5	In fixture; locate from spigot and 0·500-in-dia hole. Face, drill and tap hole in stem. Chamfer and turn relief diameter and 1·515–0·001 in stem diameter. Face boss to 1·005 $\pm$ 0·005 in for location, and form undercut	Turret lathe
6	View	View bench
7	In jig; locate from spigot and 0·500-in-dia hole. Drill 2 holes $\frac{3}{8}$ in dia	Sensitive drill
8	In fixture; locate from stem and 0·500-in-dia hole. Finish grind stem diameter and flange face	External grinder
9	Reverse; in fixture, locate from spigot and 0·500-in-dia hole. Finish grind stem diameter and boss face	External grinder
10	Remove burrs	Burr bench
11	Final view	View bench

Operation Layout for Fulcrum Pin (see fig. 4.2 on page 54)
Quantity required: 6 000 off in batches of 250 off

Chapter 5

TOOLING AND CAM LAYOUTS

Tooling layouts are used when planning the tooling for capstan, turret and automatic lathes, in order to establish the machining sequence, the equipment to be used, to check that sufficient clearances have been allowed and to provide the machine setter with setting data; in the case of automatics, the tooling layout is also used when designing the cams that are used to produce the required tool movements.

5.1. Capstan and Turret Lathes

Capstan and turret lathes are similar in function and appearance, but, in general, turret lathes are larger than capstan lathes. Both of these machines have a turret instead of a tailstock. The turret is usually hexagonal, and can be indexed from station to station, and a tool, or several tools, can be secured to each face. Most of the work, for example, drilling, boring, screwing and turning, is done from the turret because it is the most convenient way of using the machine, but the cross slide is used for plunge-cutting; the front tool post is a four-way post, and the rear post is a single post, and is used for parting-off. The main difference between capstan lathes and turret lathes is that in the case of the capstan lathe, the turret movement is limited because it is mounted on its own slide, but in the case of the turret lathe, it slides directly on the top of the bed, and its movement is limited only by the length of the bed and the tools themselves.

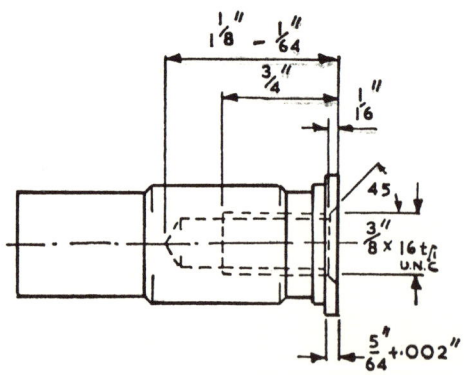

Fig. 5.1 Workpiece—Machine to Dimensions Shown

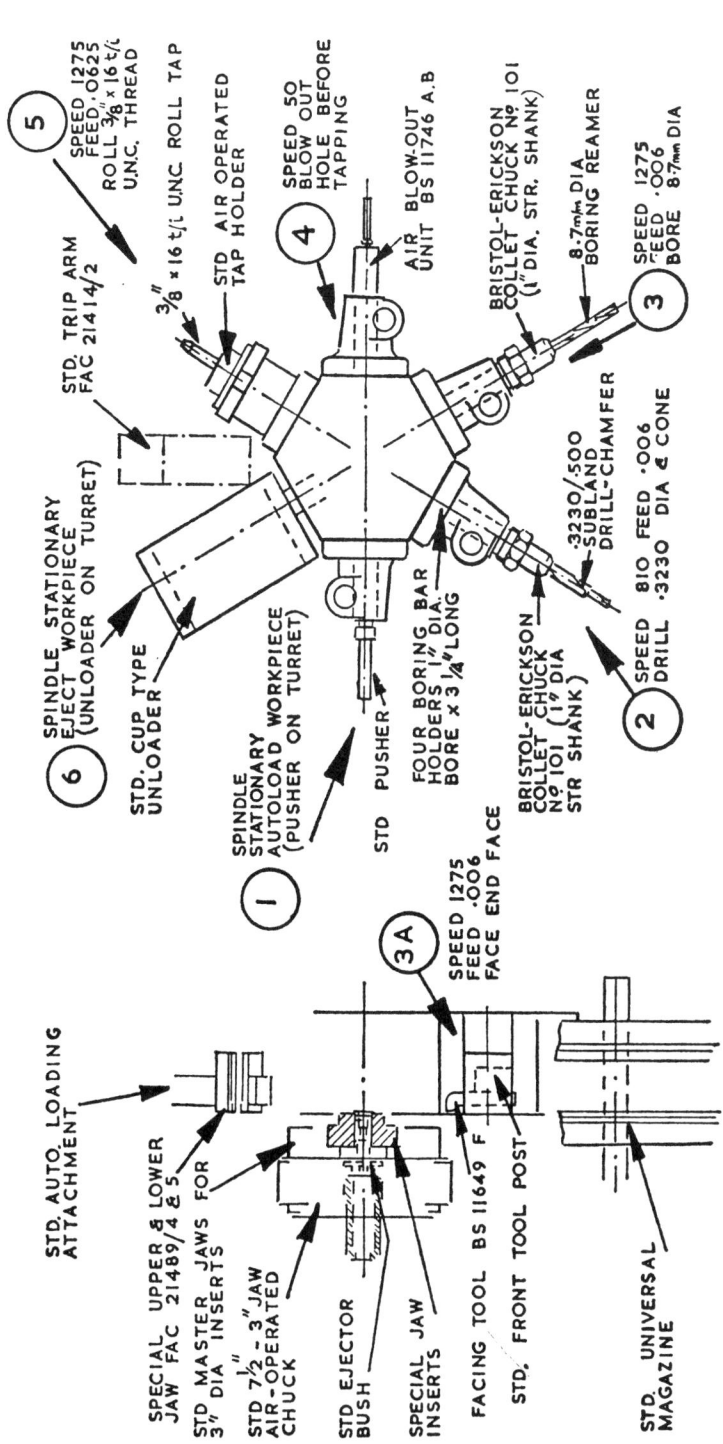

Fig. 5.2 Tooling layout

5.1.1. Fig. 5.1 shows a typical component to be produced using a capstan lathe, fig. 5.3 shows the sequence of operations involved and fig. 5.2 on page 57 shows the tooling layout using a Herbert capstan lathe, fitted with

Operation No.	Speed, r/min	Feed, in/rev	Description
1	0		Autoload workpiece, pushing from turret
2	810	0·006	Drill 0·3230 in dia. to $1\frac{1}{8}{}^{+0}_{-\frac{1}{64}}$ in dim. and form cone
3	1275	0·006	Bore 8·7 mm dia.
3A	1275	0·006	Front cross slide face end to $\frac{5}{64}{}^{+0·002}_{-0·000}$ in dim.
4	50		Blow out hole
5	1275	0·0625	Roll $\frac{3}{8}$ in × 16 T.P.I. UNC thread
6	0		Eject workpiece into cup-type unloader

Fig. 5.3 Sequence of Operations

an air-operated chuck, and standard universal loading system and ejection system.

5.2. The Automatic Screw-Type or Turret-Type Machine

This machine is an automatic version of a capstan lathe; end-cutting is done from the turret, and forming and parting-off from the cross slides.

The turret movement is controlled by the 'lead' cam, which takes the place of the capstan lathe operator working the 'star' wheel; the front and rear slides are independent, and each are operated by a cam.

The B.S.A. Automatic Screw Machines are typical, and the following example is taken from the B.S.A. book 'Tool Layout and Cam Design'; this book describes the procedure for preparing tool layouts and designing the cams, and also gives the recommended speeds, feeds, dwells, etc.; capacity charts are used to obtain details of stroke, clearances, etc. (capacity charts are discussed on page 260).

5.2.1. Tool and cam layout for screw-type automatic machines

Component to be machined

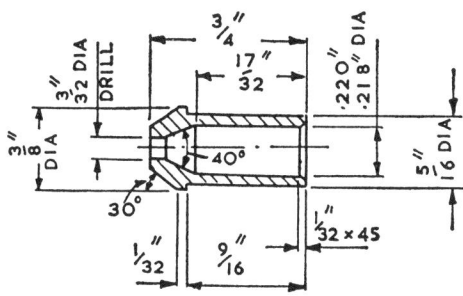

Fig. 5.4 Workpiece

From $\frac{3}{8}$-in-dia brass bar.
Machine: B.S.A. No. 48 Short Stroke Automatic Screw Machine.

Spindle speed. From data: turning, forming and cutting-off speed 600 ft/min (max). The machine can give a spindle speed of 4 995 rev/min (i.e. 490 ft/min surface speed).

Operation sequence (see fig. 5.5).

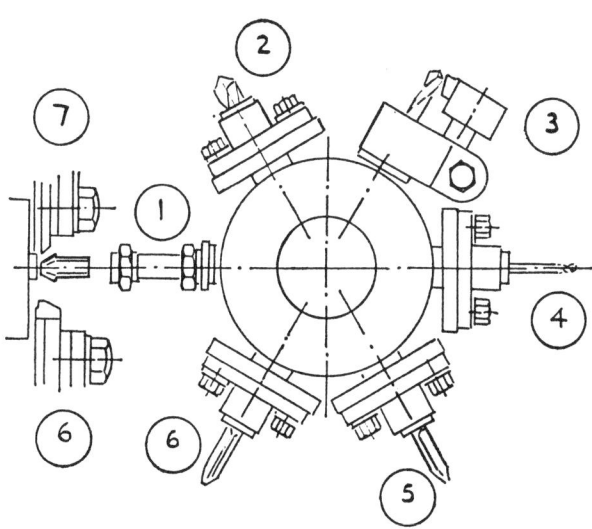

Fig. 5.5 Tool Layout

1. Feed stock to turret stop.
2. Centre drill end of bar.
3. Turn $\frac{5}{16}$-in-dia and drill 0·209-in-dia hole.
4. Drill $\frac{3}{32}$-in-dia hole.

5. Cone bottom of 0·209-in-dia hole.
*6. Ream 0·218-in-dia and cone.
7. Cut-off piece from rear cross slide.

Travel for each tool. The component to be machined is drawn ten times full size, and the various throws are obtained by scaling the drawing; an allowance of between 0·005 and 0·015 in is made to give an approach to avoid damage to the tools or to the component.

Cam design worksheet (see fig. 5.6 on page 61). The tooling and cam details are determined with the aid of a worksheet:

The column headed 'sequence of operations' is filled in first, and then the 'throw in inches' column (using the information obtained from the scale drawing of the component). The 'feed in in/rev' is filled in using the manufacturer's data, and from this the 'revolutions of workpiece for each operation' can be determined:

$$\text{Revolutions per operation} = \frac{\text{Travel or throw of tool}}{\text{Feed per revolution}}$$

The 'stock feed' and 'index turret' data is given as 'time required', but this can easily be converted into 'revolutions of workpiece' when the spindle speed is determined.

The cam blanks are supplied already divided into 100 divisions, and the cam data will need to be in terms of these divisions and not in degrees. The manufacturer's data gives allowances for 'dwell' and 'tool clearances' in terms of these divisions, and these are entered in the last column of the worksheet.

Approximate cycle time. The revolutions required for cutting and indexing operations (not including overlapped operations) are now added together; in this example the total is 516. The hundredths required for 'dwell' and 'tool clearances' are also added together; in this example the total is $7\frac{1}{2}$.

This means that $(100 - 7\frac{1}{2})$, i.e. $92\frac{1}{2}$ hundredths, are available for 516 revolutions.

The approximate revolutions required to make one component

$$= \frac{516}{92\frac{1}{2}} \times 100 = 558.$$

Actual cycle time. The manufacturer's data indicates that when the spindle speed is 4 995 rev/min, the nearest 'revolutions per piece' available using the change gears will be either 541 or 583; in this example the higher one is used. This gives a cycle time of 7 seconds, and a few extra revolutions that can be used to ease the feeds or correct the clearance allowances. At this point the tool clearances are checked; the precise technique is given in the

* The forming operation from the front cross slide is overlapped on the drilling and reaming operations.

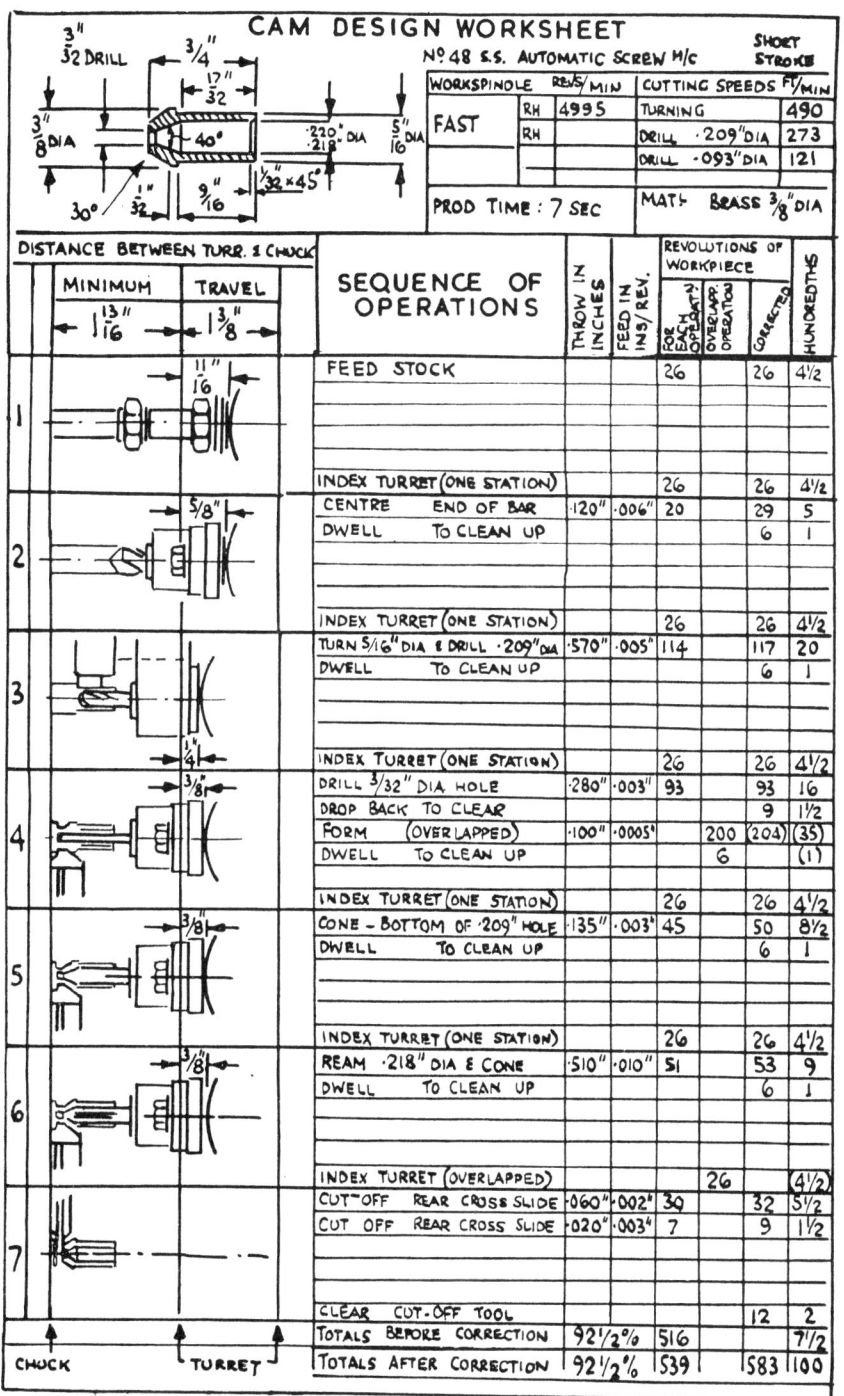

Fig. 5.6 Cam Design Worksheet

manufacturer's data book. In this example the initial allowances are assumed to be acceptable.

Correction of 'Revolutions for each Operation'

It will be recalled that the approximate number of revolutions required was 558, but that the machine could give 583 revolutions to complete one workpiece; it is now necessary to correct the initial figures.

The number of revolutions available for cutting and indexing is:

$92\frac{1}{2}\%$ of $583 = 539$ revolutions (*note:* the approximation was 516)

The number of additional revolutions is, therefore, $539 - 516 = 23$.

These additional revolutions are spread over the cycle as required, and the column headed 'corrected' is filled in.

As a check, the 'dwell' and 'clearance' hundredths are converted into revolutions of workpiece, and entered in the 'corrected' column; the figures in this column must total 583.

Finally, the corrected revolutions of workpiece are converted into hundredths. In this example:

$$\frac{583 \text{ total revolutions of workpiece}}{100} = 5\cdot 83 \text{ revolutions per hundredth}$$

$$\frac{29 \text{ revolutions of workpiece}}{5\cdot 83} = 5 \text{ hundredths}$$

The worksheet is not complete, and shows the throws required and the cam hundredths for each operation (these must total 100).

Drawing the cams. The 'lead' cam will give the maximum travel when a lobe height is maximum for the cam; it is now necessary to determine the 'cut down' for each lobe. To do this, the space headed 'distance between turret and chuck' is used. A line is drawn to represent the chuck face, and two lines representing the extreme positions for the turret faces are drawn as shown on fig. 5.6 on page 61. The completed workpiece is drawn to scale at operation 7 in the parting-off position, so that its position relative to the turret is established. The workpiece is now drawn at each operation with each tool shown in the finishing position. By drawing in the tool holders, the position of the turret face for each operation can be established; the amount that this is short of the maximum position of the turret gives the 'cut down' of each lobe of the 'lead' cam.

It is usual to draw all the cam profiles on one diagram so that the complete cycle can be examined easily; the manufacturers supply drawing blanks marked out in hundredths, and of the correct cam-blank sizes.

The 'lead' cam lobes are first drawn in, starting with the 'cut down' and working back, using the required throw. For example, considering the lobe for 'Centre' at operation No. 2. Fig. 5.6 on page 61 shows that

TOOLING AND CAM LAYOUTS

the 'cut down' is $\frac{5}{8}$ in, and so the maximum lobe height for this operation will beat the '14 hundredths' position, and cut down by $\frac{5}{8}$ in; working backwards, the throw must be 0·120 in and extends over 5 hundredths, and the profile of the lift can be drawn in using the templates provided by the manufacturer. Fig. 5.7 shows the completed cams for this example, and should be studied in conjunction with the worksheet.

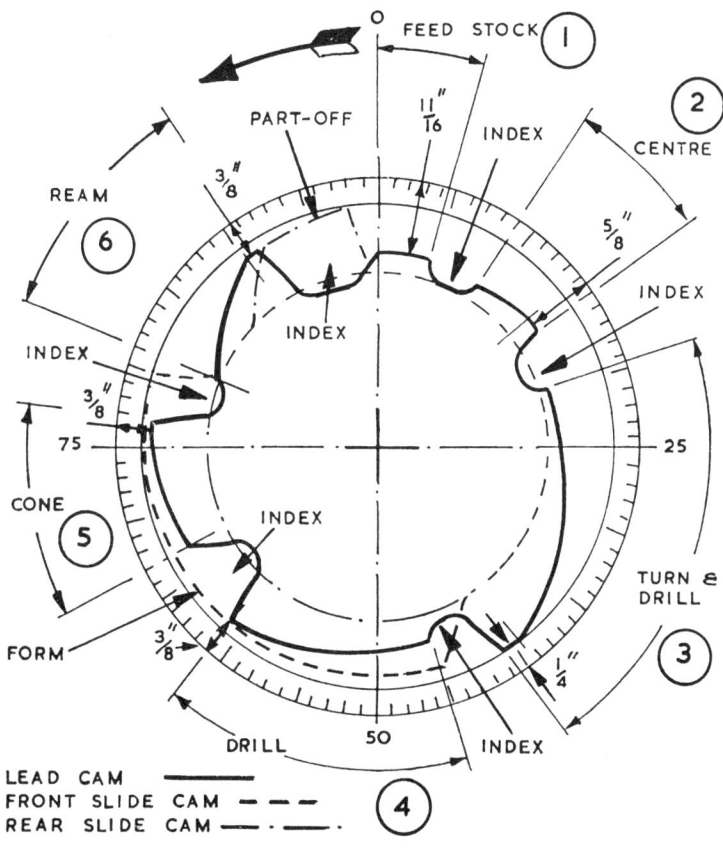

Fig. 5.7 Cam Profiles

Tool and cam layout. The complete information regarding the machine settings, speeds, feeds, tools and cams is presented in the form of a chart.

5.3. The Swiss-Type or Sliding Head Automatic Machine

The Wickman $\frac{7}{16}$-in Sliding Head Precision Automatic is typical of this type of machine, and is used for small, slender, accurate components for clocks, meters, radio equipment and precision instruments.

In this machine the material, in bar form, is gripped in the headstock spindle (which at all times revolves at high speed) and is fed slowly past five radially disposed cutting tools. These tools are made to advance radially into the workpiece for plunge-cutting, and dwell whilst the work advances when a 'feed' is required. Auxiliary attachments can be fitted for special operations.

5.3.1. Main parts:

1. **Sliding headstock**, through which the bar is passed and gripped by collet. After parting-off, the collet is released, and the spring-loaded headstock recoils over the bar stock by an amount equal to the length of the finished component plus the width of the parting-off tool, and then the collet is closed. The headstock is advanced during the operation sequence. The amount of recoil and the headstock advance is controlled by a cam; for small movements a plate cam is used, but for larger movements a bell cam must be used instead.

2. **Tool bracket.** This is close to the headstock for maximum rigidity, and it supports five tool slides (as shown in fig. 5.8) and a guide bush for

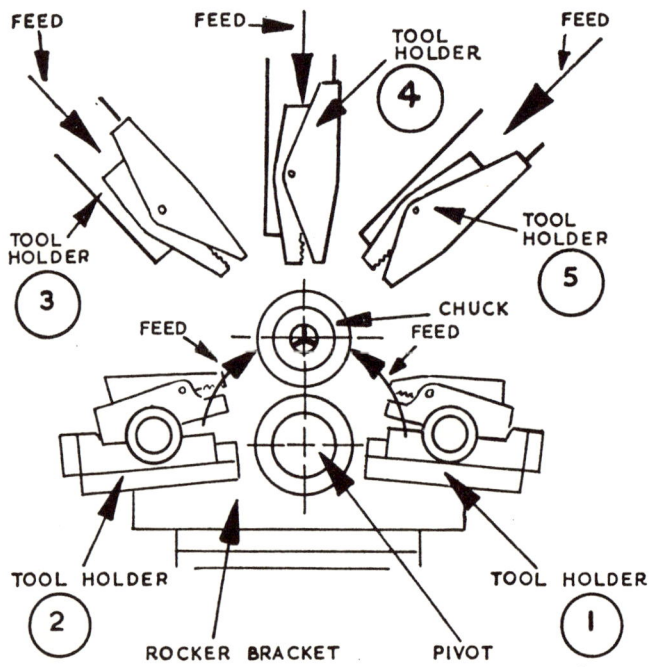

Fig. 5.8 Arrangement of Tools in Wickman $\frac{7}{16}$ in Sliding Head Automatic Machine

the bar stock. Tools 1 and 2 are carried on a rocker so that when one tool enters, the other exits; one cam is therefore used to control both of these two tools. The overhead tools (Nos. 3, 4 and 5) are controlled by three independent cams. All the five tool posts are adjustable using micrometer devices.

TOOLING AND CAM LAYOUTS

3. **Camshaft**, controlling and synchronising bar-stock and cutting-tool movements.

4. **Auxiliary attachments** for performing operations such as drilling, tapping, screwing, screwslotting, etc.

5.3.2. Tool and cam layout for the Wickman $\frac{7}{16}$-in Sliding Head Precision Automatic Machine. (This example is based upon an example given in 'Cam and Tool Design' published by Messrs Wickman Limited. The 'machine data' referred to in this example is taken from that publication.)

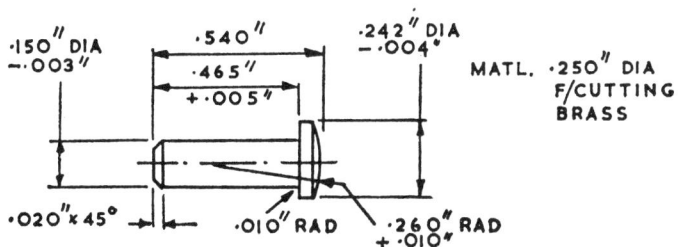

Fig. 5.9 Workpiece

Component to be produced

Tool layout. The workpiece is drawn ten times full size (see fig. 5.10 on page 66); the drawing is dimensioned to show mean sizes, and the diameters are dimensioned as radii. This drawing is used to indicate the cutting tools at the start and the end of each operation, and the movements that are required.

General plan

Operation 1. Bar feed.
Operation 2. Turn 0·150–0·003 in dia.
Operation 3. Turn 0·242–0·004 in dia.
Operation 4. Part-from radius on head and chamfer the next component.
Operation 5. Complete-form radius on head and part-off.

Consider the tools to be used

Tool 1 to be used for the two turning operations (note that this tool is least suitable for forming-in, because its in-feed is by spring-action against the cam). The other operations are forming-in and can be done by any of the remaining tools, and the clearance between them is the main consideration; in general, tools that are opposite each other should be used for successive operations. The tooling plan is then as follows:

Operation 1. Tool No. 4 used as a bar stop.
Operation 2. Tool No. 1.
Operation 3. Tool No. 1.

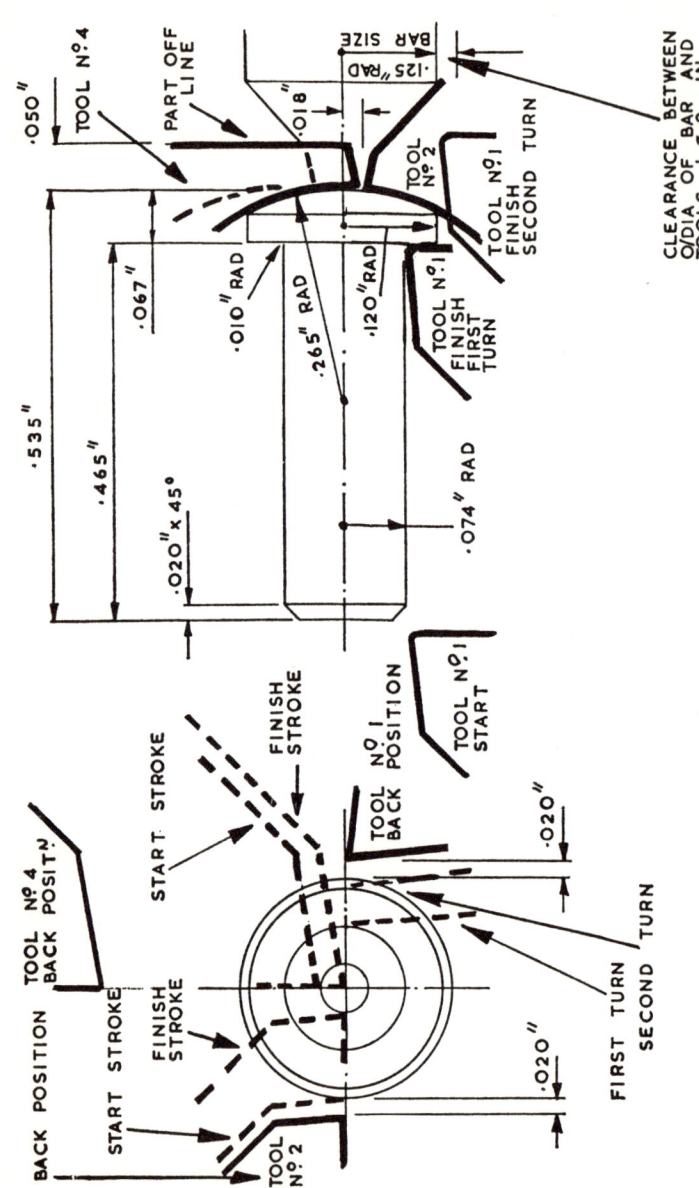

Fig. 5.10 Tooling layout for pin

TOOLING AND CAM LAYOUTS

Operation 4. Tool No. 2 (this is opposite to Tool No. 1, and the entry of this tool can take place as Tool No. 2 exits).

Operation 5. Tool No. 4 (this tool is well clear of Tools Nos. 1 and 2). After parting-off this tool will remain in position to act as a bar stop.

Preliminary tool and cam layout (see pages 68–69)

A layout chart is now produced listing all the movements required to produce the component, together with details of the cam lifts and falls, the workpiece revolutions and cam degrees. Initially it is possible to fill in details of the idle 'time' (expressed in cam degrees) and the productive 'time' (expressed in revolutions of the workpiece). The data used is obtained from the maker's handbook.

It will be seen that some operations or part of operations overlap each other and are not included in the 'time' assessment; it is necessary to determine the details of these operations because cams are required for them.

The 'dwell' at the end of certain cutting operations is introduced to ensure that the cutting is complete and is accurate. Machine data shows that about 10 revolutions of the spindle is best for brass; initially, this is taken to be 10° of cam rotation, but may need to be adjusted when the complete tooling layout is completed (See operations 8 and 16).

The ratio of the leverage that connects the cam follower to the tool slide must be taken into account when each cam is designed; the ratio for each of the tools to be used in this example is shown in the 'cam details' columns of the layout.

The total number of 'idle degrees of cam movement' (provisional at this stage) will be seen to be 88, and that total number of 'productive spindle revolutions' will be seen to be 280; this means that 272° of cam movement are available for 280 revolutions of the spindle (the camshaft rotates once during the production of one component).

The choice of 10° of cam movement to produce a dwell of 10 revolutions of the spindle can now be verified:

$$10° \text{ of cam} = \frac{280}{272} \times 10 = 10 \cdot 3 \text{ revolutions of work spindle}$$

∴ 10° of cam rotation is acceptable.

The workspindle revolutions can now be expressed as cam degrees. For example, considering operation 7:

117 revolutions of workspindle

$$= \frac{272}{280} \times 117 = 113 \cdot 6° \text{ of cam rotation (say 114)}$$

114 is now entered in the 'productive cam degrees' column.

Preliminary Tool and Cam Layout

	Operation	Notes	Cam details	Idle 'time' in cam degrees	Productive 'time' Spindle revolutions	Productive 'time' Cam degrees	Total angular displacement of cam
1	Open chuck	From machine data: 10°		10			10
2	Bar feed	Collet moves back component length plus parting-tool width. From diagram: 0·050 + 0·535 in = 0·585 in. From machine data: 45° required for 1-in-movement ∴ 26° required for 0·585-in-movement	Headstock cam: 0·585 in fall in 26°	26			36
	Tool No. 4 in place as a length stop						
3	Close chuck	From machine data: 15°		15			51
4	Exit Tool No. 4	Tool movement = 0°018 in + 0·125 in + clearance ≑ $\tfrac{5}{32}$ in. From machine data: just over 6°	No. 4 cam: (2¼:1 ratio) cam fall = $\tfrac{5}{32}$ in × 2¼ = $\tfrac{22}{64}$ in in 6°+	6			57
5	Enter Tool No. 1 (to turn 0·074 in radius)	From diagram: tool starts 0·020 in clear. Movement = 0·020 + (0·125 − 0·074) = 0·071 in. From machine data this requires 3° (from 54° to 57°)	Rocker cam: (3:1 ratio) cam fall = 0·071 in × 3 = 0·213 in in 3°. * Not idle—overlap	—			—
6	Pause	To steady tool—allow 2° (usual allowance)		2			59
7	Turn 0·074 in radius (length 0·468 in)	Feeds table gives 0·002–0·010 in for brass—say 0·004 in per rev. for good finish. 'Time' = $\tfrac{0.468}{0.004}$ in = 117 revolutions	Headstock cam: 0·468-in lift		117	114	173
8	Dwell	To ensure good finish and relieve tool spring allow 10°—to be checked later		10			183
9	Exit Tool No. 1	Tool No. 1 turns head diameter as well as body diameter. This operation is to change from 0·074 to 0·120 in radius. Tool moves 0·046 in in 3° (from data)	Rocker cam: (3:1 ratio) cam lift = 0·046 in × 3 = 0·138 in in 3°	3			186

10	Pause	To steady the tool—2° usual allowance		2			188
11	Turn	Distance = 0·585 − 0·468 in = 0·067 in Including 0·050 in parting-off = 0·117 in Feed = 0·005 in 'Time' = $\frac{0\cdot117}{0\cdot005}$ in = 23 revolutions			23	22	210
12	Pause	For complete cutting and finish allow 2°		2			212
13 & 14	Exit Tool No. 1 Enter Tool No. 2	Tool No. 1: from 0·120 to 0·145 in radius = 0·025 Tool No. 2: from 0·020 to 0·005 in clear, i.e. 0·015 in. Total movement = 0·025 + 0·015 in = 0·040 in	Rocker cam: (3:1 ratio) cam lift = 0·040 in × 3 = 0·120 in in 2° (2° from data)	2			214
15	Form-in. Tool No. 2	Part-form head radius and chamfer next component. From practice allow 0·092 in stroke and 0·0012 in feed 'Time' = $\frac{0\cdot092}{0\cdot0012}$ in = 76 revolutions	Rocker cam: (3:1 ratio) cam lift = 0·092 in × 3 = 0·276 in		76	74	288
16	Dwell	For complete cutting and finish allow 10°—to be checked later		10			298
17	Exit Tool No. 2	To return to back position Movement = 0·092 + 0·015 in = 0·107 in * This is an overlapping operation		—			—
18	Enter Tool No. 4	Overlaps with operations 15 and 16		—			—
19	Part-off. Tool No. 4	Complete 'end' of radius and part-off Tool movement 0·064 × 0·001 in feed 'Time' = $\frac{0\cdot064}{0\cdot001}$ in = 64 revolutions	No. 4 cam: (2½:1 ratio) cam lift = 0·064 in × 2½ = 0·160 in		64	62	360
			Totals	88	280	272	360

Each of the spindle revolutions are converted to cam degrees, and adjusted if necessary so that the total is 272.

Finally, a running total is entered in the last column to assist in the cam design, and as a check.

Spindle speed and time cycle

The surface speed for cutting brass is recommended as 450–550 ft/min. From the machine data, the nearest spindle speed available is 7 124 rev/min (this gives a cutting speed of 466 ft/min). Details of the machine change gears to give this are shown in the machine handbook.

The number of spindle revolutions to produce one component

$$= \frac{280}{272} \times 360 = 370\cdot8$$

$\therefore$ The time to produce one component $= \dfrac{370\cdot8 \times 60}{7\ 124} = 3\cdot12$ seconds

The nearest time cycle that the machine can give to this is 3·13 seconds.

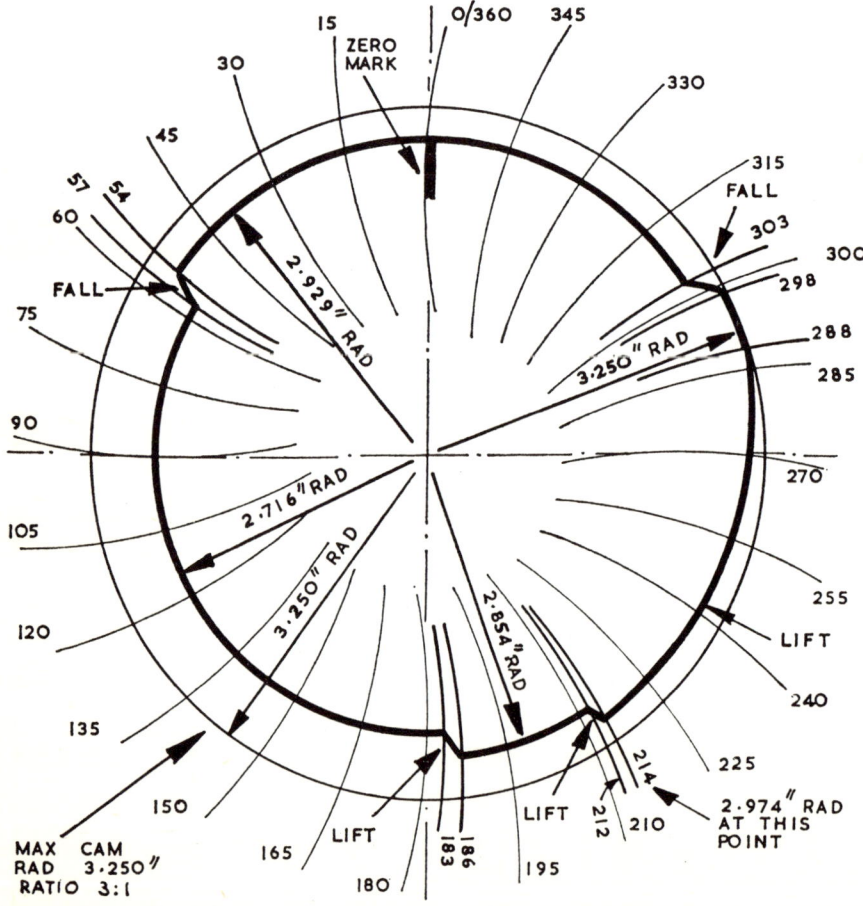

Fig. 5.11 Rocker Cam

TOOLING AND CAM LAYOUTS

Cam design

Cams are required for (*a*) the rocker tools (Nos. 1 and 2), (*b*) Tool No. 4 and (*c*) the headstock.

The Rocker Cam (Tools Nos. 1 and 2)

Note: The machine data gives the maximum cam radius to be 3·250 in. When designing this cam, start at the point where the cam radius is maximum. Take details of the cam lift and fall and angles from the preliminary cam layout (see pages 68–69). A cam lift causes Tool No. 2 to enter and Tool No. 1 to exit.

Operation	Notes	Cam details		
		Deg.	Radius	Range
17 Exit Tool No. 2	Cam falls 0·321 in from 3·250 to 2·929 in radius	5	To 2·929 in radius	298–303
18–4	No movement of Tools Nos. 1 and 2	111	2·929 in radius cam dwell	303–54
5 Enter Tool No. 1	Cam falls 0·213 in from 2·929 to 2·716 in radius	3	To 2·716 in radius	54–57
6–8	No movement of Tools Nos. 1 and 2	126	2·716 in radius cam dwell	57–183
9 Exit Tool No. 1	Cam lifts 0·138 in from 2·716 to 2·854 in radius	3	To 2·854 in radius	183–186
10–12	No movement of Tools Nos. 1 and 2	26	2·854 in radius cam dwell	186–212
13 Exit Tool No. 1 & 14 Enter Tool No. 2	Cam lifts 0·120 in from 2·854 to 2·974 in radius	2	To 2·974 in radius	212–214
15 Enter Tool No. 2	Cam lifts 0·276 in from 2·974 to 3·250 in radius	74	To 3·250 in radius	214–288
16	No movement of Tools No. 1 and 2	10	3·250 in radius cam dwell	288–298

The Rocker Cam is drawn from this information (see fig. 5.11); the follower is of the radial-arm type, and so the lifts and falls are considered along arcs. The follower can be a roller or a toe (or point) type; the cam shown is for a toe follower. The profile shape for lifts and falls is drawn using templates supplied by the machine manufacturers.

Tool No. 4 Cam

Note: The machine data gives the maximum cam radius to be 2·500 in. Again, start at the point where the cam radius is maximum. As there are few changes in the cam profile, there is no need to use a table to lay out the information.

Cam Data (from the preliminary cam layout):

Maximum position 0°–51° (cam dwell).
Lift of 0·160 in—from 2·340 in to 2·500 in radius between 298° and 360°.
Approach to 298° and fall from 51° using manufacturer's templates.

The cam profile is drawn from this information (see fig. 5.12).

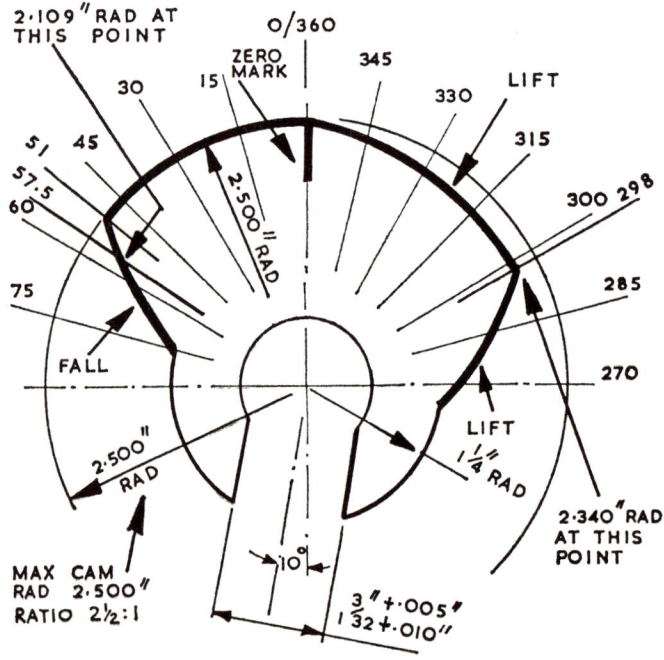

Fig. 5.12 No. 4 Cam

The Headstock Cam

Note: The machine data gives the maximum cam radius to be 3·625 in. Start at the point where the cam radius is maximum (i.e. after operation 1). Take details of the cam lifts and falls and the angles, from the preliminary cam layout (pages 68–9). A cam lift produces 'feed out'. The follower is of the radial-arm type, and can be roller or toe type; the cam shown is for toe-type follower.

Operation	Notes	Cam details		
		Deg.	Radius	Range
2 Bar feed	Cam falls 0·585 in from 3·625 to 3·040 in radius	26	To 3·040 in radius	10–36
4–6	No headstock movement	23	3·040 in radius cam dwell	36–59
7 Turn 0·468 in length	Bar advances. Cam lift 0·468 in from 3·040 to 3·508 in	114	To 3·508 in radius	59–173
8 Dwell: 10° 9 Exit Tool No. 1: 3° 10 Pause: 2°	Cam dwell	15	3·508 in radius cam dwell	173–188
11 Turn 0·117 in length	Bar advances. Cam lift 0·117 in from 3·508 to 3·625 in	22	To 3·625 in radius	188–210
12–1	Cam dwell	160	3·625 in radius cam dwell	210–10

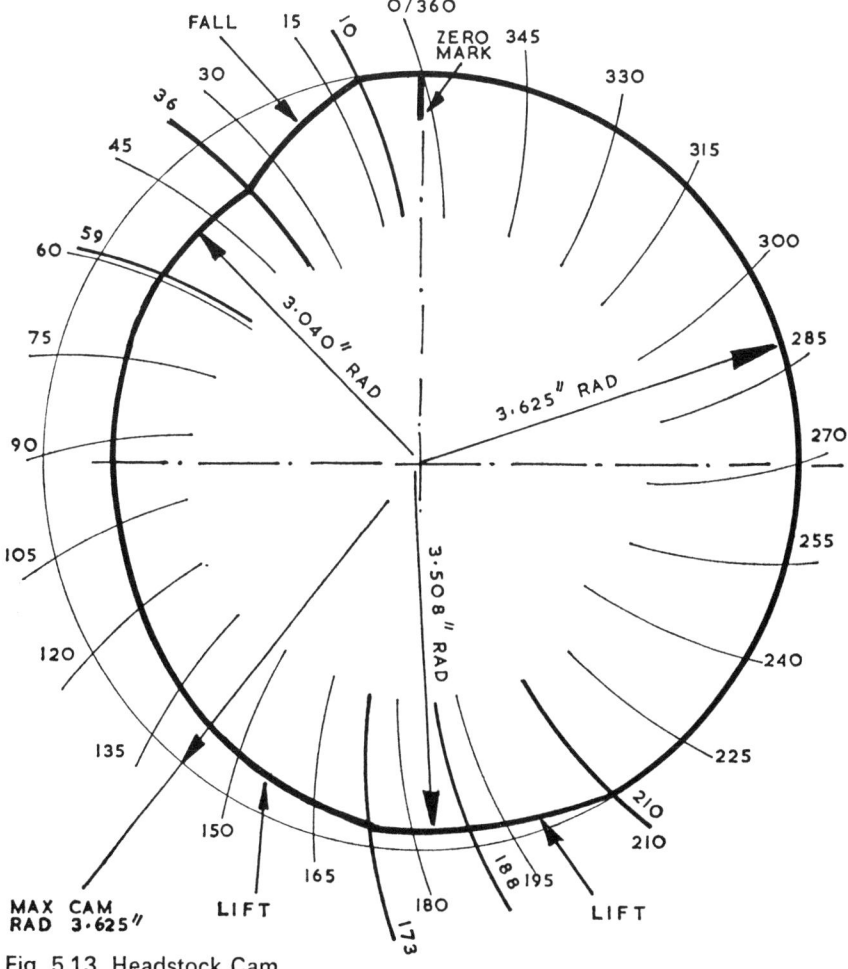

Fig. 5.13 Headstock Cam

Tool and cam layout. The preliminary tool and cam layout is used to compile the final layout. This gives details of the operations, the cam degrees and the spindle revolutions, and also details of the tools and their settings, and of the machine change wheels to produce the required time cycle.

5.4. Multi-Spindle Automatic Machines

The production rate of automatics can be increased by having several spindles on one machine, so that a number of workpieces are machined simultaneously. In these machines the headstock contains a number of spindles, and the entire headstock is indexed from station to station, so that at any time a number of workpieces are being machined, and each workpiece is at a different stage in its production. Multi-spindle machines are available for chuck-work and for bar-work.

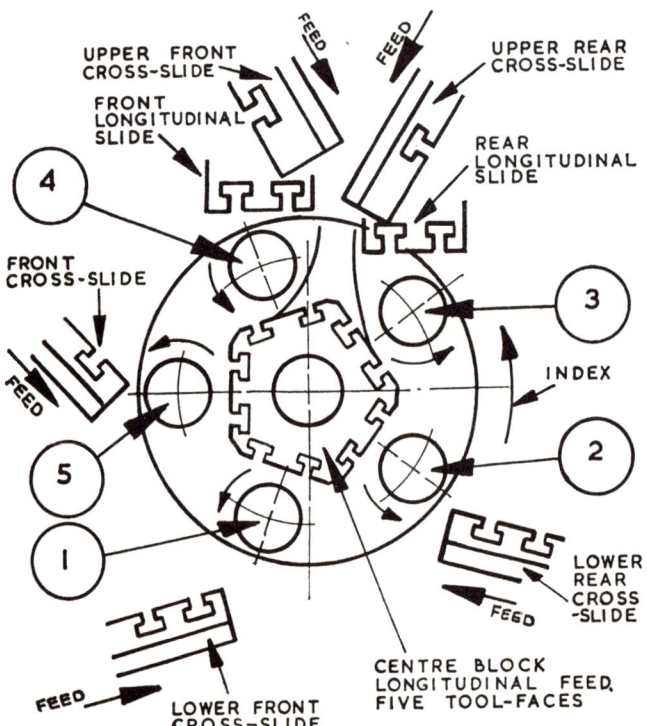

Fig. 5.14 Arrangement of Tools in Wickman 5-Spindle Automatic Machine

The Wickman 5-Spindle Automatic is typical of this type of machine, and the tool arrangement of this type of machine is illustrated in fig. 5.14.

The cross slides are wide and very rigid; the two lower slides are particularly rigid and are used for the heavy primary operations usually allocated to these stations; the front cross slide is used, in bar-work, for parting-off.

TOOLING AND CAM LAYOUTS

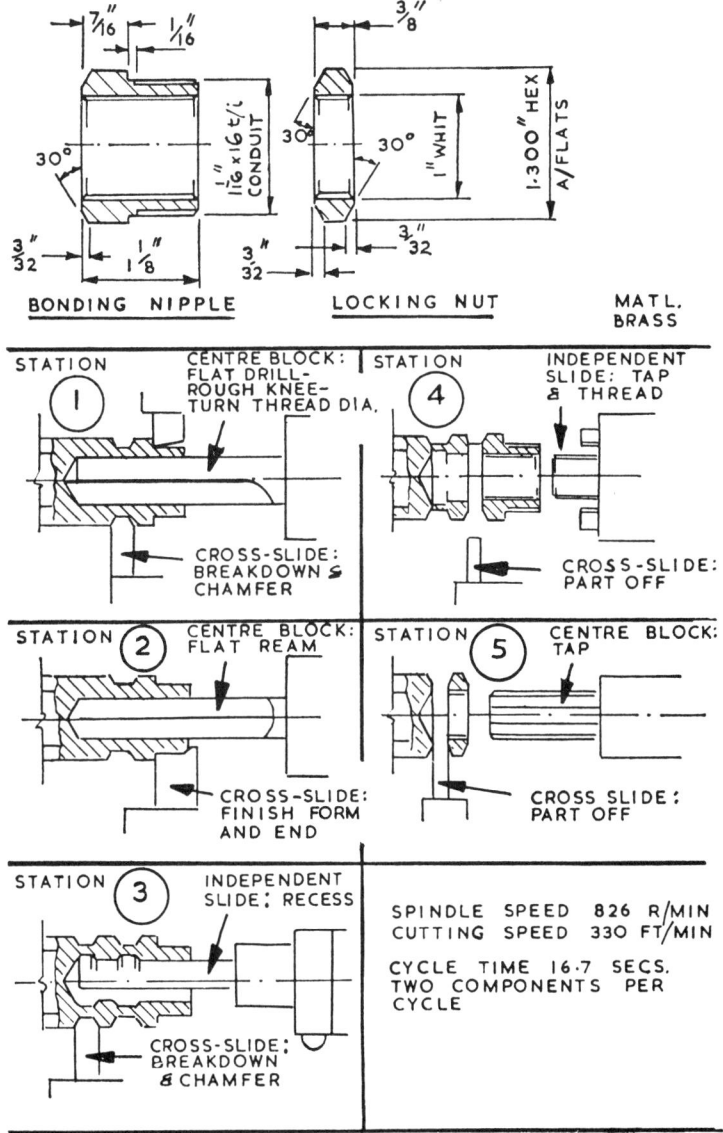

Fig. 5.15 Tooling Layout for Wickman 5-Spindle Automatic Machine

Longitudinal feeds are done by the centre block, and the two independent longitudinal slides at stations 3 and 4. The centre block has five faces, and variation of feed lengths from one station to another is obtained by variation of the tool positions along these faces.

The machine is provided with a simple stroke adjustment and changeable cams are dispensed with, so that automatic machining can be applied to short-run production.

The cycle time is governed by the machining time at the station at which most work is done, and therefore, when planning the sequence of operations, the work should, if possible, be evenly distributed between the five stations. When the time for the 'longest' station has been determined, the time for the other stations will be adjusted accordingly.

Fig. 5.15 on page 75 shows a tooling layout for a five-spindle automatic machine. The material is in the form of hexagonal bar, and one pair of components is produced during one machine cycle.

It will be seen that the heavy work is done at stations 1 and 2, and that parting-off is done at stations 4 and 5.

The complete tooling layout would include details of the tools and settings as well as the operation sequence.

References

'Turret Lathe Work', catalogues and other literature published by Alfred Herbert Ltd.
'Laying-out Tools and Designing Cams', published by B.S.A. Tools Ltd.
'Cam and Tool Design', catalogues and operator's manuals published by Wickman Ltd.

Chapter 6
LOCATION

6.1. Introduction

When a workpiece is placed upon a machine table, spindle or other holding device it must be correctly positioned relative to the table or cutter feed before it is clamped. It is the duty of the location system to provide the necessary constraint so that the workpiece is correctly located; it is also necessary to foolproof the location system so that there is no chance of incorrect location.

6.2. The Six Degrees of Freedom

Fig. 6.1 illustrates a body that is completely free in space. When in this condition the body has six degrees of freedom. Consider the freedom of the body with respect to the three mutually perpendicular axes X–X, Y–Y and Z–Z.

The body can:

1. Move along axis Y–Y
2. Move along axis X–X
3. Move along axis Z–Z

These are the three freedoms of translation.

The body can also:

4. Rotate about axis Y–Y
5. Rotate about axis X–X
6. Rotate about axis Z–Z

These are the three freedoms of rotation.

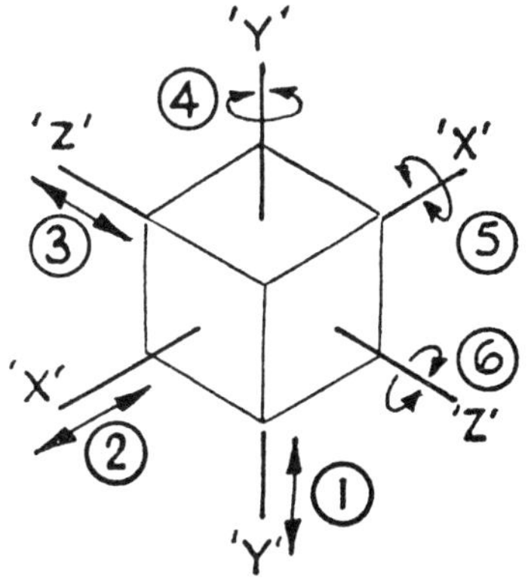

The total number of freedoms is six.

Fig. 6.1 The Six Degrees of Freedom of a Body in Space

It is the duty of the location system to eliminate as many of these freedoms as it is necessary for the operation to be performed with the required accuracy.

6.3. The Extent of Location and the Choice of Location System

The location system used for a particular operation will depend upon the operation to be performed and upon the shape of the workpiece before that operation.

6.3.1. Fig. 6.2 illustrates three stages in the machining of a part; when stage 1 is completed the part is symmetrical about the *X–X* axis, and so it does not need to be positioned about that axis when drilling hole *A* at stage 2. It must, however, be positioned so that the axis of hole *A* will be parallel to the *X–X* axis. When hole *A* has been produced the part is no longer symmetrical about the *X–X* axis and so it must be completely constrained before slot *B* is machined at stage 3.

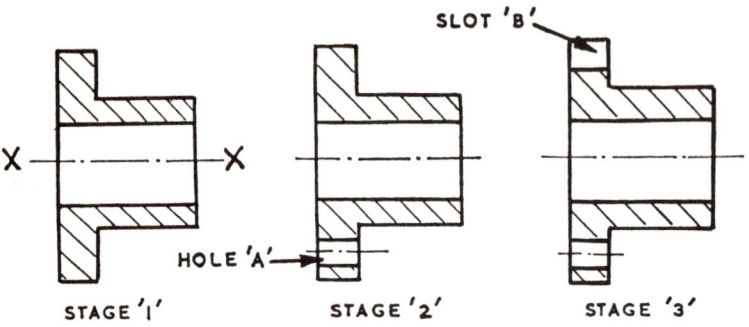

Fig. 6.2 Three Stages in the Machining of a Part

6.3.2. When there is a choice of location points the most effective location system must be selected; the choice will depend upon such features as simplicity of the system, accuracy of the location, ease of location and effect of wear upon the accuracy of the system. When a cylindrical hole or shaft is available as a location point it should be used if possible because cylindrical location eliminates all the freedoms except that of rotation about that of the cylinder (see page 86).

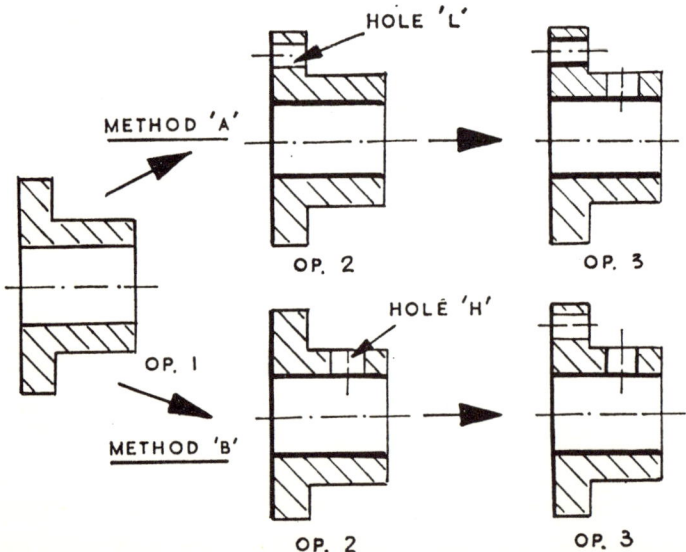

Fig. 6.3 Two Methods of Machining a Part (Locations Shown in Heavy Lines)

LOCATION

Fig. 6.3 shows two methods of machining a part. After operation 1 has been completed that part is symmetrical about the axis of the bore. At this point there is a choice of method; either hole L or hole H can be machined at operation 2. In order to compare these two alternatives it is necessary to consider the location for operation 3. When method A is used the part can be positioned for operation 3 by locating it from the bore and from hole L, and when method B is used it can be positioned by locating it from the bore and hole H. Method A is the better method because the locators will be parallel and easy to see during location; method B is not a good method because the two locators will not be parallel, and so the locator that positions the part about the axis of its bore must be retractable and will wear during use; in addition to this objection, location will be awkward because the operator will not be able to see the location in hole H being done.

6.4. Redundant Location

A redundant location is said to be present when more than one locator is used to constrain a freedom; redundant location must be avoided.

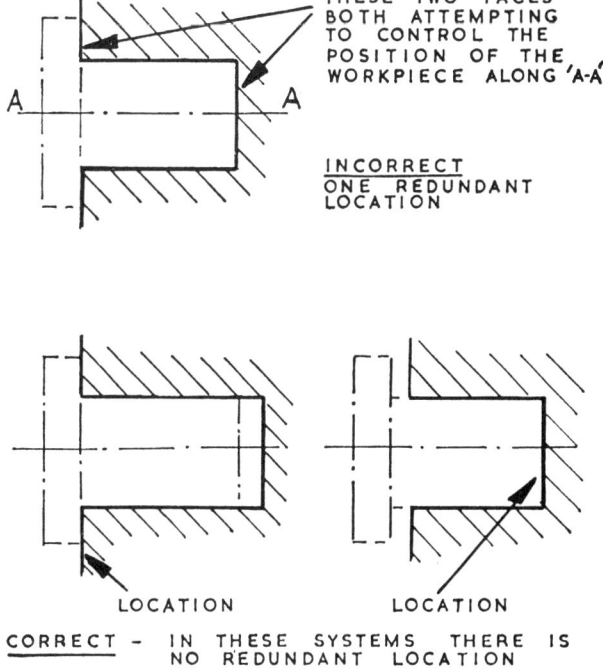

Fig. 6.4 Redundant Location and How It Can Be Avoided

Fig. 6.4 shows a location system that includes a redundant location. The duty of the location system is to position the workpiece relative to axis A–A and to control its position along A–A; in the upper illustration two faces are attempting to control the position of the workpiece along A–A—this would only be successful if the location system and the workpiece were without any error, and so only one of these faces must be used for location, as shown in the lower illustrations. The choice of face used depends upon the dimensional requirements of the workpiece, and whether the workpiece is to be supported under the flange.

Fig. 6.5 illustrates another system that includes a redundant location.

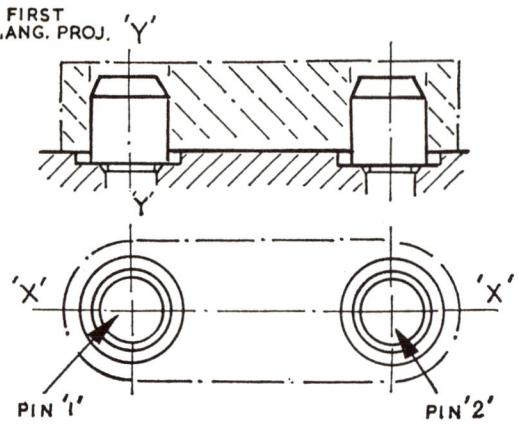

Fig. 6.5 A Location System with Redundant Location

The duty of this system is to control the position of the workpiece using the two holes as location points. Pin 1 will, on its own, constrain all the freedoms except that of rotation about axis $Y-Y$, and it is intended that pin 2 controls this freedom. But as this pin is full form, it too is attempting to control movement along $X-X$, thus introducing a redundant location. Pin 2 must be specially shaped so that it will not influence any freedom except that of rotation about $Y-Y$—fig. 6.28 on page 88 illustrates the solution to this problem.

Fig. 6.6 also illustrates a system that includes redundant location. In this example the workpiece is located from a cylindrical feature and the two fixed flatted locators are attempting to control freedom of rotation about the axis of the cylinder. Two redundant locations are present in this example because the cylindrical location and the flatted locators are both attempting to fix the position of the cylinder axis, and at the same time the two flatted locators are competing for the control of rotation. The workpiece should be located from the cylinder and *one* flat, and to avoid redundant location the flatted locator should be adjustable or alternatively a pin locator be used to control the rotation.

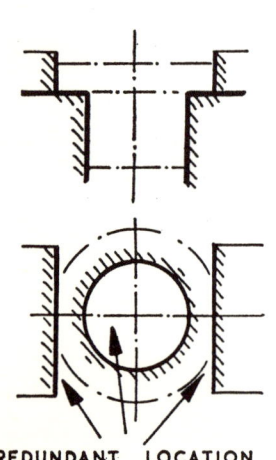

Fig. 6.6 A Location System with Two Redundant Locations

LOCATION 81

It might be argued that redundant location can be avoided by simply introducing a large clearance between the workpiece location point and the locator; this may certainly eliminate redundant location, but it will result in incomplete location, and should be used only for an approximate 'location'.

6.5. Foolproofing

The location system should be such that it will be difficult for the operator to incorrectly load the workpiece into the location. Fig. 6.7 illustrates a

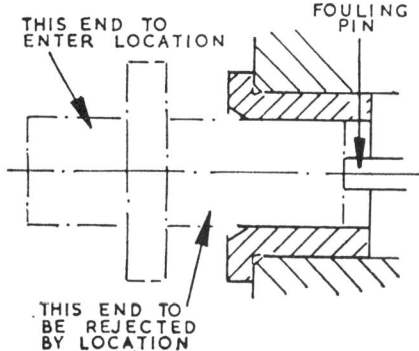

Fig. 6.7 Example of Foolproofing

fouling pin that indicates to the operator that the long end of the workpiece has been put into the location instead of the short end. In this example the foolproofing can be made absolute by designing the clamping system so that the workpiece cannot be clamped when the flange is not

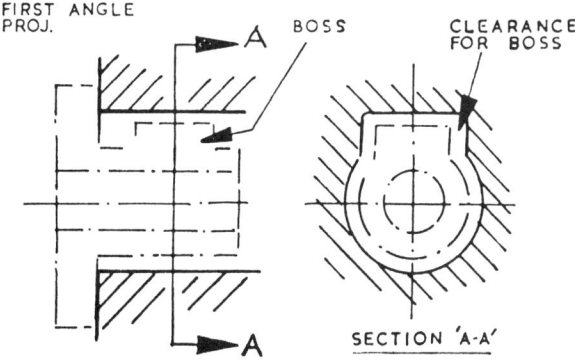

Fig. 6.8 Example of Foolproofing

seated on the face of the bush. Fig. 6.8 illustrates part of a piece of equipment that ensures correct loading relative to the boss by having a clearance for the boss. Fig. 6.9 illustrates a workpiece that has two holes of the same diameter and equi-distant from the principal locator, so that either will be accepted by the second locator. In this example the flange shape demands that the workpiece be accepted in one position only, and so this flange shape is used for foolproofing, by introducing a fouling pin that will not allow the workpiece to be incorrectly loaded. Very often the product designer will assist the production engineer by changing the size or position of holes that will allow incorrect loading, and at the same time foolproof the actual assembly of the product.

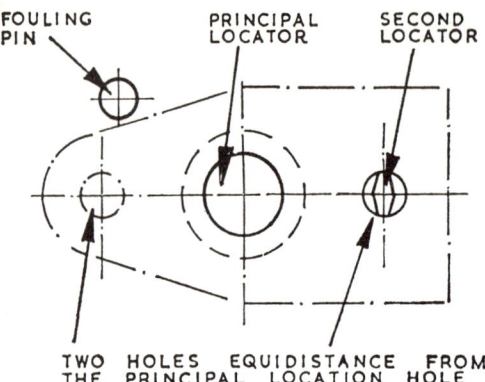

Fig. 6.9 Example of Foolproofing

6.6. The Six-Point Location Principle

Fig. 6.10 illustrates the six-point location principle; in this example the workpiece, shown as a cube, is to be positioned in a corner and retained there by the clamping force acting towards the corner as shown.

Three pads, 1, 2 and 3, are required on the base; these three pads will constrain the workpiece along $Y-Y$, about $Z-Z$ and about $X-X$.

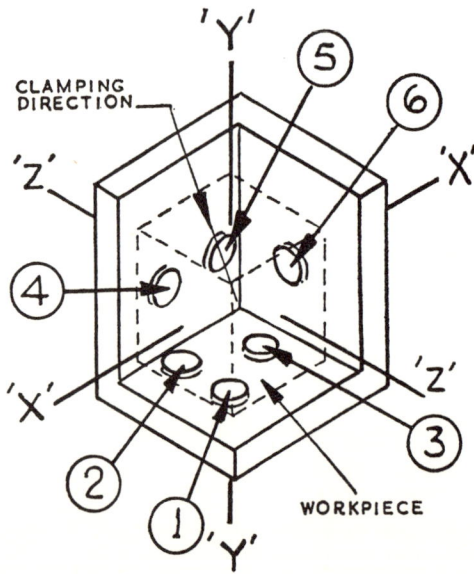

Fig. 6.10 The Six-Point Location Principle

LOCATION 83

The two pads 4 and 5 will constrain the workpiece along Z–Z and about Y–Y.

Pad 6 will constrain the workpiece along X–X. These six pads, positioned as shown, will therefore fully constrain the workpiece.

At the early stages in the machining of a workpiece the surfaces may be uneven and the pads may not contact it properly; this can be overcome by making some of the locators adjustable as shown in figs. 6.12, 6.13 and 6.14.

6.7. Location Devices

Location is obtained by a combination of location from plane surfaces, profiles and cylinders; location devices used in conjunction with each of these types of surface will now be considered.

6.7.1. Location from plane surface. Fig. 6.11 shows a simple pad made from casehardened steel and located in the body of the jig or fixture.

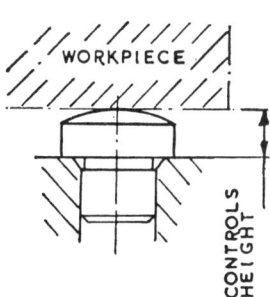

Fig. 6.11 Location Pad

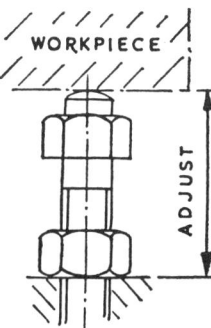

Fig. 6.12 Adjustable Location Pad

'Point' contact is obtained by making the location surface of spherical radius shape. Several locators of this type would be used to produce a location system. When locating from an uneven surface some of the locators will need to be made adjustable; adjustable support points would also be used when plenty of support is necessary and where, if the additional support be made fixed, it would produce redundant location. A very simple adjustable support is shown in fig. 6.12. A more elaborate arrangement is shown in fig. 6.13; in this system the support point can be in a recess or similarly difficult place for the operator to reach, because the adjustment point is some way from the support point. The support can be spring-loaded so that it adjusts itself to the uneven surface; the pin being locked in position as shown in fig. 6.14.

When designing locators care must be taken to ensure that swarf, burrs and dirt will not interfere with the function of the location system. In fig. 6.15 a generous undercut is provided for this purpose, and in fig. 6.16 the same effect is obtained by using pads.

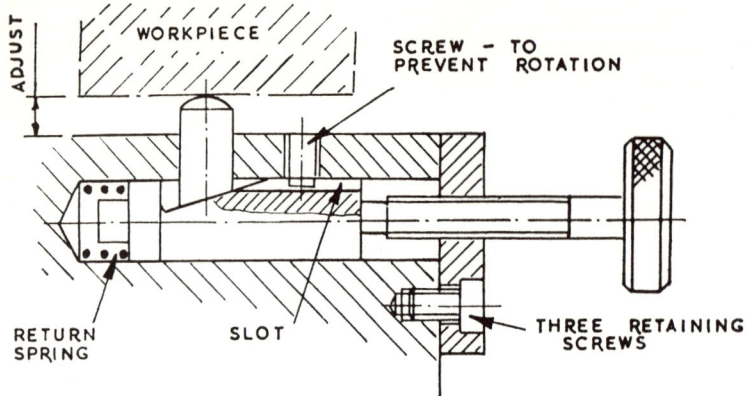

Fig. 6.13 Adjustable Location Pad

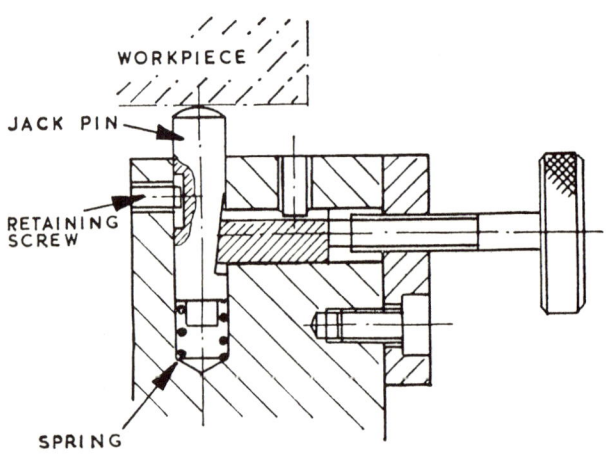

Fig. 6.14 Spring-Loaded Adjustable Pad

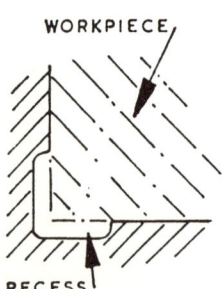

Fig. 6.15 Recess to Prevent Mal-Location

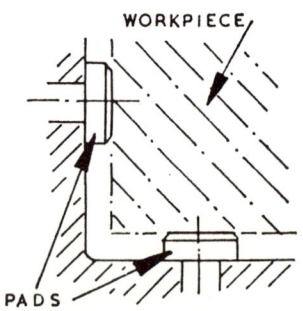

Fig. 6.16 Arrangement of Pads to Prevent Mal-Location

LOCATION

6.7.2. Location from a profile. When, for example, a group of holes is to be drilled in a 'best' position relative to a cast profile, the component can best be positioned on the base of the jig or fixture by using a sighting plate as shown in fig. 6.17.

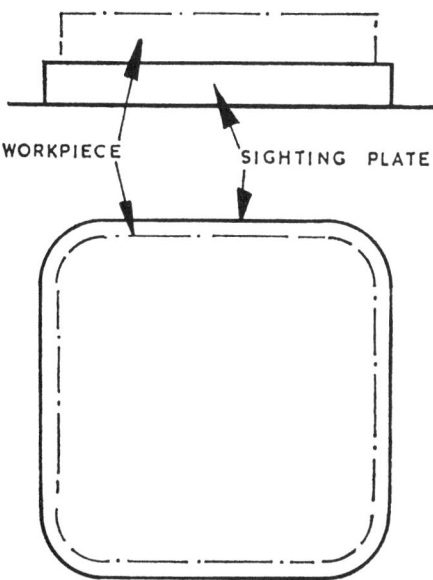

Fig. 6.17 Location by Sighting Plate

At the early machining stages of a workpiece it may be positioned from a profile by means of a series of pins that form the location system as shown in figs. 6.18 and 6.19; the arrangement shown in fig. 6.19 produces a form of 'vee' location. When there is a large amount of variation from batch to batch the pins should be made adjustable if the location system is to be reasonably reliable; fig. 6.20 shows a typical adjustable locator of this type.

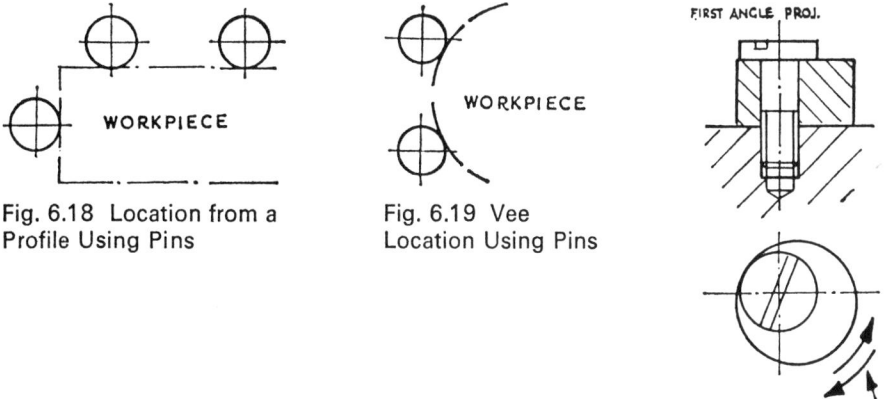

Fig. 6.18 Location from a Profile Using Pins

Fig. 6.19 Vee Location Using Pins

Fig. 6.20 Adjustable Location Pin

6.7.3. Location from a cylinder. Location from a cylinder is the most effective form of location because a cylinder, mounted on a base, will control all freedoms except that of rotation about its own axis. A suitable cylindrical location point can be found in very many workpieces partly because a cylindrical location will be used to locate it upon assembly and

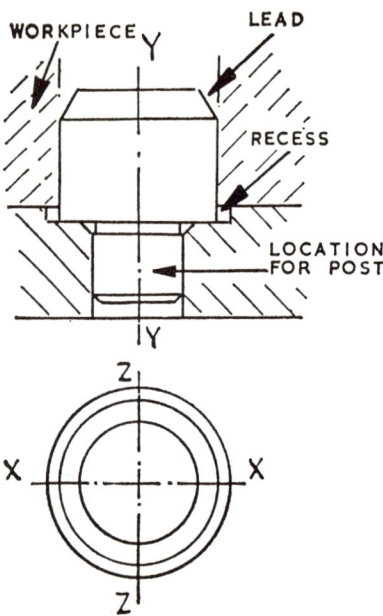

Fig. 6.21 Location Post

partly because cylindrical features will be present where shafts are involved. Fig. 6.21 shows a location post as used to locate a workpiece from a bore or hole; as already stated, the arrangement shown will control all freedoms other than rotation about the Y–Y axis. The post shown is located in the base, and it is shaped so that it seats correctly. It will be seen that the base surface is recessed so that dirt, etc., can fall into it and not prevent accurate seating of the workpiece. The post should be kept as short as possible,

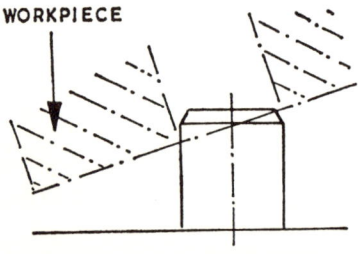

Fig 6.22 Jamming Caused by a Long Location Post and Short Lead

Fig. 6.23 Short Locator and Long Lead Does Not Produce Jamming

and a lead provided as shown. Fig. 6.22 shows the ill-effect of a long post with a short lead; the workpiece can easily be tipped upon loading and jammed as shown; when the post is short and the lead is long, jamming will not occur (see fig. 6.23). When a rather fragile workpiece must be supported over a long length a long post may be required; when designing

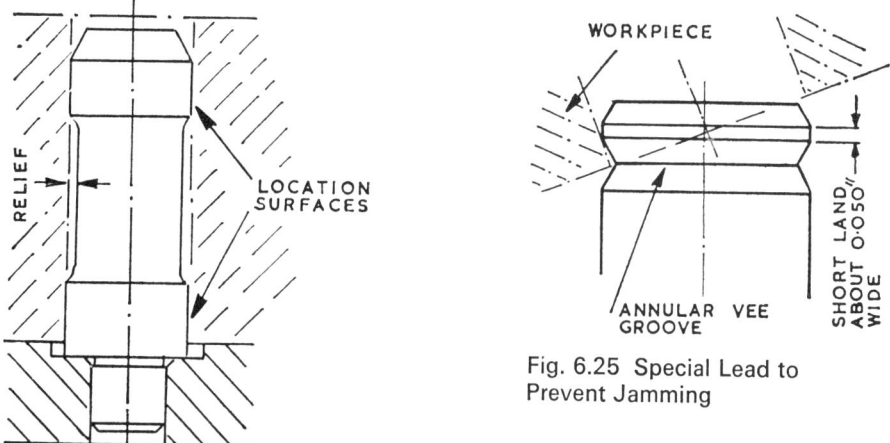

Fig. 6.24 Long Location Post

Fig. 6.25 Special Lead to Prevent Jamming

such a post, care must be taken to prevent jamming and to prevent 'over-location'. Fig. 6.24 shows a long locator, and it will be seen that the lead is very generous and that the post is relieved slightly so that location only occurs at the extreme ends. Fig. 6.25 shows how the jamming problem can be eased by using a specially-shaped lead; it will be seen that the recess will prevent the jamming from taking place.

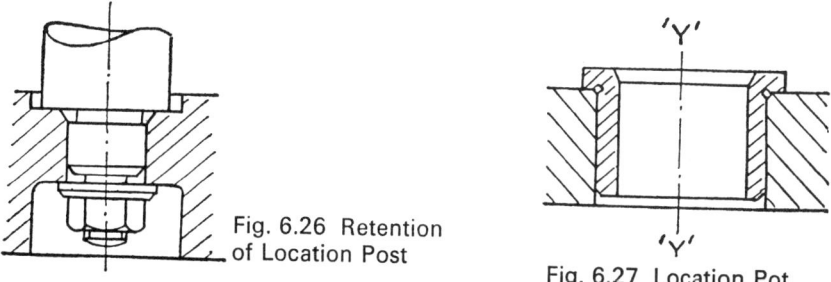

Fig. 6.26 Retention of Location Post

Fig. 6.27 Location Pot

Sometimes the workpiece is clamped using a clamp that is placed on the location post as shown in fig. 8.24 on page 125. Unless the clamping force is very light, the location post must be retained by nut, set-screw, etc., otherwise it will be pulled out by the clamping force. Fig. 6.26 illustrates a typical retention method.

Fig. 6.27 shows pot location for locating from a cylindrical shaft or similar shape. The same principles apply here as to post location.

It has already been explained that one cylindrical locator will constrain all freedoms except that of rotation about its own axis; a second locator is required to produce complete constraint. Very often the second location point is another cylindrical feature, for example, a dowel hole; the second locator is therefore of cylindrical shape also. Care must be taken to avoid redundant location, and the second locator must be shaped so that it

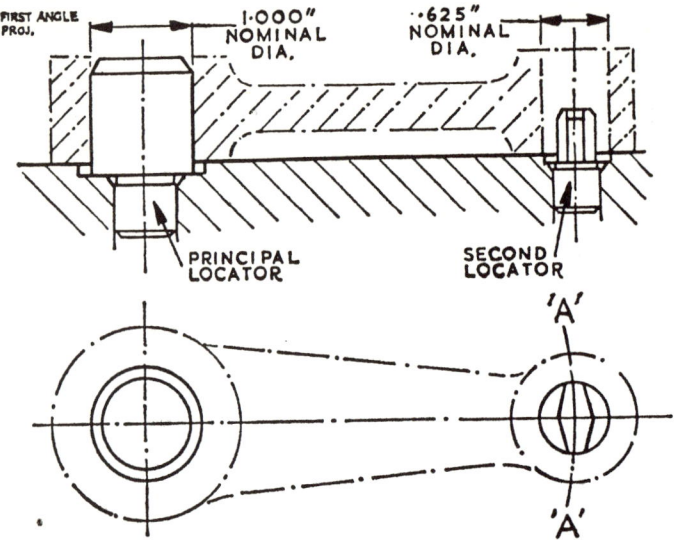

Fig. 6.28 Location from Two Cylindrical Holes (Dimensioned to Indicate Proportions)

controls only the rotation about the principal locator; figs. 6.28 and 6.28 (a) illustrate this point. The principal cylindrical locator should be longer than the second locator so that the workpiece can be positioned on this locator before it is engaged with the second locator.

When several features are available for use as the second location point, the accuracy may well be the deciding factor. Fig. 6.29 shows that if two

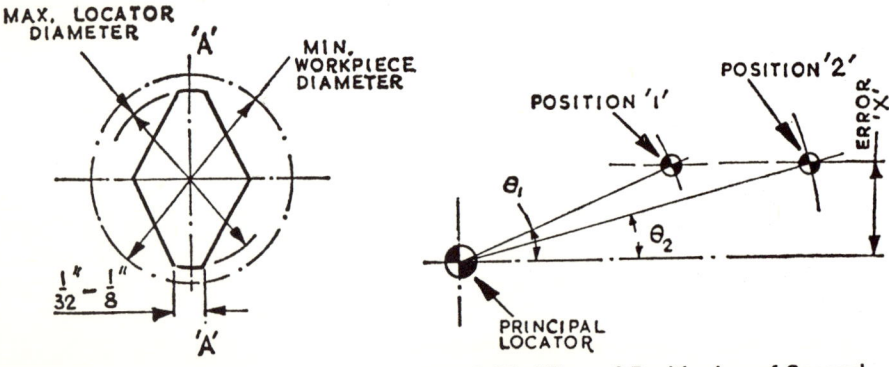

Fig. 6.28 (a) Enlarged View of Diamond-Shaped Pin

Fig. 6.29 Effect of Positioning of Second Locator upon Accuracy of Location

LOCATION

location points 'position 1' and 'position 2' are available, and if they both have a positional error of X, the locator at 'position 2' will produce a more accurate rotational accuracy than the locator at 'position 1'. Sometimes it is necessary to position the workpiece from a conical feature; the disadvantage of using conical location is that location along the cone axis will depend upon the accuracy of the cone itself, and so when conical location must be used some form of adjustment is usually incorporated to allow for this. Conical location is also used when it is necessary to locate from the axis of a cylindrical feature, but when the diameter of that feature is not particularly accurate; a typical example being a hole drilled in a cast circular boss, and where the hole must be drilled in the centre of the boss

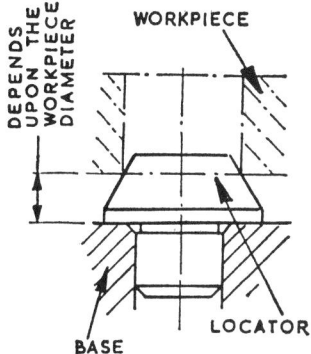
Fig. 6.30 Conical Location

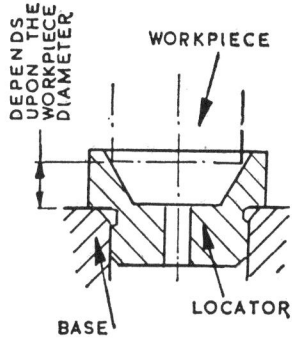
Fig. 6.31 Conical Location

for good appearance. Figs. 6.30 and 6.31 show location from cylindrical features using conical locators; it will be seen that location relative to the axis of the feature is obtained at the expense of position along the axis.

A more complicated conical location system but one which produces accuracy along the axis of the location hole is shown in fig. 6.32; the locator is

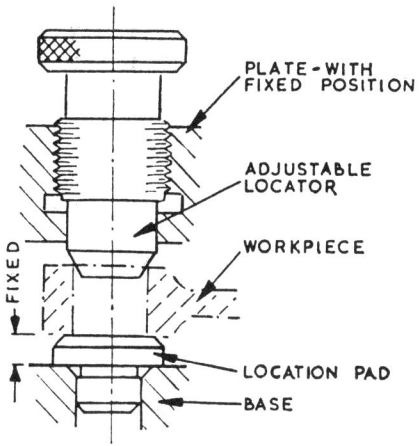
Fig. 6.32 Adjustable Conical Locator

adjustable to compensate for the variation. It must be noted that a screw thread is unsuitable for control of axis position, and so the locator axis is controlled by a plain diameter.

6.7.4. Vee location. This type of location can be used for location from a cylinder or from a part-cylindrical profile. The vee locator can be fixed as shown in fig. 6.33 or sliding. A vee location system can consist of two

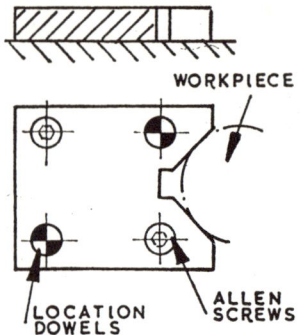

Fig. 6.33 Fixed Vee Locator

fixed vees for an approximate location, or of one fixed and one sliding vee, when the effect of profile variation can be compensated. A vee location can be used as the second locator in conjunction with a cylindrical locator.

Figs. 6.34 and 6.35 show two arrangements for sliding vee location—the

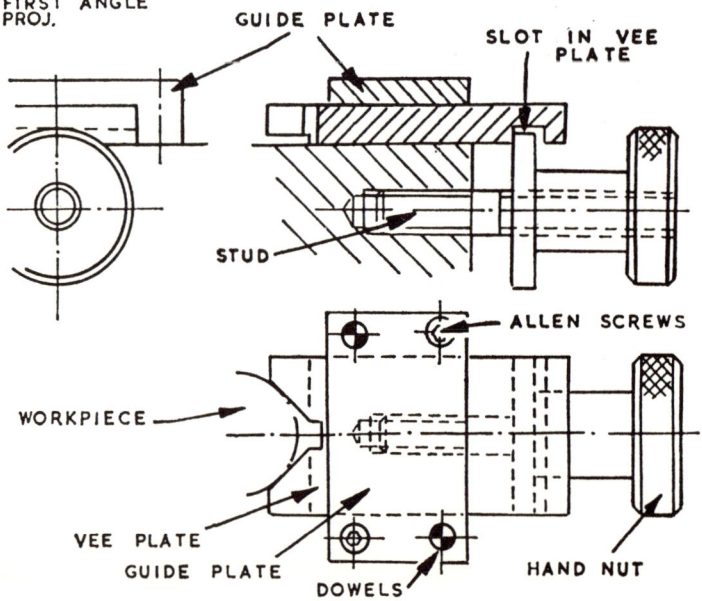

Fig. 6.34 Sliding Vee Location

LOCATION

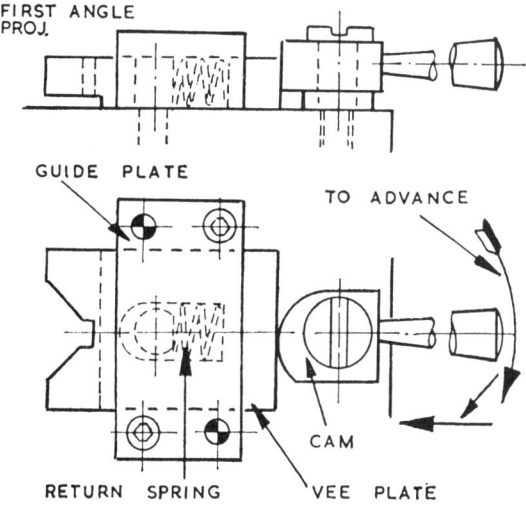

Fig. 6.35 Sliding Vee Location

cam-operated system shown in fig. 6.35 being the more rapid to operate. A vee location system that includes a sliding vee can be made to produce a small downward clamping force by inclining the sides of the vee as shown; the vee plate should be thicker than the workpiece (see fig. 6.36).

When a workpiece is seated in a fixed vee locator its position will depend upon its diameter; care must therefore be taken to ensure that the use of a

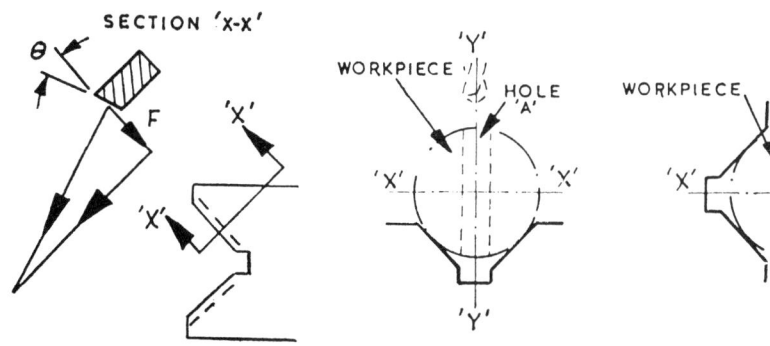

Fig. 6.36 Vee to Produce Small Clamping Force

Fig. 6.37 Correct Application of Vee Location

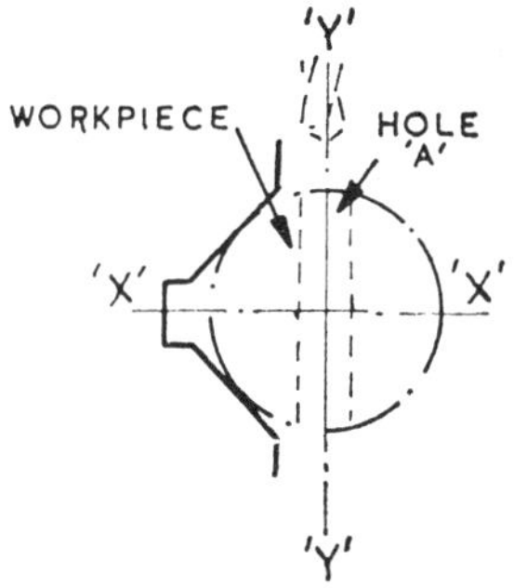

Fig. 6.38 Incorrect Application of Vee Location

vee will not cause errors due to component variation. In the arrangement shown in fig. 6.37, variation of workpiece size will produce variation of the position of workpiece centre along Y–Y but not along X–X; this can be used when drilling hole A with complete success. The arrangement shown in fig. 6.38 is unsuitable because variation of workpiece size will cause variation of the position of the workpiece centre along X–X but not along Y–Y.

6.8. Workpiece Ejectors

If the workpiece is awkward to grasp, heavy, or when the coolant is likely to produce a 'suction effect', some form of ejection will be needed. The position of ejectors is usually considered in relation to the position of the locators, and typical ejectors are therefore included in this chapter.

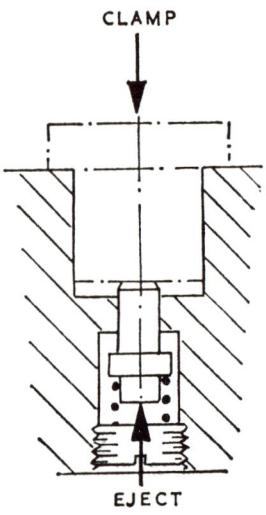

Fig. 6.39 Spring-Operated Ejector

Fig. 6.39 shows a simple spring-loaded ejector; the clamp presses the ejector down against the spring, which operates the ejector when the clamping pressure is removed. Fig. 6.40 shows a trigger-operated ejector; the

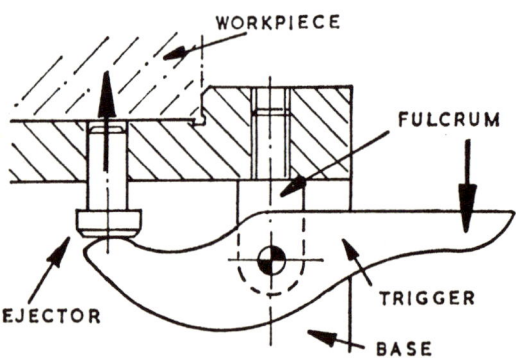

Fig. 6.40 Trigger-Type Ejector

trigger is made out of balance so that it allows the ejector to drop below the location face when the finger pressure is withdrawn; the amount of the trigger movement is restricted so that the ejector will not fall out of position.

Fig. 6.41 shows a push-operated ejector; the plunger returns under the action of the compression spring, and the ejector returns under the action of the light spring. This arrangement can be extended to include several ejectors operated by one plunger as shown in fig. 6.42.

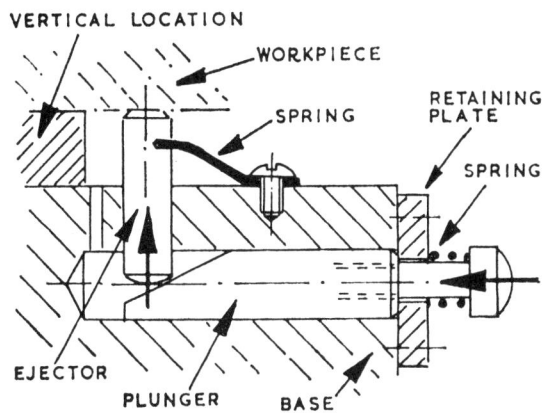

Fig. 6.41 Push-Operated Ejector System

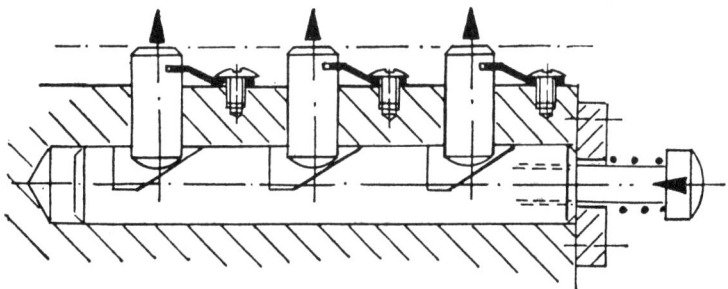

Fig. 6.42 Push-Type Ejector System Adapted for Several Ejectors

Chapter 7

CLAMPING

7.1. Introduction

The clamping system must constrain the workpiece against the cutting forces without damaging it or causing inaccuracies.

7.1.1. Requirements of the clamping system. The clamps must be positioned so that the clamping forces act on supported or rigid parts of the workpiece (see fig. 7.1). The reaction to the clamping and cutting forces must be taken by the main frame of the jig or fixture. Care must be taken to ensure that the clamps can be operated in safety, as quickly as possible and with minimum effort on the part of the operator. The clamps must not be loosened by the vibration caused by the cutting action. The clamping forces must be regulated so that they are adequate, and yet do not cause damage to the workpiece; the force can be regulated by design of the clamp. When the force is exerted by hand nut, the size of the nut can be designed to give the required force.

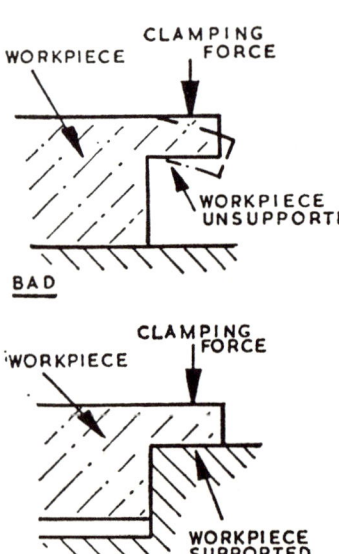

Fig. 7.1 Workpiece Must Be Supported at Point of Clamping

7.1.2. Lever systems

Taking moments about R_A (in fig. 7.2):

$$R_B L - W(L - X) = 0$$

$$R_B = \frac{W}{L}(L - X)$$

When $X = \frac{L}{2}$, $R_A = R_B = \frac{W}{2}$

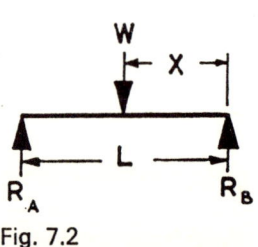

Fig. 7.2

CLAMPING

Taking moments about F (in fig. 7.3):

$$WX - R(L - X) = 0$$

$$R = \frac{WX}{L - X}$$

When $X = \frac{L}{2}$, $R = W$

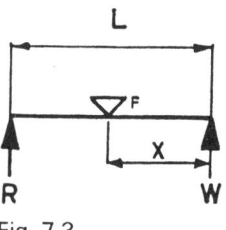
Fig. 7.3

Taking moments about F (in fig. 7.4):

$$WX - RL = 0$$

$$R = \frac{WX}{L}$$

When $X = \frac{L}{2}$, $R = \frac{W}{2}$

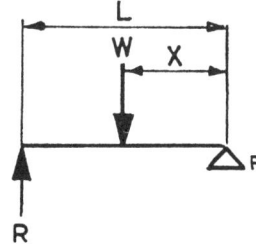
Fig. 7.4

Taking moments about F (in fig. 7.5):

$$RX - WL = 0$$

$$R = \frac{WL}{X}$$

When $X = \frac{L}{2}$, $R = 2W$

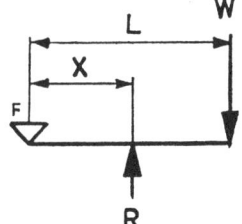
Fig. 7.5

7.2. Plate Clamps

These are simple levers; the clamping force being increased as the stud is placed nearer the toe-end. Figs. 7.6–7.14 show some typical plate clamps.

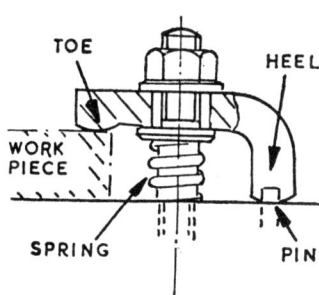

Fig. 7.6 Simple Plate Clamp

Fig. 7.6 shows a simple plate clamp made from a bent plate. The 'toe' and the 'heel' are shaped so that the clamp will seat correctly upon the workpiece and the base, even when the height of the workpiece varies. A spring is placed between the clamp plate and the base so that the clamp can be operated easily, and also to allow the swarf to be cleaned away without holding the clamp up by hand; the washer between the clamp plate and the spring prevents the latter from entering the hole in the clamp plate. In this example the clamp is rotated about the stud axis to allow the workpiece to be removed, and the pin in the base prevents the clamp plate from rotating during the clamping operation.

Fig. 7.7 shows a flat plate clamp; in this system the clamp is supported by a heel pin. This clamp is also rotated about the stud axis to remove the workpiece and held against the rotational forces by the heel pin engaging in the plate.

Fig. 7.7 Plate Clamp with Heel Pin

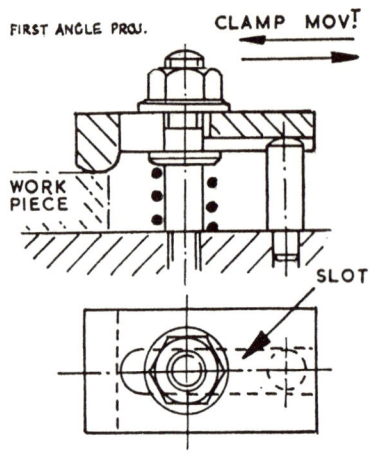

Fig. 7.8 Plate Clamp with Heel Pin (Slotted Type)

Fig. 7.8 also shows a plate clamp with heel pin, but in this arrangement the heel pin engages in a slot in the clamp plate to produce a better clamp location.

Large variations in workpiece height can be compensated for by having an adjustable heel pin as shown in fig. 7.9.

Fig. 7.9 Plate Clamp with Adjustable Heel Pin

Fig. 7.10 Application of Spherical Washers

Smaller variations in workpiece height can be allowed for by using a pair of spherical washers so that the clamp plate can be at an angle to the stud axis and still permit the nut to clamp it correctly (see fig. 7.10).

As an alternative to the slot shown in the previous examples, the plate can be stepped as shown in fig. 7.11 so that it can be rotated to remove the workpiece.

CLAMPING

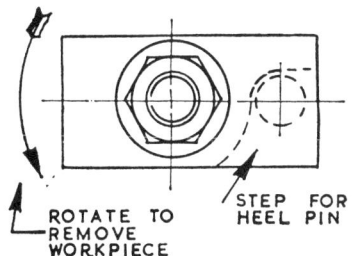
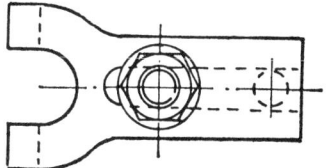

Fig. 7.11 Alternative Form of Plate Clamp (Swinging Type)

Fig. 7.12 Two-Point Clamp

The clamp can be shaped as shown in fig. 7.12 to suit the workpiece.

Large workpieces can be clamped using 'spider' clamps as shown in fig. 7.13; this clamp can be made up from a cylindrical boss, to which is welded the three arms made up from 'Tee' section for good stiffness with minimum weight.

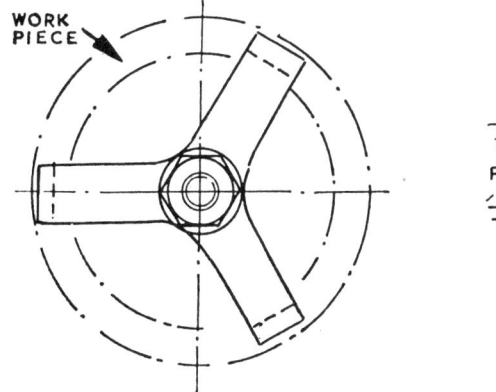

Fig. 7.13 Spider Clamp

Fig. 7.14 Edge Clamp

Fig. 7.14 shows a wedge-type edge clamp used when the only horizontal surface is that to be machined, and so a vertical face must be used for clamping.

7.2.1. The clamps shown so far have all been secured by hexagonal nut and spanner; this is acceptable when a high clamping force is necessary as when milling and when turning, but a hand-operated nut is better if it permits the required force to be applied. The hand nut should be large enough to allow the operator to grip it in comfort. The hand nut can be machined from bar, but the manufacturers of 'unit tooling parts' market small cast hand nuts of the type shown in figs. 7.15 and 7.16.

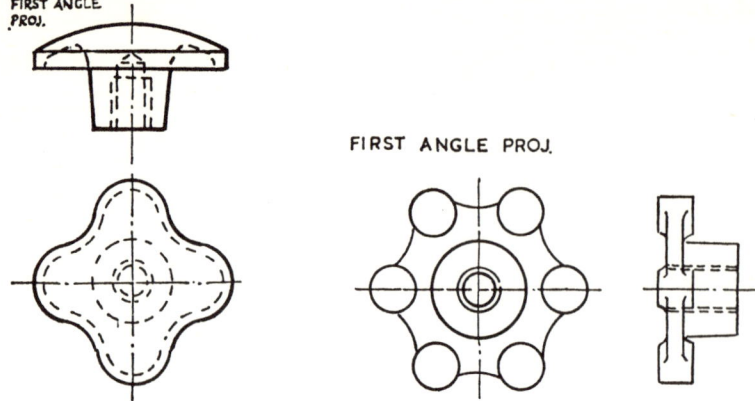

Fig. 7.15 Cast Hand Nut   Fig. 7.16 Cast Hand Nut

7.3. Pivoted and Latch-Type Clamps

Figs. 7.17 and 7.18 show two typical pivoted clamps; these clamps allow more space around the cutting area because the clamping screws are set some distance away from the point of clamping.

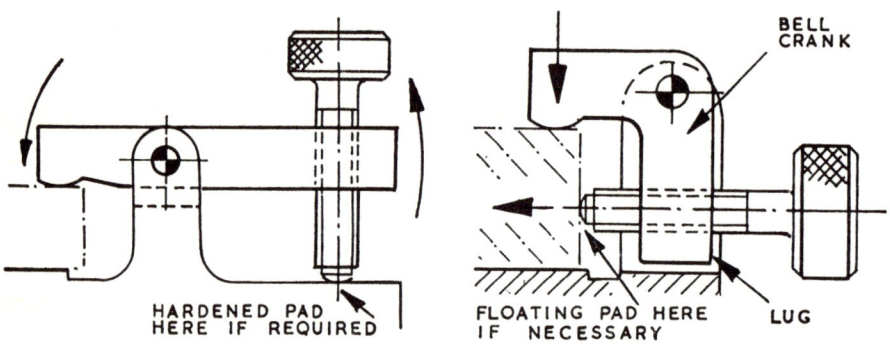

Fig. 7.17 Pivoted Clamp   Fig. 7.18 Pivoted Clamp

Fig. 7.19 shows a latch-type clamp; to release the workpiece, the nut is slackened off and the bolt swung away as shown, to allow the hinged clamp to be swung away as shown until face X on the clamp meets face X on the body. The clamping pad is shaped so that variation of workpiece height will not prevent correct clamp seating. The quarter-turn thumb screw shown in fig. 7.20 can be used instead of a swinging bolt.

The latch-type clamp can be adapted to produce two-way clamping as shown in fig. 7.21. The swinging bolt is attached to a link member that is in turn attached to the body. When the clamping nut is tightened it causes this link member to rotate about the fixed pin so that a horizontal clamping force is applied to the workpiece. When this clamping is completed, further tightening will produce a vertical clamping force.

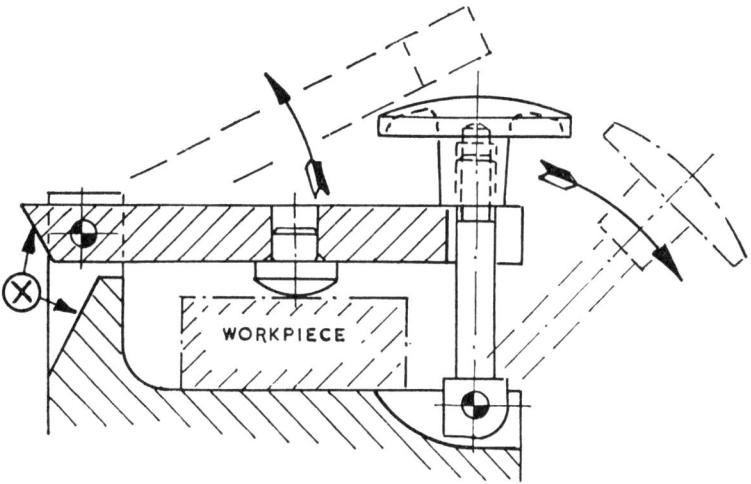

Fig. 7.19 Latch-Type Clamp

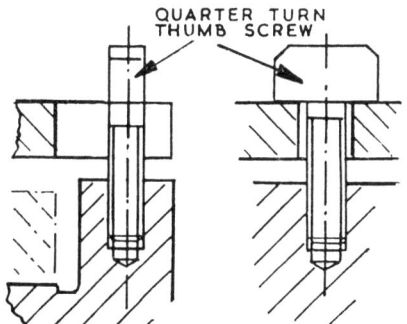

Fig. 7.20 Quarter-Turn Thumb Screw Applied to a Latch-Type Clamp

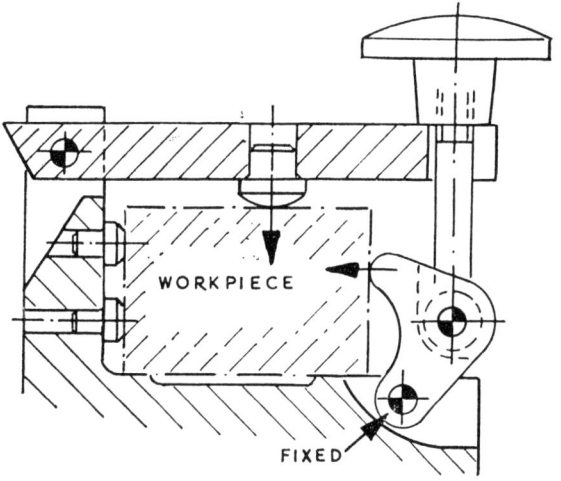

Fig. 7.21 Two-Way Clamp

7.4. Direct Clamping

The button clamp shown in fig. 7.22 is a simple direct clamping device that can be swung away to allow the workpiece to be removed. When the clamping strap must be removed completely one of the arrangements shown in fig. 7.23 can be used.

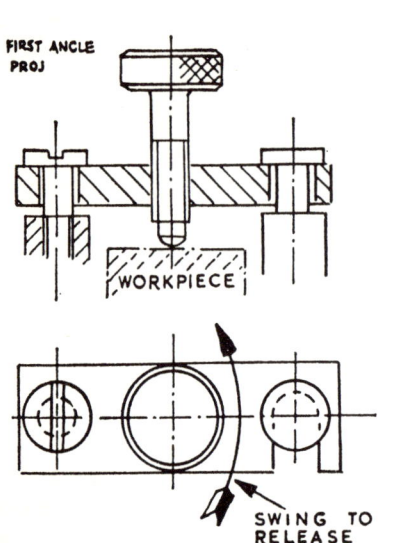

Fig. 7.22 Button Clamp

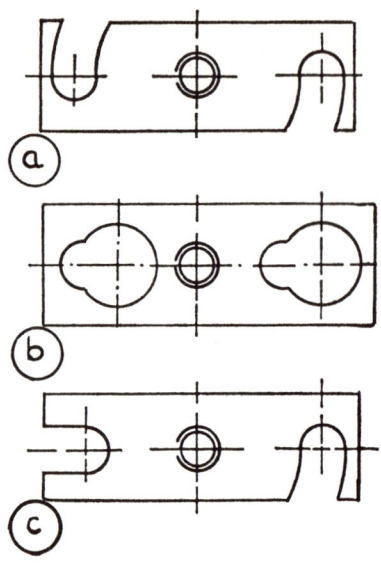

Fig. 7.23 Alternative Forms of Button Clamp

Fig. 7.24 shows another form of direct clamping device; the clamp body is supported by a post, and can be swung away to allow the workpiece to be removed.

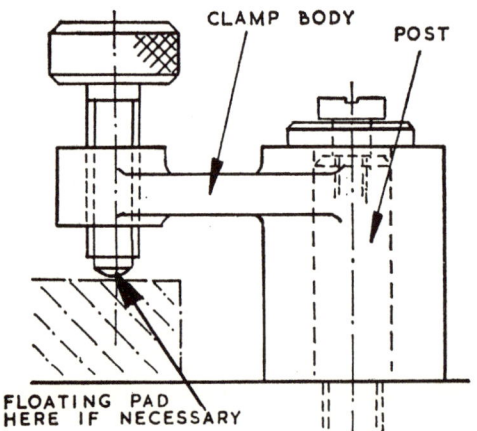

Fig. 7.24 Swinging Clamp

CLAMPING

7.4.1. Floating pads. When the clamping screw contacts the workpiece directly it may damage it. When it is necessary to protect the workpiece a floating pad can be fitted to the end of the clamping screw as shown in fig. 7.25. The pad will be stationary when the screw is finally clamping the

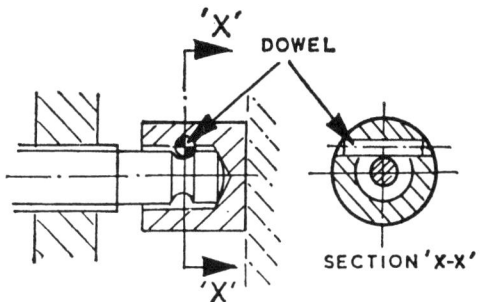

Fig. 7.25 Floating Pad

workpiece—the end of the screw pressing upon the bottom of the hole in the pad. The dowel is used only to prevent the pad from falling from the screw when released. The pad will adjust itself to accommodate variation in workpiece flatness.

7.4.2. Direct clamping using a post. A workpiece with a bore can be clamped directly by using a post, washer and nut. This arrangement can use a drill plate as part of the clamping system as shown in fig. 8.24 on page 125. As

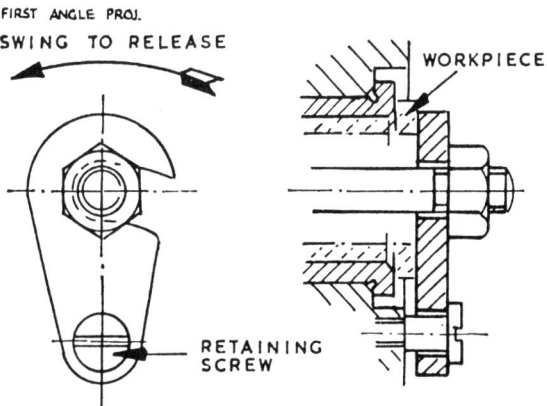

Fig. 7.26 Application of Captive Washer

removal of the nut is time-consuming, the clamping system should allow it to remain on the post when the workpiece is removed. When the bore of the workpiece is large enough to allow the nut to pass through it, the nut can remain on the post, and a washer inserted between it and the workpiece to allow the clamping to be done; fig. 7.26 shows an arrangement using a captive washer that can be swung away so that the workpiece can be removed; the slot in the washer is positioned so that as the nut is rotated the washer is rotated about the retaining screw to keep it engaged. When a captive washer cannot be incorporated a 'Cee' washer is used as

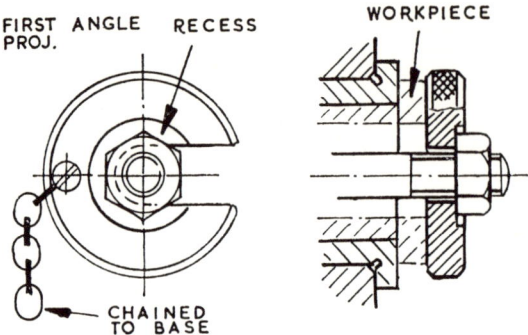

Fig. 7.27 Application of 'Cee' Washer

shown in fig. 7.27. This type of washer is often chained to the base so that it can be located quickly. When the post is horizontal the washer is recessed so that the nut prevents it from falling away when the slot is in the vertical position.

7.4.3. Quick-action nuts. When the bore of the workpiece is too small to allow the workpiece to be removed with the nut still on the post, the latter must be removed. To reduce the time required to remove the nut a quick-action nut is used. The nut shown in fig. 7.28 is typical of the type marketed by the manufacturers of 'unit tooling parts'; this nut has a threaded bore and a second, plain bore

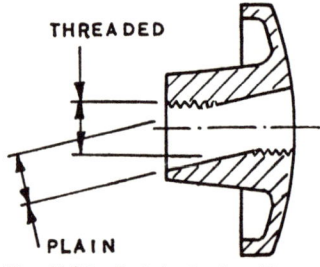

Fig. 7.28 Quick-Action Nut

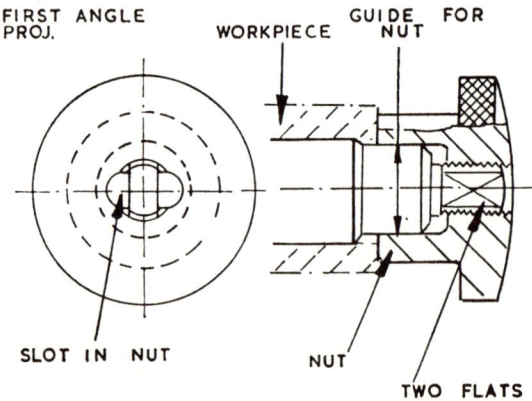

Fig. 7.29 Quick-Action Nut

CLAMPING   103

machined at a small angle to it to produce an elongated-hole effect. When placing the nut in position it can be inclined so that the plain hole (slightly larger than the thread size) can be passed over the post; the nut is then inclined so that the threaded portion engages with the screwed part of the post to produce the clamping effect. An alternative system is to machine two flats on the threaded part of the post, and to machine an elongated hole in the nut as shown in fig. 7.29; the nut can be positioned by aligning the slot in the nut with the flats on the post, and then rotating the nut to produce the clamping.

7.4. Hook Bolts

When there is little room available, a hook bolt can be used as shown in fig. 7.30; this bolt can be rotated to free the workpiece. Fig. 7.31 shows a hook bolt operated by an eccentric; this allows very rapid clamping.

7.5. Clamping Plates

Fig. 7.32 shows a clamping plate that is secured by two swing bolts. This system is awkward to use but is useful when the shape of the workpiece

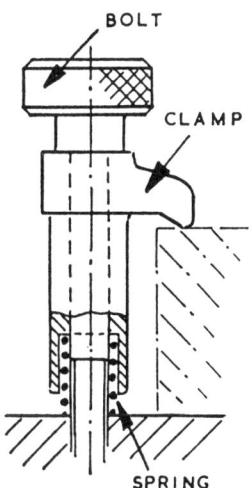

Fig. 7.30 Hook Bolt

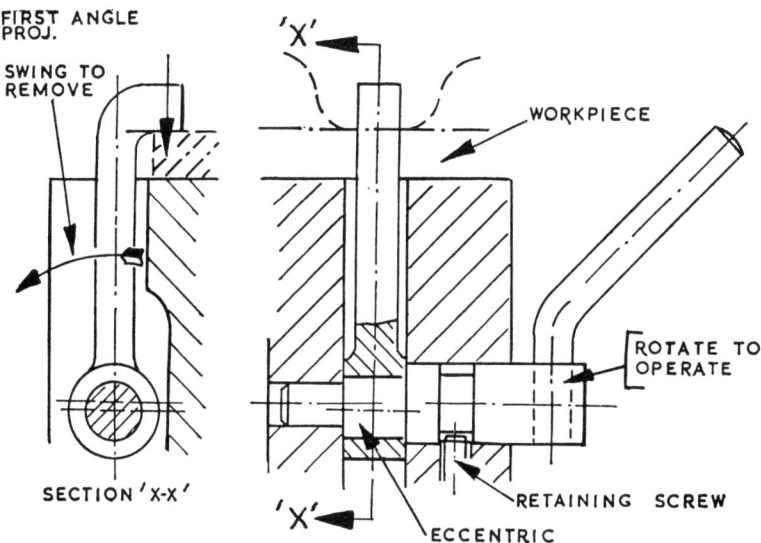

Fig. 7.31 Hook Bolt Operated by an Eccentric

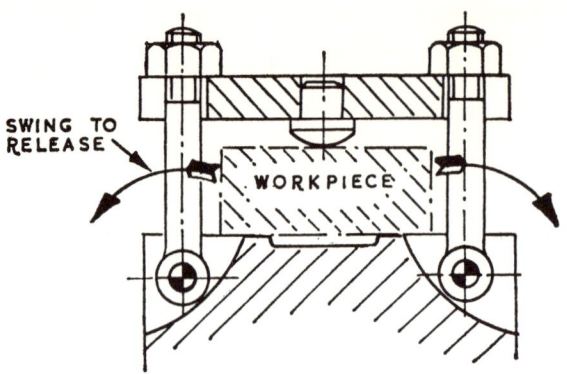

Fig. 7.32 Clamping with a Clamp Plate

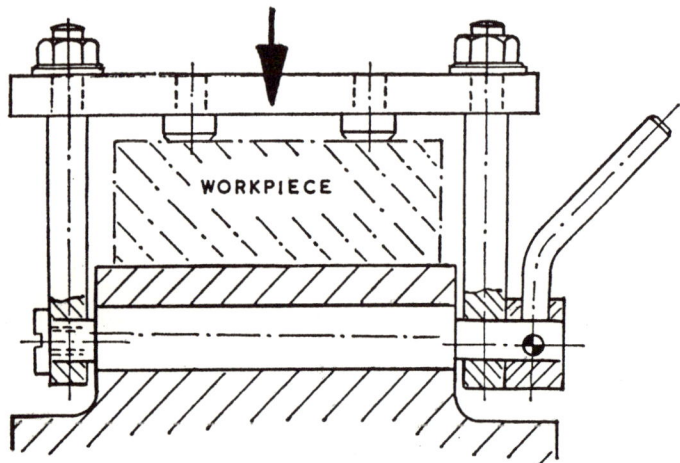

Fig. 7.33 Clamping Plate Operated by an Eccentric

does not permit another clamping system to be used. The arrangement shown in fig. 7.33 is an eccentric-operated version of the clamping plate system.

7.6. Clamping More Than One Workpiece

A plate clamp can secure no more than two workpieces at once owing to variation in workpiece size. In the arrangement shown in fig. 7.34 the clamp can adjust itself to accommodate two workpieces so that both are secured, but if more than two workpieces are introduced, the clamp will only sit on two of the workpieces so that the others will be free as shown in fig. 7.35.

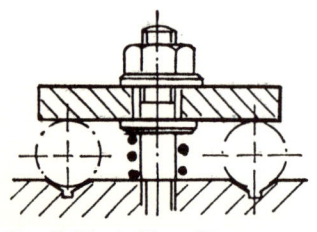

Fig. 7.34 A Plate Clamp Securing Two Workpieces

CLAMPING 105

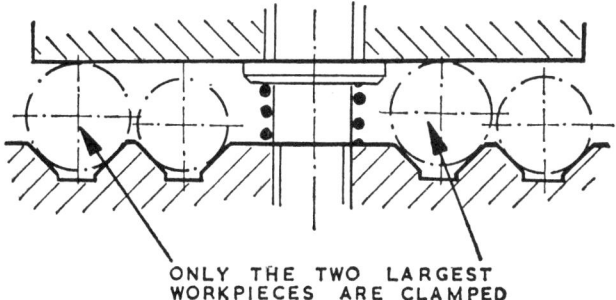

Fig. 7.35 Effect of Attempting to Secure More Than Two Workpieces with a Plate Clamp

7.6.1. Equalising clamps. When clamping at two places on an uneven surface, or when clamping two workpieces whose heights are likely to vary, an equalising clamp as shown in fig. 7.36 can be used. This system can be extended as shown in fig. 7.37 to enable several workpieces to be clamped at once.

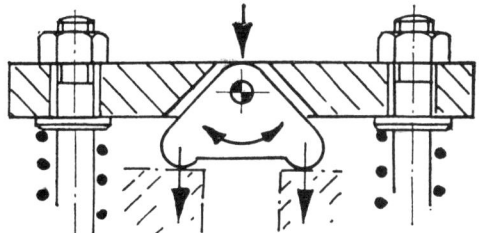

Fig. 7.36 Clamping Two Workpieces with an Equalising Clamp

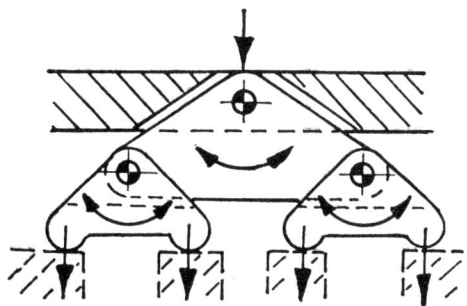

Fig. 7.37 Clamping Four Workpieces with an Equalising Clamp

7.7. Differential Clamping

Fig. 7.38 shows a differential clamping system. This arrangement is used because it is required to clamp the workpiece at the sides of the cast boss, the position and size of which can vary. The clamping force must be the same on each side of the boss to avoid straining the workpiece, which must be held rigidly. In the clamping system shown, the actuating plate is moved towards the workpiece by the clamping screw, and it operates the two clamps by wedge action; variations in the position of the workpiece boss are allowed for, because as soon as one of the clamps touches the workpiece, further clamping causes the actuating plate to move at right angles to the screw movement (see 'float' on fig. 7.38) so that continued screw movement only moves the other clamp. When both clamps are touching the workpiece, further movement of the screw produces equal clamping forces on the sides of the workpiece.

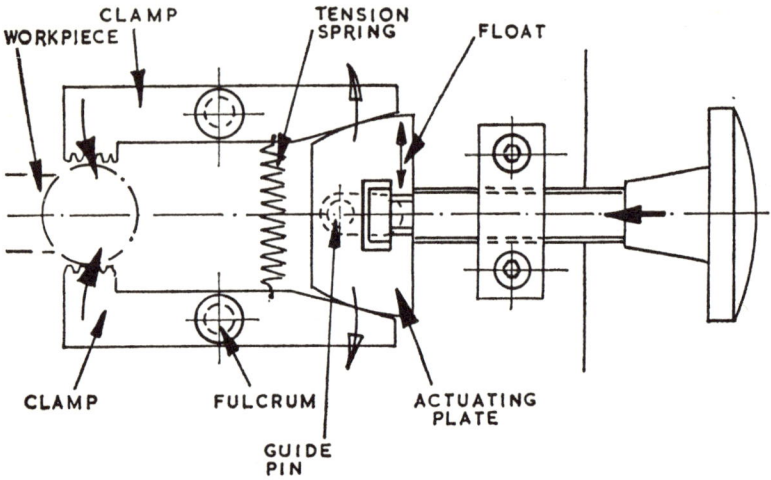

Fig. 7.38 Differential Clamp

If it is necessary to withdraw the clamps in a direction parallel to the axis of the clamping screw, the fulcrum holes in the clamps can be elongated.

7.8. Cam-Actuated Clamping

Cam-actuated clamps are very rapid to operate, but care must be taken to ensure that the cutting action will not loosen the clamp; milling is particularly inclined to cause this loosening. The cam must be arranged so that the clamping action is a natural one to perform, and if possible, a continuation of the clamp-positioning movement. Components for cam-actuated clamps are marketed by the manufacturers of 'unit tooling parts'.

Fig. 7.39 shows a cam plate used to secure a latch-type clamp or jig. It will be seen that the tightening action is in the same direction as that to close the latch.

Cams can be used to secure plate clamps; for example, the system shown in fig. 7.40 is a cam-actuated version of the clamp shown in fig. 7.8 on page 96, and that shown in fig. 7.41 the cam-actuated version of that shown in fig. 7.17 on page 98.

CLAMPING

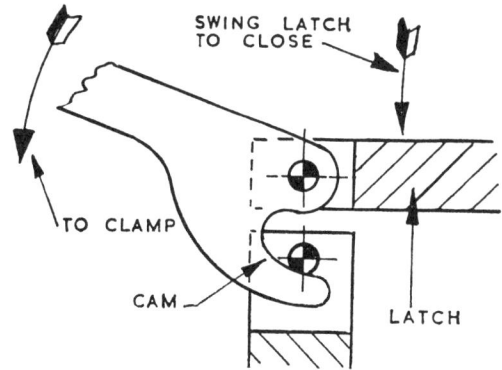

Fig. 7.39 Cam Plate Used to Secure a Latch

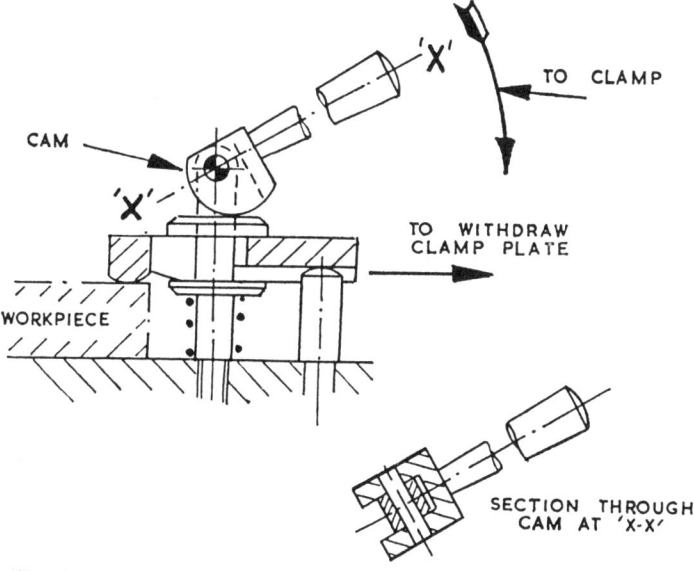

Fig. 7.40 Cam-Actuated Plate Clamp

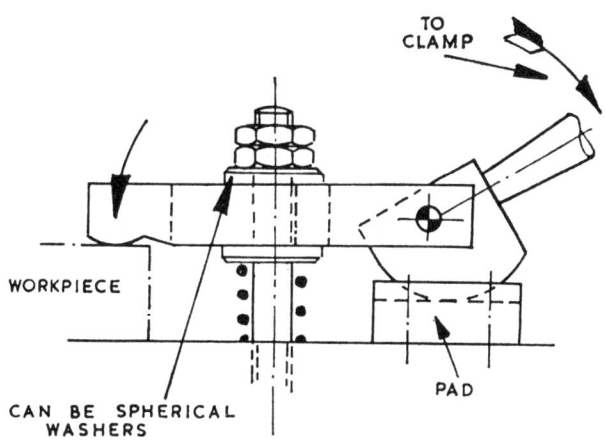

Fig. 7.41 Cam-Actuated Pivot Clamp

JIG AND TOOL DESIGN

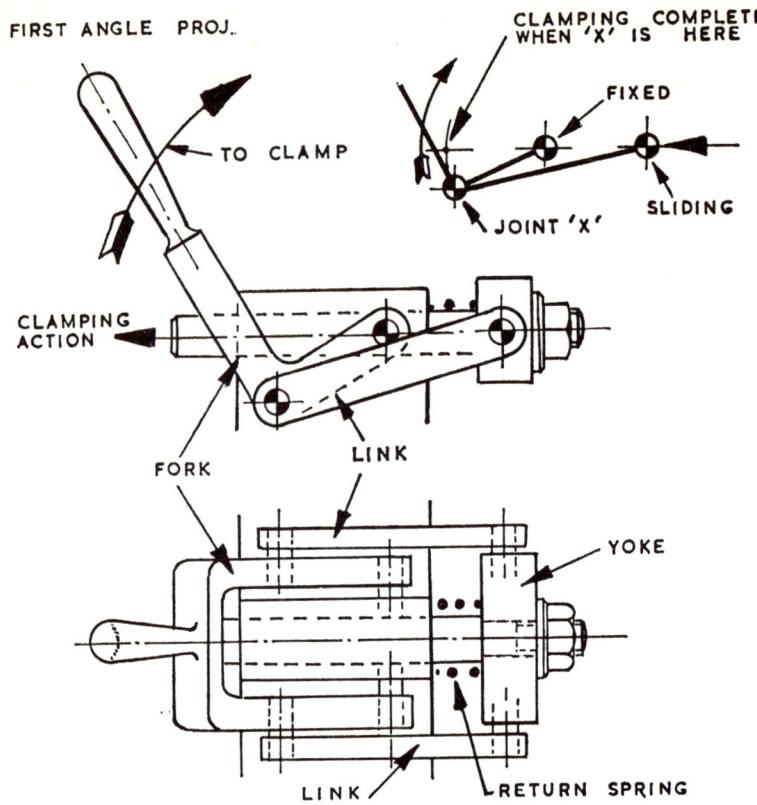

Fig. 7.42 Toggle Clamp

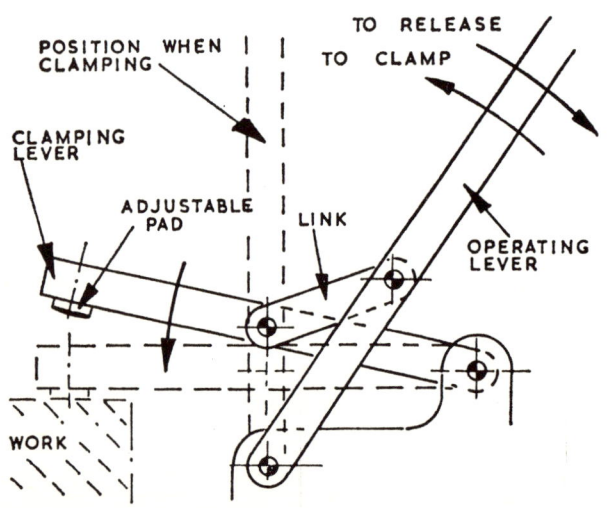

Fig. 7.43 Toggle Clamp

7.9. Toggle Clamps

Toggle clamps are very rapid to operate and give a secure clamping action. Fig. 7.42 shows the toggle part of a plunger-operated clamp; the linkage diagram shows the relative movements, and indicates that when the linkage is in the clamping position a large angular movement is necessary to unlock the clamp. Fig. 7.43 shows another toggle clamping system, in which the clamping lever is quickly moved clear of the workpiece.

7.10. Pneumatic Clamping

Compressed air is used extensively for clamping, and for operating location and indexing devices. The advantages claimed for pneumatic clamping are that there is less wear on the clamps and associated parts, there is less tendency for damage to occur to the workpiece, that the operation is more rapid and that the clamping pressure can be controlled more accurately.

7.10.1. The basic features of a pneumatic system are a control valve, a reducing valve to control the pressure and a cylinder unit to operate the mechanical part of the arrangement. An air-flow regulator and pressure gauge may also be included in the system, and several cylinders may be operated by one system. The air being tapped from air-line, or supplied by a small compressor.

7.10.2. Fig. 7.44 shows a simple single-acting cylinder system; in this system the air operates the push rod against the compression spring and

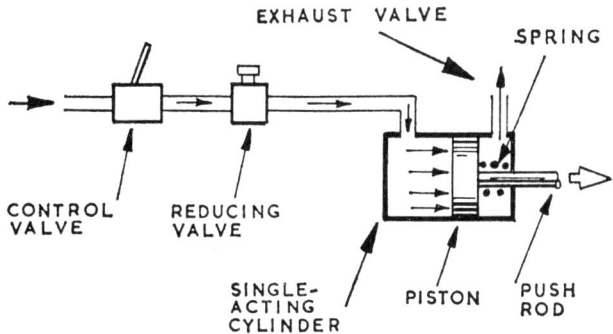

Fig. 7.44 Pneumatic System with Single-Acting Cylinder

the air is expelled to atmosphere through the exhaust valve. When the control-valve lever is operated, the supply of air is cut off and the piston returned by the action of the spring. The control-valve lever can be made to operate by the action of a trip, actuated by the table.

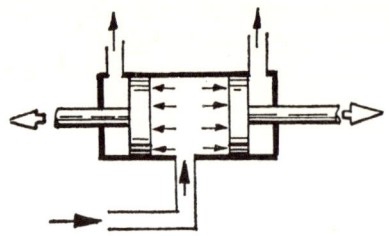

Fig. 7.45 Single-Acting Twin Piston Cylinder

Fig. 7.45 shows a single-acting twin-piston cylinder that can be used to operate two clamps and produce equal clamping forces.

7.10.3. Fig. 7.46 shows a system using a double-acting cylinder so that air is used to control piston movement in both directions. The control valve reverses the pressure direction.

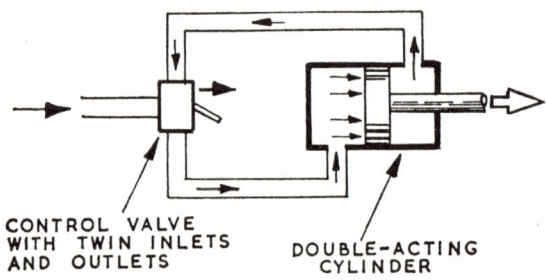

Fig. 7.46 Pneumatic System with Double-Acting Cylinder

7.10.4. The push rod can be used to clamp the workpiece directly or to operate a rack and pinion, wedge or toggle clamp. Fig. 7.47 shows a typical clamp.

To safeguard against accidents following air-pressure failure, a non-

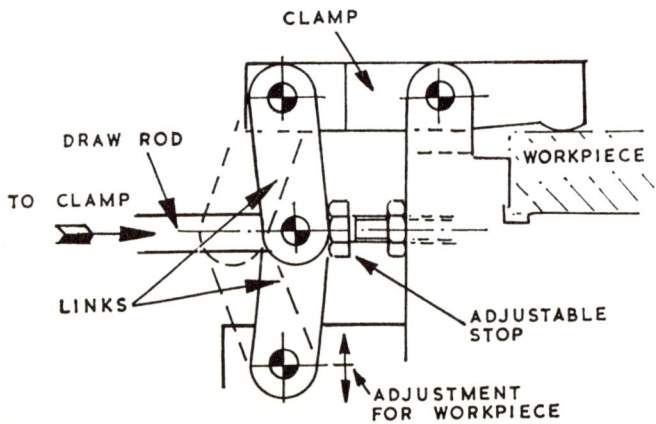

Fig. 7.47 Air-Operated Clamp

CLAMPING

return valve can be incorporated into the system, or as an alternative, the clamping force can be produced by powerful spring, and the air used to release the clamp as shown in figs. 7.48 and 7.49. Care must be taken to ensure that the operator cannot place his fingers between the clamp and the workpiece; the clamp–workpiece clearances should therefore be made as

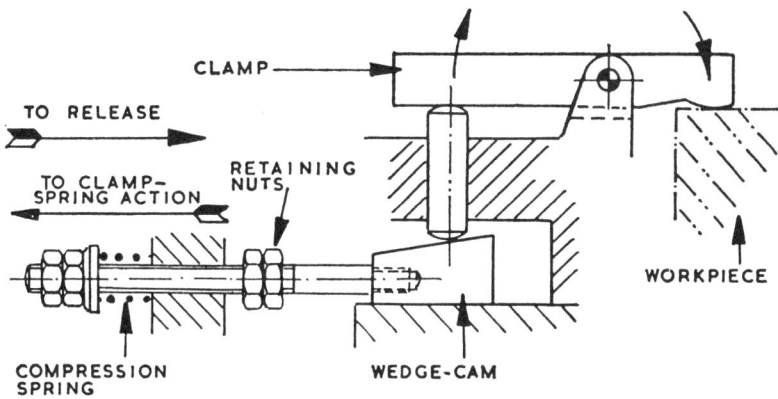

Fig. 7.48 Air-Operated Clamp (Clamping by Spring, Release by Air)

small as possible, and if further safeguard is necessary, the valves arranged so that he must be operating both valves with his hands before clamping can take place.

7.11. Hydraulic Clamping

Hydraulic pressure can be used to clamp the workpiece or to operate the equipment. In general, the principles are the same as those associated with pneumatic operation; the pressure oil being either supplied by the machine itself or by pump.

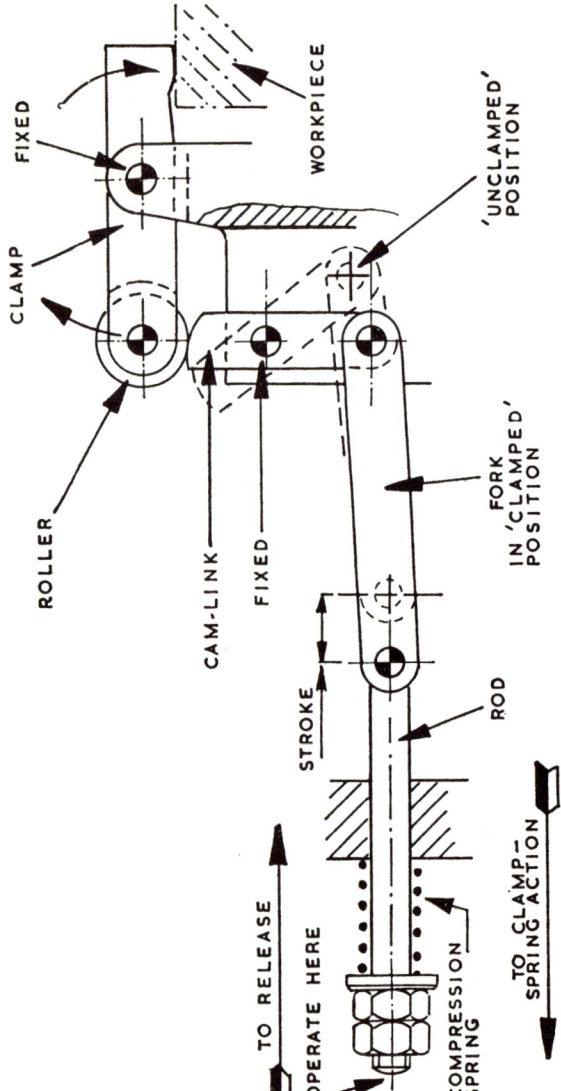

Fig. 7.49 Air-Operated Clamp (Clamping by Spring, Release by Air)

Chapter 8

DRILLING JIGS

8.1. Introduction

A **jig** was originally so called to distinguish it from a **fixture**—this latter type of equipment being fixed to the machine-tool table, chuck or faceplate. The only equipment which does not normally need to be fixed to the machine tool to counteract the effect of cutting forces, machine movement or gravity is that used for drilling and associated operations. Another characteristic of operations performed on a drilling machine is that the cutter can be guided by the equipment whilst cutting is taking place—this cannot be applied to other machining operations. Thus the term **jig** has, in machine-shop use, come to be applied to equipment that incorporates means of guiding the tool during cutting—although in some cases this equipment may be bolted to the machine table during the machining operation.

8.1.1. A drilling jig thus incorporates means of locating and clamping the workpiece, and also means of controlling the cutter during cutting.

8.2. Location

The location of the component is obtained by applying the principles laid down in Chapter 6; but because drilling is a comparatively short operation the location system must be such that positioning and clamping takes up as little time as possible. Drilling is very rarely done as a first operation and so there will be suitable machined surfaces from which to locate the workpiece. Components that are drilled at an early stage in the machining process may not have sufficient machined surfaces to produce complete location, and location may need to be partly from machined surfaces and partly from cast or forged features; locators that position the workpiece from these latter features may need to be adjustable to allow variation in workpiece size from batch to batch. Drill plates may also be positioned from cast features by employing 'sighting' plates (see page 85).

8.3. Clamping

Usually the clamping force required to hold the workpiece will be small, and so hand nuts, cams and similar clamping features can be used. Hand-operated clamps are usually preferred to spanner-operated clamps because time can be lost in locating a spanner, and shifting its position on the nut during tightening.

8.4. Handling the Jig

Some drill jigs have to be inverted between loading and drilling, moved about on the machine table if a fixed-spindle drilling machine is used or, in the case of box jigs, turned over from one 'station' to the next. These jigs should incorporate handles, lifting points, trunnions or similar handling aids, and be made as light as possible; care should be taken to ensure that the jig will not foul parts of the machine when moved about on the machine table.

8.5. Feet

Drill jigs should be provided with feet to ensure correct seating on the machine table. There should be **four** feet so that the jig will be unstable if pieces of swarf are under one or more of the feet; it must be emphasised that the requirements of the feet of a jig are different from those of a tripod; the tripod must allow stability when placed on an uneven seating plane, but the feet of a jig must ensure that the axes of the drill bushes are vertical to prevent incorrect work or broken tools.

8.5.1. The feet should be large enough to 'bridge' any slots in the machine table, and also be far enough apart to ensure that the cutting forces act within the figure enclosed by the feet.

8.5.2. Fig. 8.1 shows a foot incorporated into the base casting, fig. 8.2 shows a typical separate foot which is an interference fit in the base and fig. 8.3 shows a base that is machined from stock to incorporate four feet.

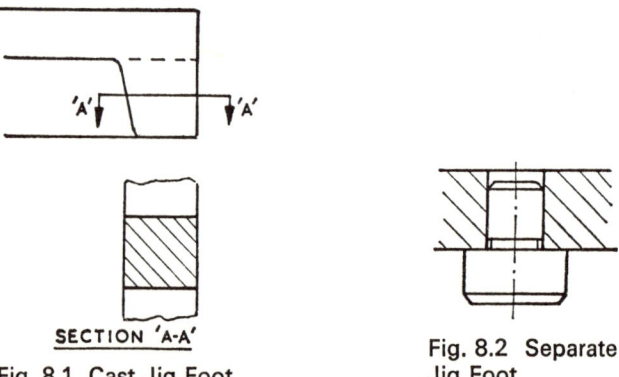

Fig. 8.1 Cast Jig Foot

Fig. 8.2 Separate Jig Foot

8.5.3. Long feet are used for table (or turnover) jigs and are as shown in fig. 8.4. In this example the foot bolt is retained by a foot nut which serves as a supporting foot when the jig is inverted for loading. The foot bolt and foot nut are tightened by applying a spanner to the hexagons

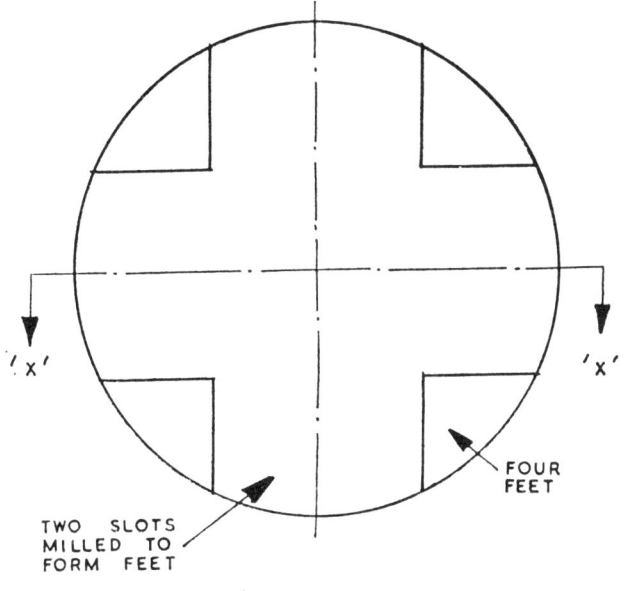

SECTION 'X-X'

Fig. 8.3 Machined Jig Foot

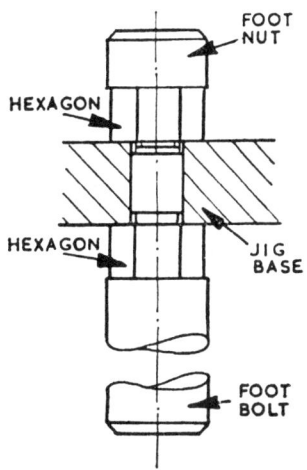

Fig. 8.4 Foot Nut and Bolt Assembly

shown; these hexagons are so positioned to prevent errors resulting from the strain that would be produced if the hexagons were at the far ends of the foot bolt and foot nut. The ends of separate feet are ground after assembly for accuracy.

8.5.4. Jig feet are marketed by the manufacturers who specialise in jig and fixture details.

8.6. Controlling the Cutters

The cutters are guided by means of holes in the drill plate, which is located relative to the workpiece, or by holes in the jig body. It is usual to bush these holes, but unbushed holes may occasionally be used for small runs or if it is impossible to make room for drill bushes. These bushes are of case-hardened and ground steel, and shaped to provide a lead for the drill when entering the bush. Recommended sizes for drill bushes are contained in British Standard B.S. 1098:1953; and drill bushes to these standards are marketed by the manufacturers of jig and fixture details.

8.6.1. Fig. 8.5 shows a typical headed drill bush as used when the hole depth must be controlled. It will be seen that a good seating for the bush is

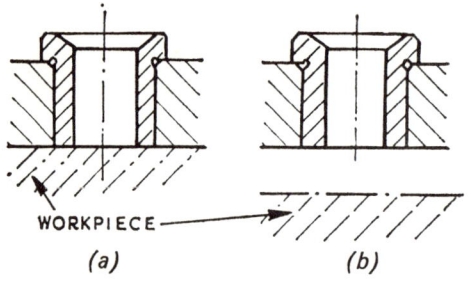

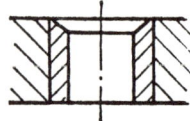

Fig. 8.5 Headed Drill Bushes

Fig. 8.6 Headless Drill Bush

obtained by chamfering the hole in the drill plate and undercutting the bush under the head. As shown in fig. 8.5, the bush must either touch the workpiece (fig. 8.5 (*a*)), so that the swarf can only escape via the bush, or be far enough away (fig. 8.5 (*b*)) to allow it to escape between the workpiece and the drill plate. Fig. 8.6 shows a headless drill bush used when the depth of the hole does not have to be controlled.

8.6.2. Special bushes are used for awkward workpieces. Fig. 8.7 shows a bush used to prevent drill run caused by a sloping surface, and fig. 8.8 shows a drill-bush arrangement designed to give adequate guidance when drilling a hole in a face that is some distance from the drill plate. When such a drill bush is particularly long its bore is relieved (as in fig. 8.8) so that only the end near the workpiece controls the drill.

DRILLING JIGS   117

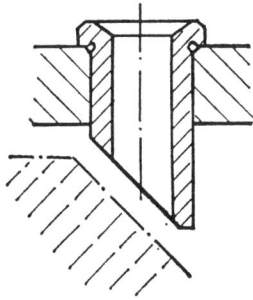

Fig. 8.7 Shaped Drill Bush

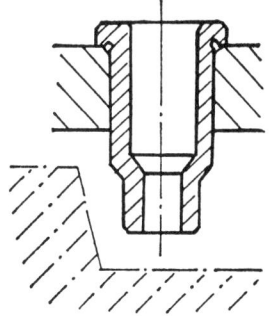

Fig. 8.8 Extended Drill Bush

8.6.3. When two or more holes are to cut on the same axis, as when drilling followed by reaming, slip renewable bushes are used (usually referred to as 'slip bushes'). A typical arrangement is shown in fig. 8.9; a separate 'slip bush' is used for each operation, and is located in the liner bush, which is a press fit in the drill plate. The slip bush must not rotate or run up the drill during cutting, and is therefore retained by a retaining screw; the slip bush

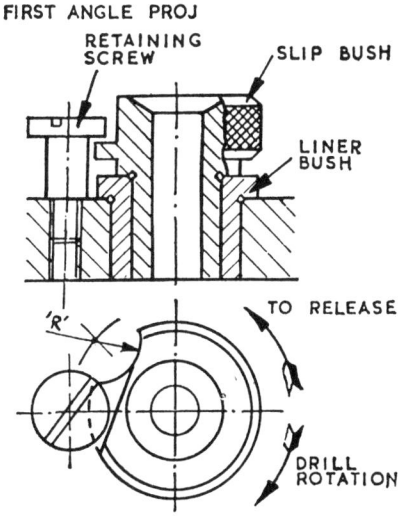

Fig. 8.9 Slip Renewable Bush Assembly

is shaped so that it can be released quickly without removing the screw. Fig. 8.9 (a) and (b) shows two typical slip bushes.

If a larger cutter is used in conjunction with smaller cutters, as when drilling and reaming followed by spotfacing, the largest cutter (in this example the spotfacing cutter) is located in the liner bush.

It is usual to supply one bush for each tool and to transfer it from one liner to the next; it is more convenient to do this than to have more bushes

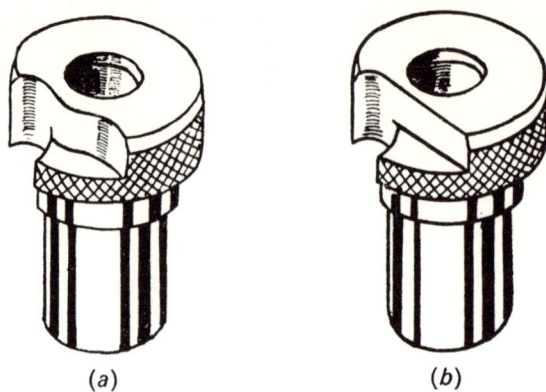

Fig. 8.9 Slip Renewable Bush

because it is easier to select the required bush when the numbers are kept small. The bushes should be marked with the size and the tool, e.g. '0·750″ REAM', and shaped or colour-coded for rapid identification.

8.6.4. A slip bush may be used when only one cutting tool is involved if a long bush is required for adequate support (as quoted in para. 8.6.2) but when it needs to be removed to allow the workpiece to be loaded and unloaded; fig. 8.27 on page 126 shows a typical example of this.

8.6.5. A fixed renewable bush as shown in fig. 8.10 is similar to a slip renewable bush but can only be taken out by removing the retaining screw.

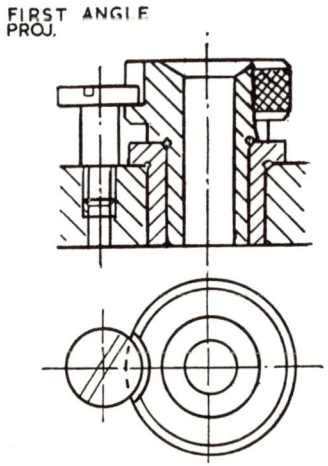

Fig. 8.10 Fixed Renewable Bush

This type of bush is used instead of a fixed drill bush if it needs to be replaced frequently because of wear.

DRILLING JIGS

8.6.6. A group of holes may be drilled using a bush with the required number of holes in it if they are too close together to permit a bush to be used for each hole. When these holes are to be drilled and then reamed, slip bushes must be used, and arrangements made to ensure that the bush is located radially—see fig. 8.11.

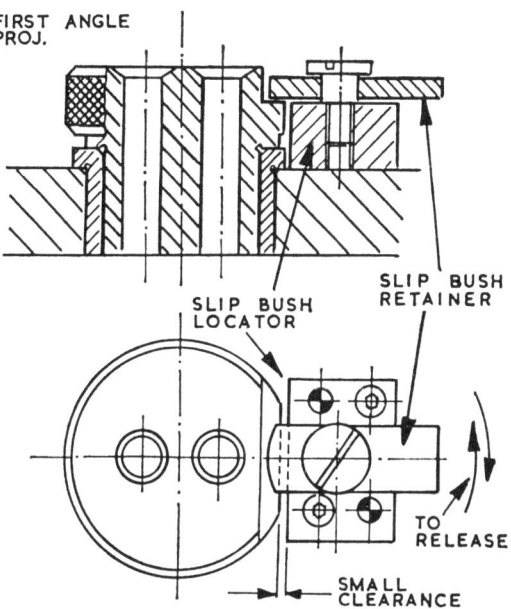

Fig. 8.11 Special Slip Renewable Bush Assembly

8.6.7. A drill bush can be used to locate the workpiece. Fig. 8.12 shows a bush that centralises a workpiece boss beneath the bush; this arrangement is unsuitable if the hole must be positioned accurately relative to other holes drilled during the same operation, but the more elaborate arrangement shown in fig. 8.13 allows greater accuracy.

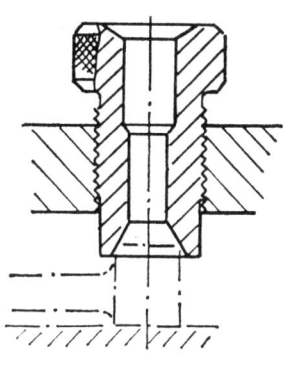

Fig. 8.12 Locating Bush

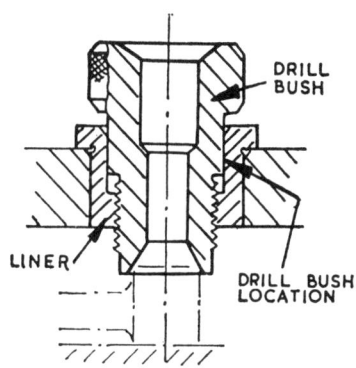

Fig. 8.13 Locating Bush

8.6.8. A drill bush can be used to provide a light clamping in the region of drilling; fig. 8.14 shows a clamping bush with a floating pad to prevent damage to the workpiece surface.

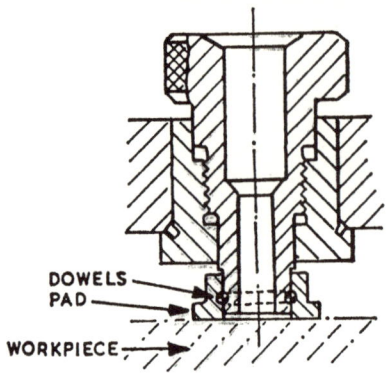

Fig. 8.14 Clamping Bush

8.7. Control of Depth

The hole depth can be controlled by holding the cutter in a special socket incorporating a stop nut which is set with the aid of a special setting gauge. Figs. 8.15 and 8.16 show such a socket; fig. 8.15 illustrates how the hole

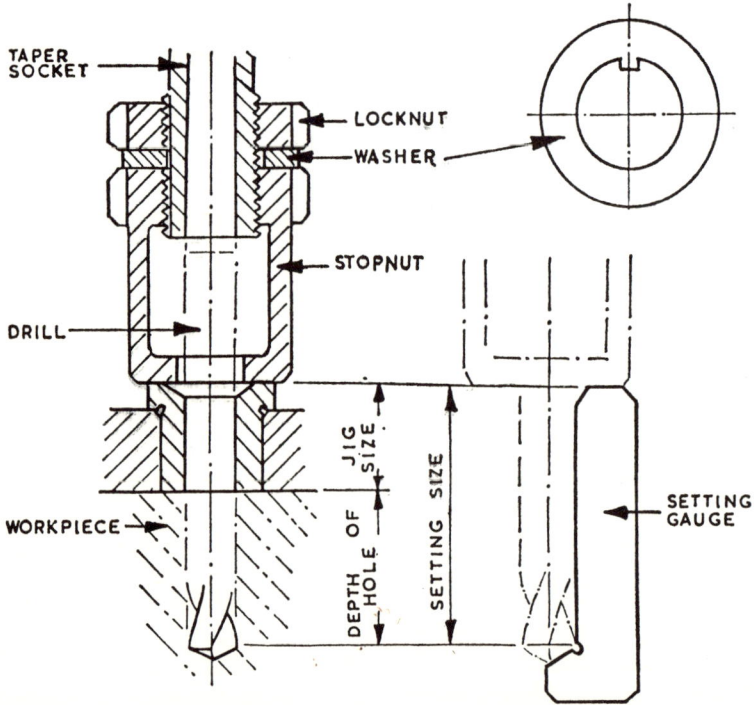

Fig. 8.15 Drill Stop Assembly and Setting Gauge

DRILLING JIGS

depth, drill length, position of the drill plate and the bush height must be taken into account when calculating the setting size. A simpler system as shown in fig. 8.17 can be used if economic considerations do not permit the more elaborate system to be used.

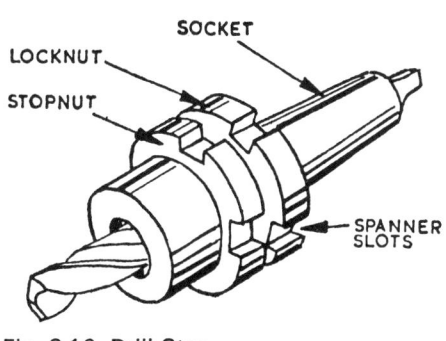

Fig. 8.16 Drill Stop

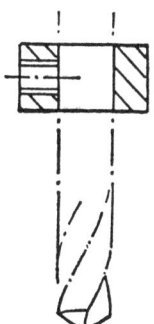

Fig. 8.17 Simple Drill Stop

8.8. Burr Grooves

When a drill starts to enter the workpiece a small 'piling up' of workpiece material occurs to produce a 'minor' burr, and when it breaks through, some of the material is pushed out to produce a 'major' burr (fig. 8.18). These burrs must be taken into account if the workpiece cannot be removed in a direction parallel to the hole axis (fig. 8.19).

8.9. Drill Over-Run

Drilling jigs for use when drilling through-holes must incorporate clearances to allow the drill to over-run after breaking through. When the

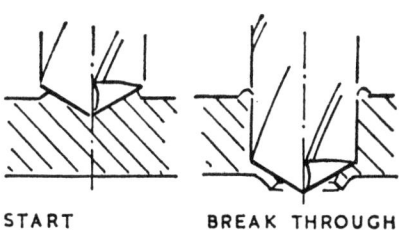

START   BREAK THROUGH

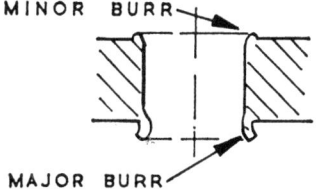

Fig. 8.18 Burr Formation

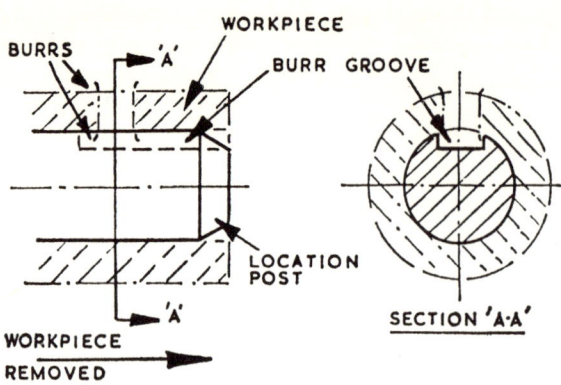

Fig. 8.19 Burr Groove

drill plate is to be removed when loading and unloading, a location system must be incorporated to ensure that the drill bushes line up with these clearances.

8.10. Handling Clearances

Care must be taken to ensure that there is adequate room for the operator's hands when using the equipment. Clearances should satisfy the conditions when both the workpiece and the equipment are on maximum metal limits —this is of particular importance when designing box jigs.

8.11. Swarf and Cutting-Fluid Clearances

Particular attention must be paid to the removal of swarf and cutting fluid; exit ports should be introduced into the body of a pot jig and a box jig, provision should be made for the brushing away of swarf and pockets where swarf and cutting fluid can accumulate should be eliminated.

8.12. Typical Drilling Jigs

Drilling jigs can be very broadly classified as follows:

    Plate jigs and channel jigs
    Solid jigs
    Post jigs
    Pot jigs
    Sandwich jigs
    Nutcracker jigs
    Table jigs (also called turnover jigs and open jigs)
    Latch jigs
    Box jigs
    Trunnion jigs

The following examples will illustrate the principal features of the more common types of drilling jig.

DRILLING JIGS

8.12.1. Plate jigs and channel jigs. These consist simply of a drill plate positioned on the workpiece by locators or by sighting, and clamped in position; heavy workpieces may even not be clamped to a base plate, but seated directly upon the machine table. The plate jig shown in fig. 8.20 is the simplest form of plate jig, and the channel jig shown in fig. 8.21 is a slightly more elaborate form of plate jig. Large workpieces may be located and

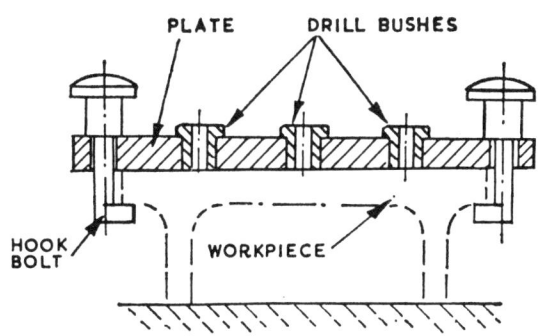

Fig. 8.20 Plate Jig

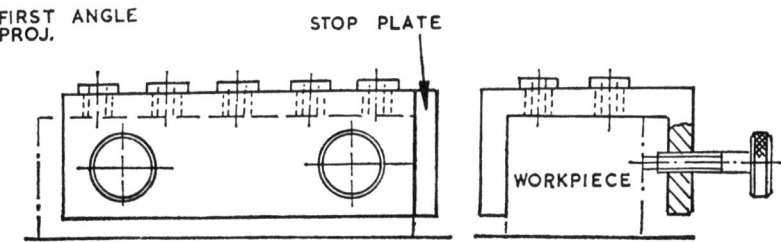

Fig. 8.21 Channel Jig

clamped to a base that can be adapted to suit several operations, and local plate jigs located and clamped to the facings to be drilled as shown in fig. 8.22.

8.12.2. Solid jigs. Small workpieces can be drilled using a jig, the body of which is machined from a solid block of steel as shown in fig. 8.23. This example also illustrates the application of burr grooves.

8.12.3. Post jigs. Post jigs are used for location from a bore. The post should be as short as possible to facilitate loading, but if it must be long to support the workpiece or to locate the drill plate it must be relieved so that location is only at its extreme ends; large posts should be bored for lightening purposes.

Fig. 8.24 shows a jig with a long location post so that it can locate the drill plate. The drill plate clamps the workpiece and is fitted with a captive

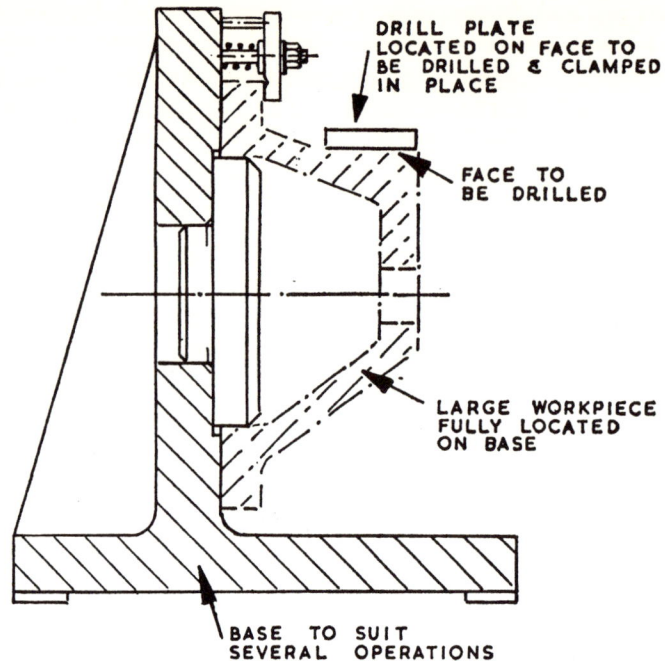

Fig. 8.22 Base and Local Drill Plate

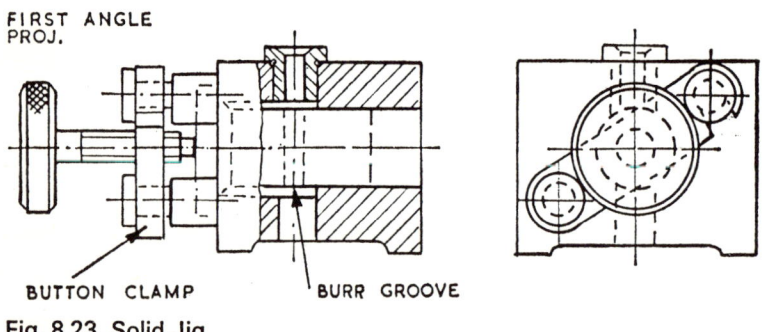

Fig. 8.23 Solid Jig

'Cee' (or swing) washer so that it can be removed when the hand nut is still attached to the post.

The jig shown in fig. 8.25 is designed to support the workpiece under the flange; the workpiece and the drill plate are both located at the top of the post. The drill plate is located relative to the drill clearance grooves in the support piece.

The post need not be vertical. Fig. 8.26 shows a drilling jig with a horizontal post; in this example a short locator is used to aid loading, to allow for drill over-run and to obviate the need for burr grooves. The post would need to be longer if the workpiece must be given better support, and a longer slip bush used if the drill must be guided close to the workpiece.

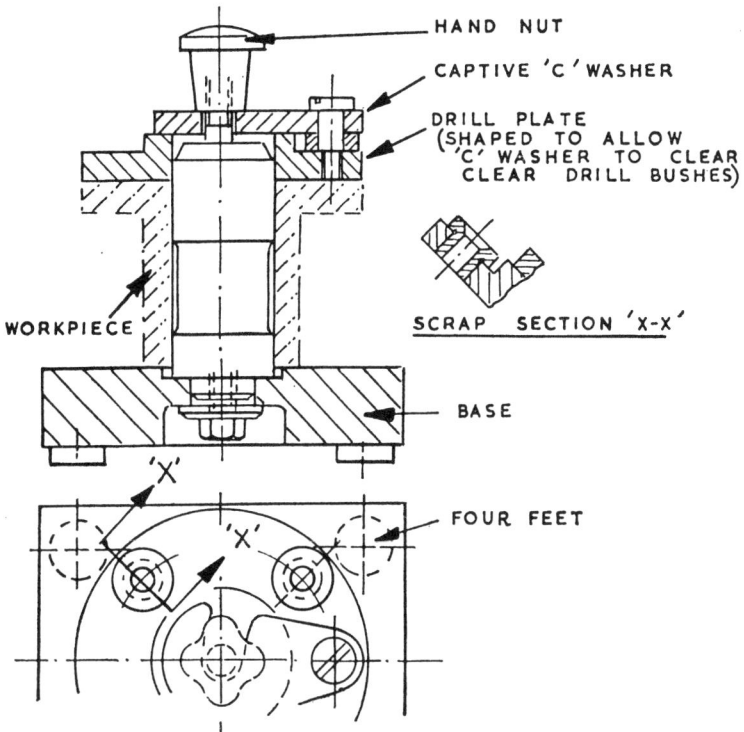

Fig. 8.24 Post Jig

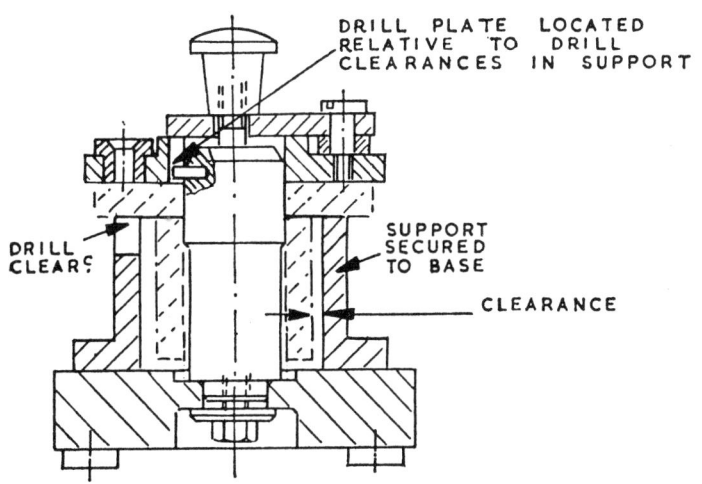

Fig. 8.25 Post Jig

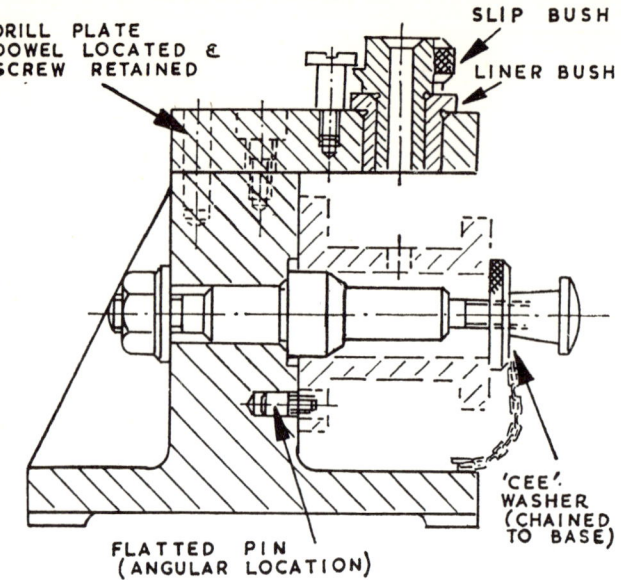

Fig. 8.26 Post Jig

The angular drilling jig shown in fig. 8.27 illustrates welded construction and the use of a quick-action nut; the latter is used because it must be removed to allow the workpiece to pass because of the smallness of the workpiece bore. The long slip bush gives good support and can be removed to allow the workpiece to pass (this type of bush is discussed in para. 8.6.4).

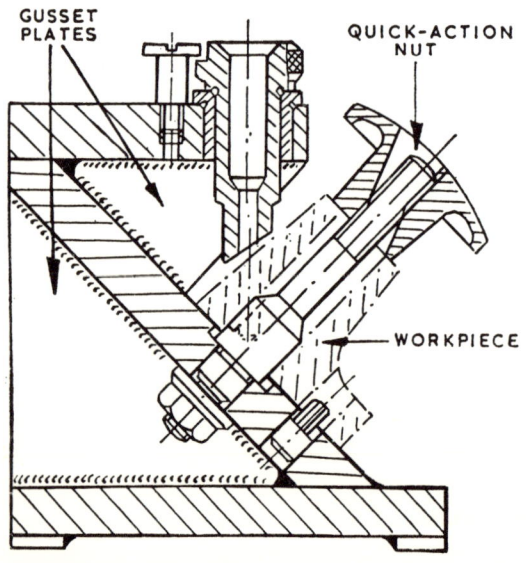

Fig. 8.27 Angular Post Jig

DRILLING JIGS 127

8.12.4. Pot jigs. Pot jigs are used when locating from an outside diameter, or to give support to the workpiece in the region of the drilling forces. When through-holes are to be drilled it is necessary to provide drill, swarf and cutting-fluid clearance; the drill plate must then be located relative to these clearances as shown in figs. 8.28 and 8.29.

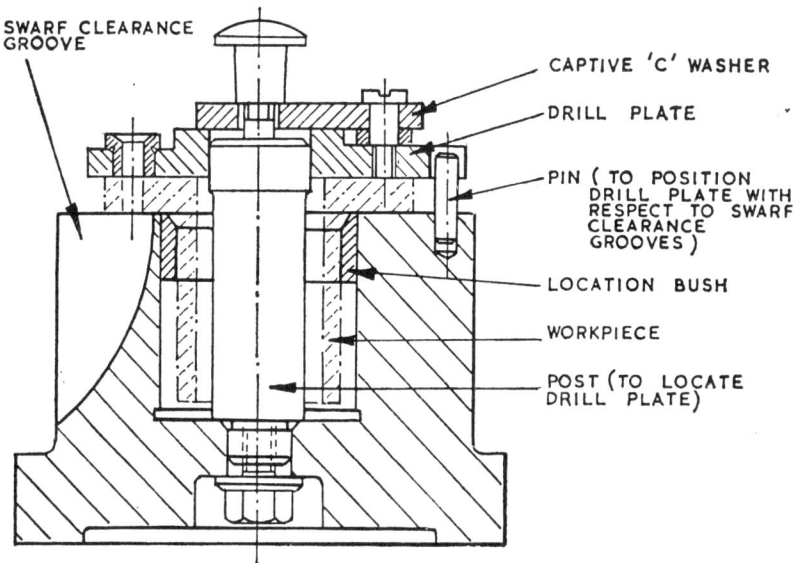

Fig. 8.28 Pot Jig

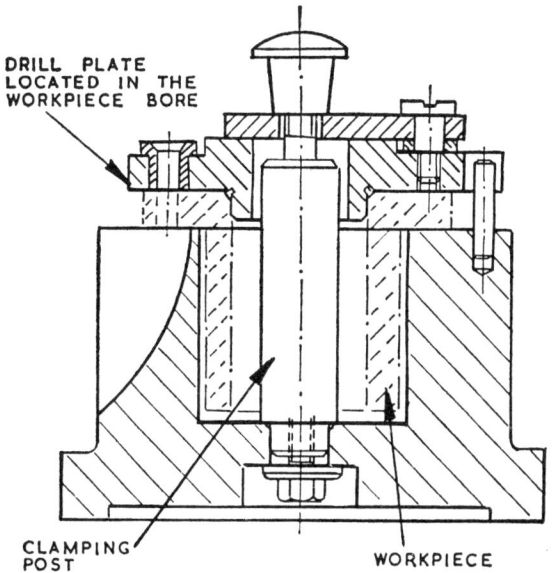

Fig. 8.29 Pot Jig

The jig shown in fig. 8.28 shows the workpiece located from its outside diameter and the drill plate located from a post, which also is used to clamp the workpiece and drill plate; the workpiece and drill-plate locations being accurately positioned relative to each other. The jig shown in fig. 8.29 provides support for the workpiece; the drill plate locates directly in the workpiece bore and the post is used only to clamp the workpiece by the drill plate.

8.12.5. Sandwich jigs. Fig. 8.30 below shows a sandwich jig; in this example the component shape demands that the drill plate is located from separate locators from those used for the workpiece; the workpiece is sandwiched between the base plate and the drill plate by two swing bolts. This example illustrates the use of 'unit tooling' parts—the base plate, drill plate, swing bolts, nuts, forks and bushes all being 'unit tooling' parts.

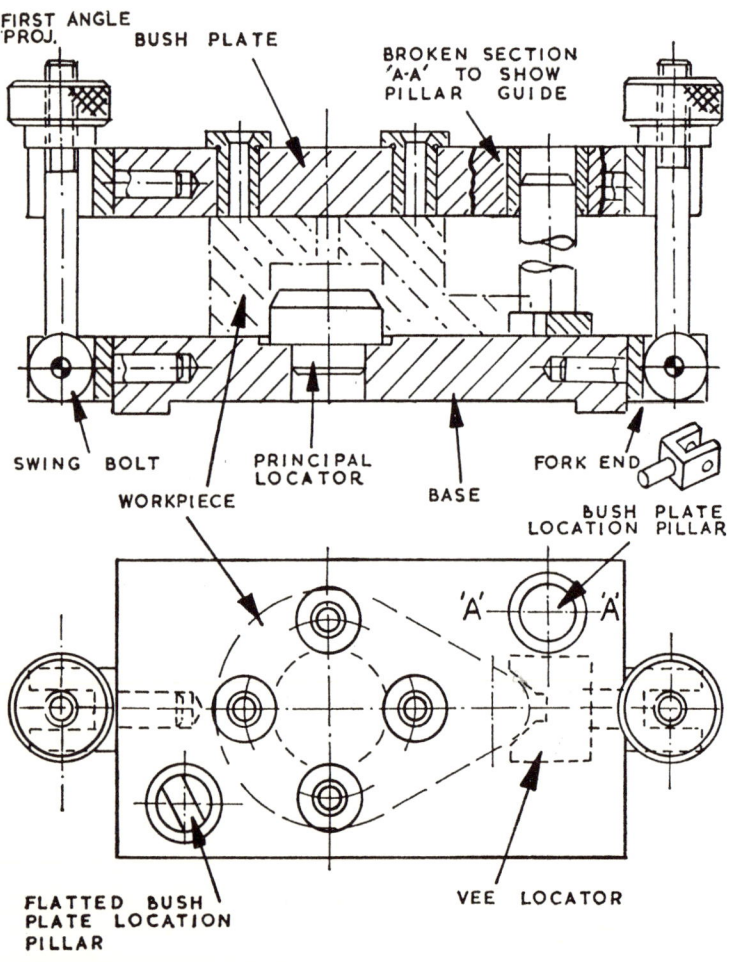

Fig. 8.30 Sandwich Jig

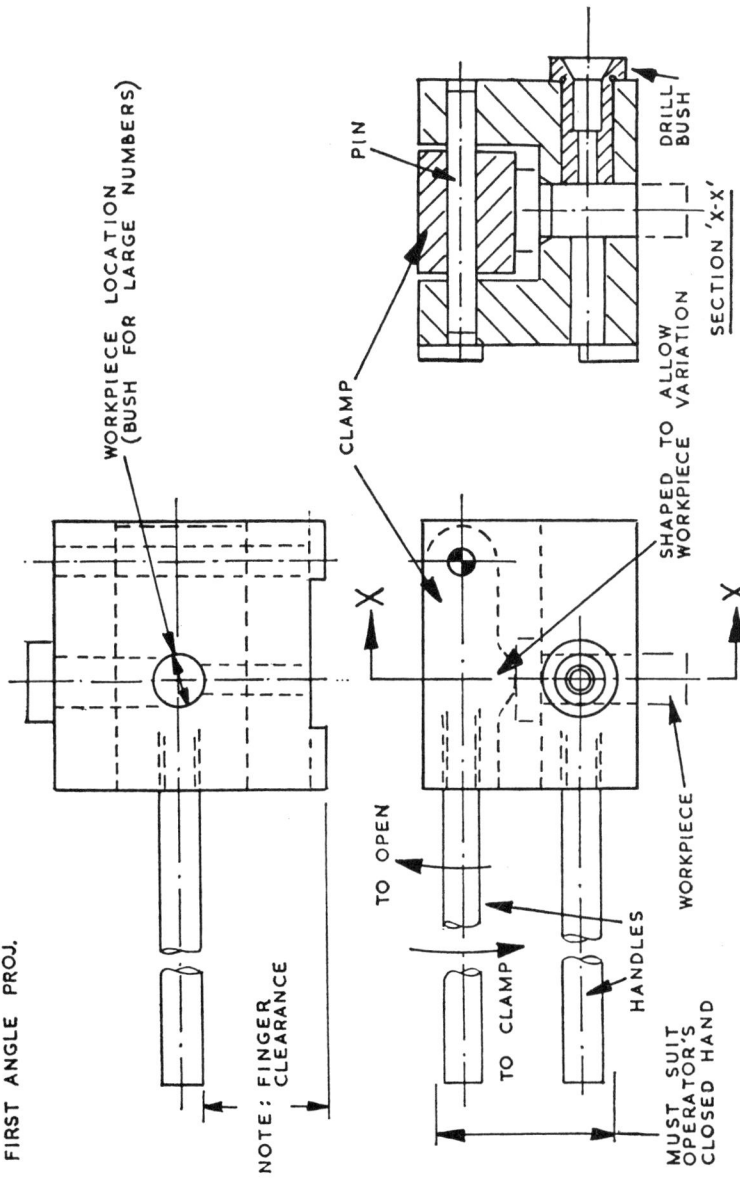

Fig. 8.31 Nutcracker Jig

8.12.6. Nutcracker jig. This type of jig is shown in fig. 8.31 on page 129; it is useful for light workpieces and enables the operator to load the jig rapidly and to hold the jig on the machine table with safety during drilling.

8.12.7. Table jigs (also called turnover jigs and open jigs). These are used when it is necessary to locate the workpiece from the face that is to be drilled. Fig. 8.32 illustrates typical components that must be so located—the location faces that must be used are shown in heavy lines.

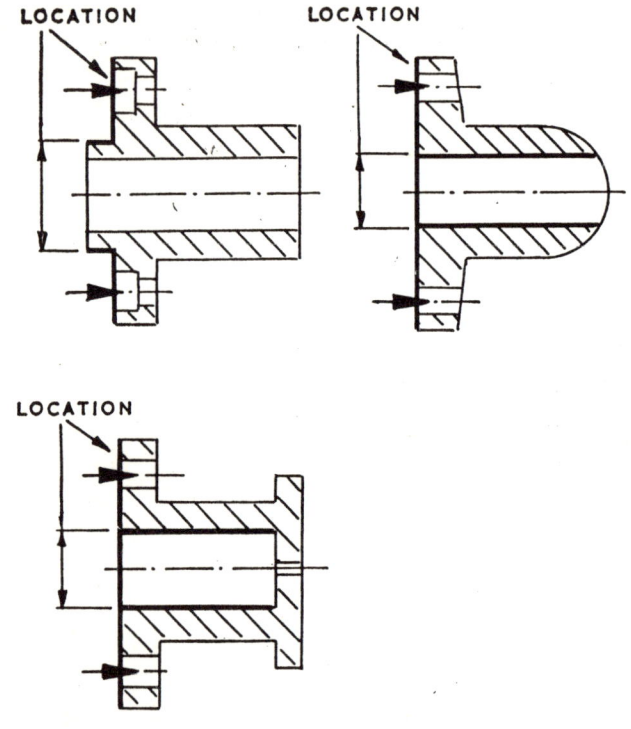

Fig. 8.32 Conditions that Require a Table Jig

Fig. 8.33 on page 131 shows a table jig; this jig is seated on the four foot nuts when the workpiece is located and clamped, and then inverted before machining is done (the jig then being in the position shown in fig. 8.33). This type of jig presents no problems in swarf and cutting-fluid disposal, and loading and clamping is easy, but the workpiece is only supported against the cutting forces by the clamps.

8.12.8. Latch jigs. Latch jigs (see fig. 8.34 on page 132) are an elaboration of the latch type of clamp shown in fig. 7.19 on page 99. The latch must be positively located (faces X and slot Y) so that the bush bores are vertical whatever the workpiece height, and is secured by nut A. The workpiece is

DRILLING JIGS

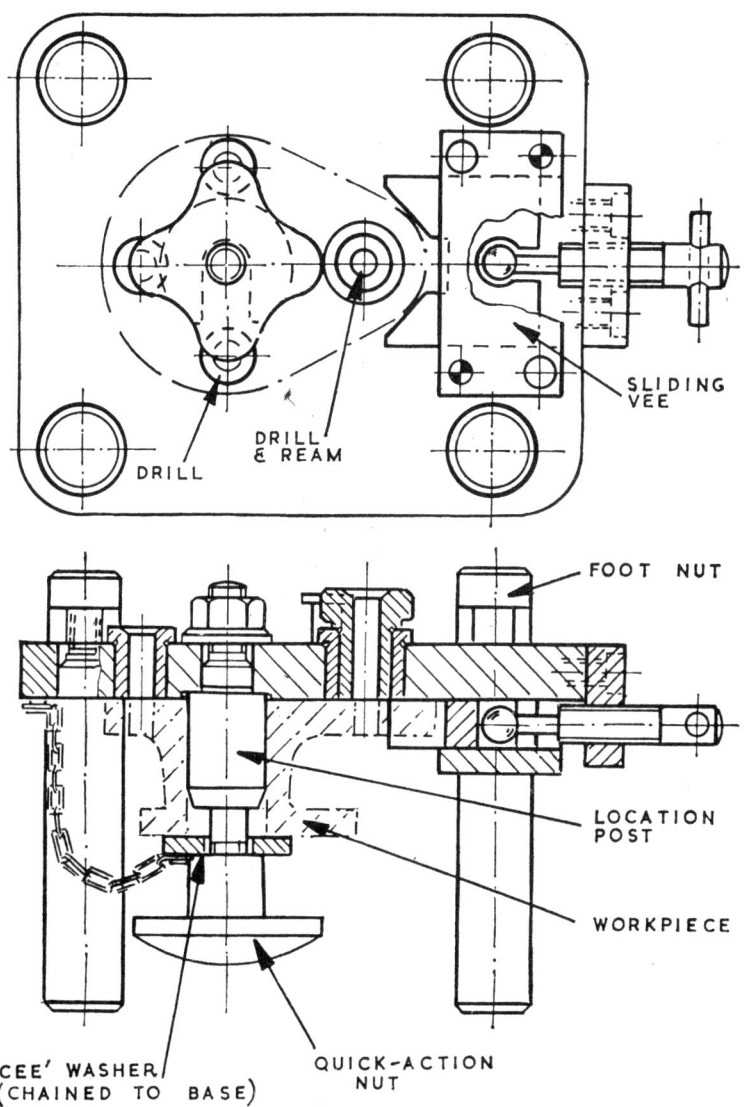

Fig. 8.33 Table Jig

in turn clamped by screw B; if the latch is used to clamp the workpiece directly, the bush axes will not necessarily be vertical owing to the workpiece variations.

8.12.9. Box jigs. When a relatively small workpiece is to have holes drilled in it in different directions, a box jig can be used so that the workpiece need

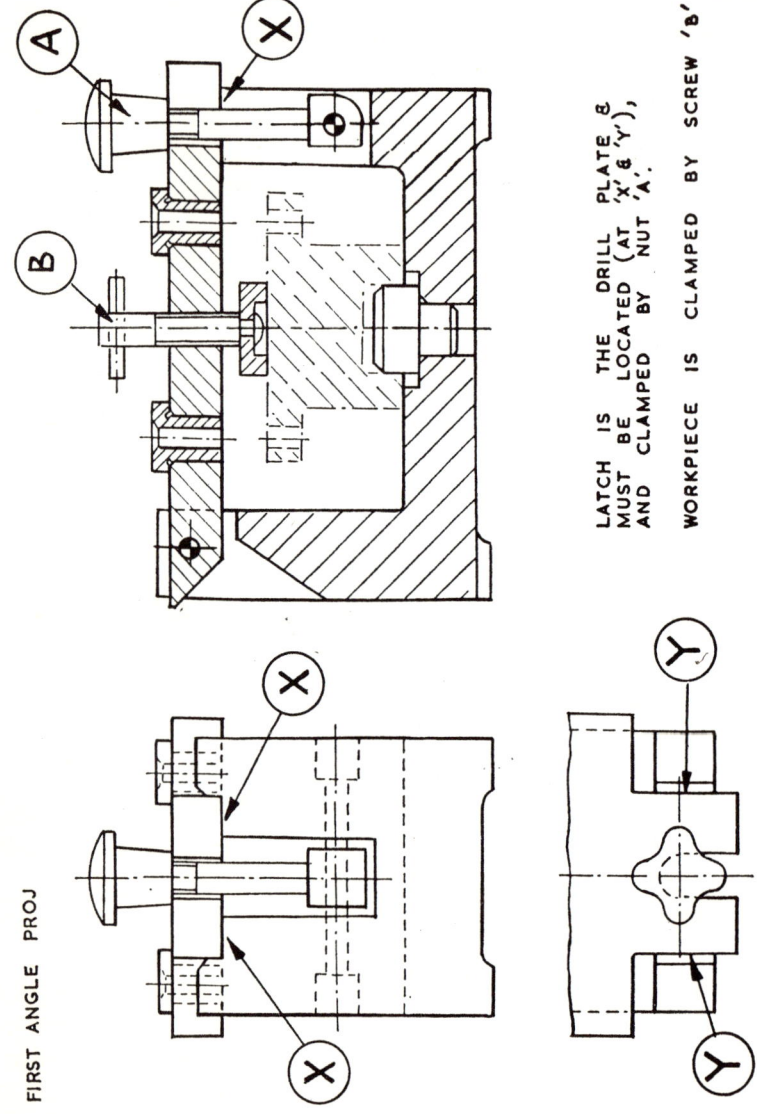

Fig. 8.34 Latch Jig

DRILLING JIGS

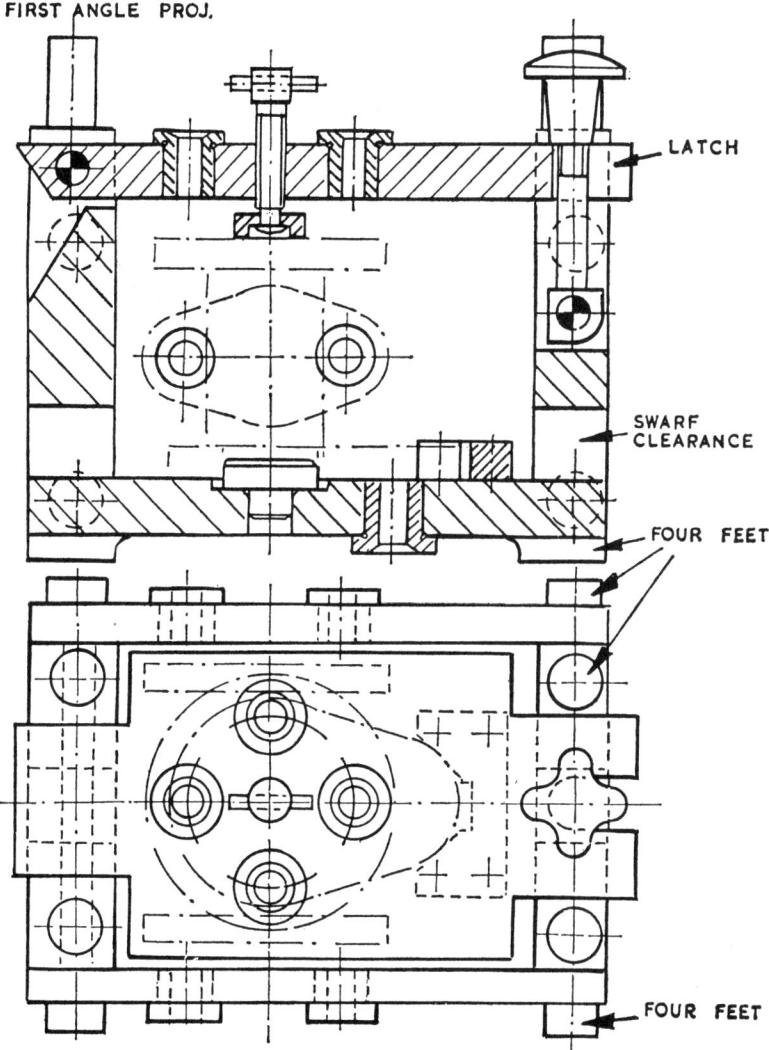

Fig. 8.35 Box Jig

only be located and clamped once. The jig is turned over and seated on each face in turn. Fig. 8.35 above shows a typical box jig. Care must be taken when designing a box jig to ensure that there is ample clearance for the operator's hands when loading and unloading, adequate swarf and cutting-fluid escape ports must be provided, and the inside must be free from recesses where accumulation of swarf and cutting fluid can occur.

8.12.10. Trunnion jigs. When a large or awkwardly-shaped workpiece is to be drilled in many directions a trunnion jig as shown in fig. 8.36 on page 134 is used; this is an extension of the box-jig principle. The workpiece is

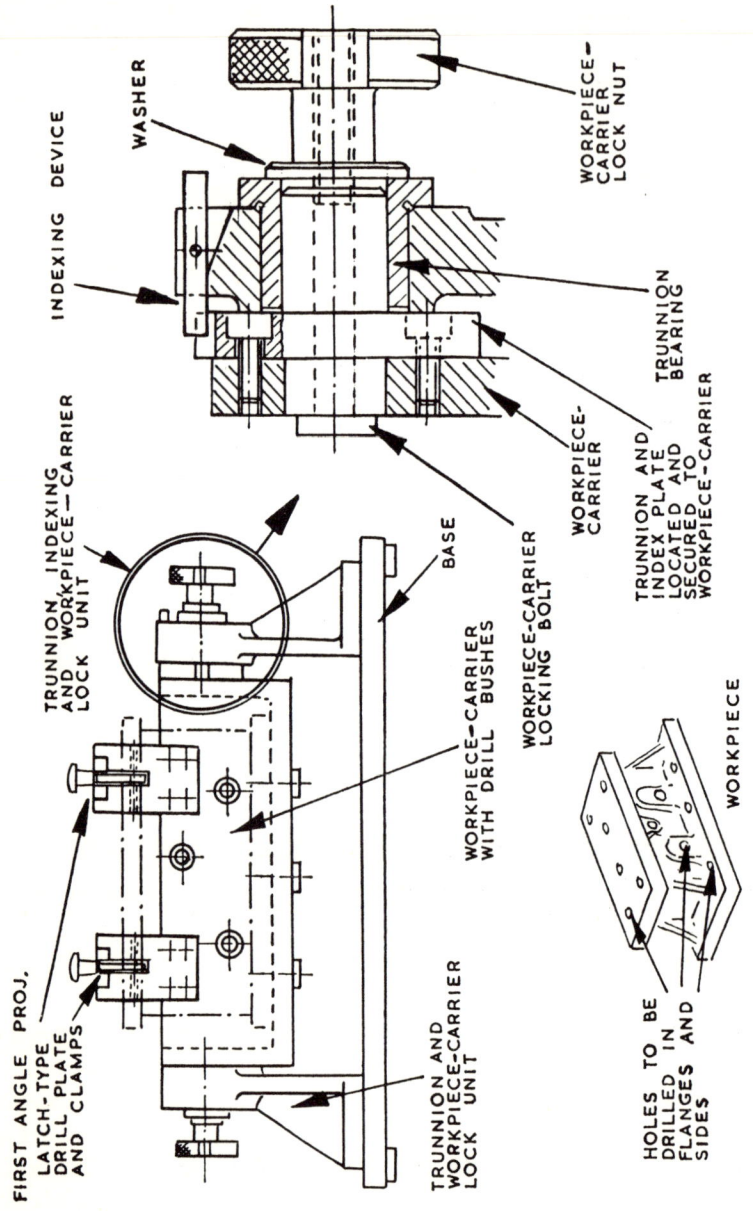

Fig. 8.36 Trunnion Jig

located and clamped to the carrier, which also carries the drill plates. The carrier is mounted on trunnions so that it can be rotated from station to station, positioned using an indexing device and secured in position before drilling. *Note:* Although not classed as an 'indexing jig', the principles associated with that type of jig and fixture apply here; for details of indexing devices see Chapter 11.

References

B.S. 122:Part 2:1964, Reamers, Countersinks and Counterbores.
B.S. 328:Part 1:1959, Twist Drills and Combined Drills and Countersinks.
B.S. 1098:1953, Jig Bushes.

Chapter 9
MILLING FIXTURES

9.1. Introduction

A milling fixture is located on the machine table and bolted in position; the workpiece is, in turn, located and clamped to the fixture. As explained in Chapter 8, a fixture does not control the cutting tools during the actual cutting, but the tools can be positioned before cutting is commenced.

9.2. Milling Methods

The fixture design depends upon the milling methods to be employed; the following examples illustrate variations of the basic milling process.

9.2.1. Straddle milling. In this method two cutters are mounted on the arbor so that two faces are machined simultaneously; the machine table is positioned relative to **one** of the cutters using a setting block (setting-block design is discussed on page 141).

9.2.2. Gang milling. Three or more cutters can be mounted on the arbor so that several faces can be machined at once; again, the machine table is positioned relative to **one** of the cutters in the gang.

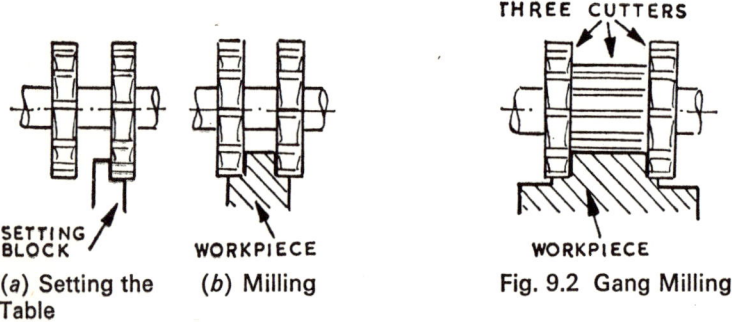

(a) Setting the Table   (b) Milling   Fig. 9.2 Gang Milling

Fig. 9.1 Straddle Milling

9.2.3. String or line milling. In this method several workpieces are mounted along the length of the machine table so that they can be machined in one pass. One cutter or a number of cutters can be used, and the workpieces can be arranged in a single line or a double line. Fig. 9.3 illustrates string milling methods.

MILLING FIXTURES

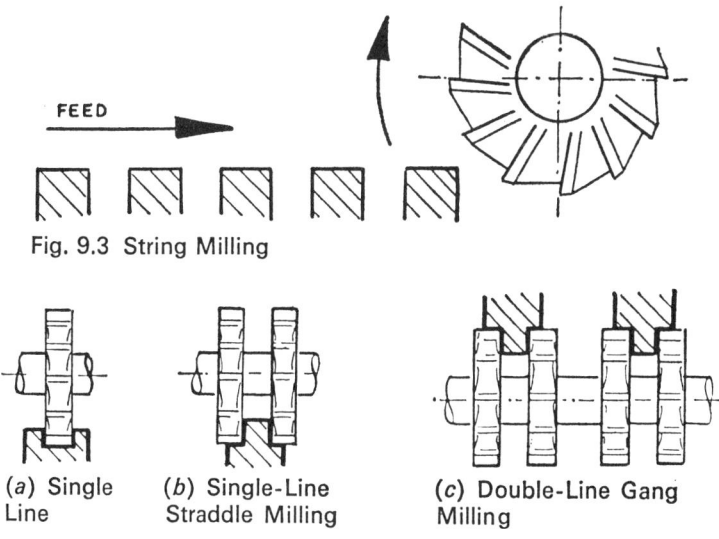

Fig. 9.3 String Milling

(a) Single Line (b) Single-Line Straddle Milling (c) Double-Line Gang Milling

Fig. 9.3 String Milling Methods

9.2.4. Pendulum milling. Pendulum milling implies that cutting takes place when the table moves to the right and also when it moves to the left. Figs. 9.4, 9.5 and 9.6 illustrate pendulum milling methods. In fig. 9.4 one workpiece is machined at a time, and is indexed between passes (the design of indexing jigs and fixtures is the subject of Chapter 11).

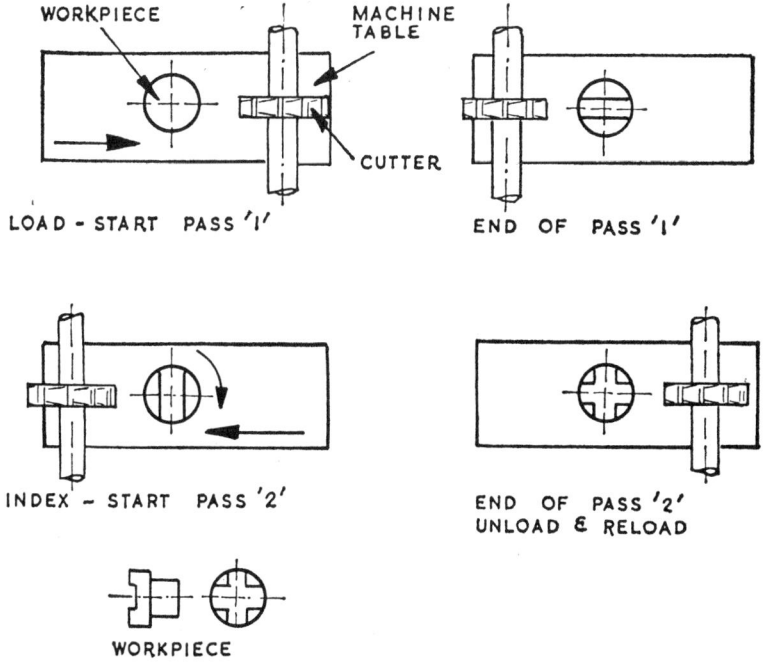

Fig. 9.4 Pendulum Milling (Single Cutter—One Workpiece Cut at a Time)

Instead of machining one workpiece at a time, two workpieces may be mounted on an indexing fixture, and indexing performed between passes, as shown in fig. 9.5. Another version of pendulum milling uses two fixtures, so that when one workpiece is being machined the other fixture can be unloaded and reloaded.

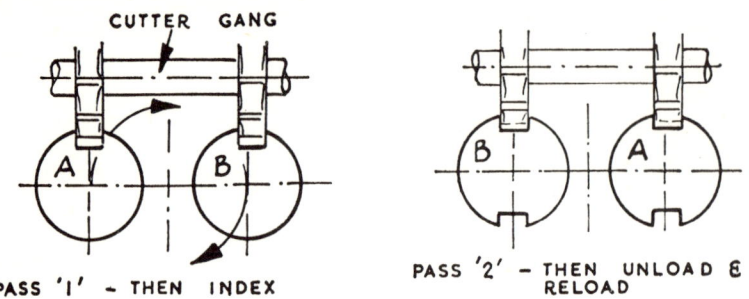

Fig. 9.5 Pendulum Milling (Two Workpieces Cut at a Time)

A feature of pendulum milling is that movement in one direction produces up-cut milling, and movement in the other direction produces down-cut milling; the workpiece and the milling machine must be suitable for machining by both methods.

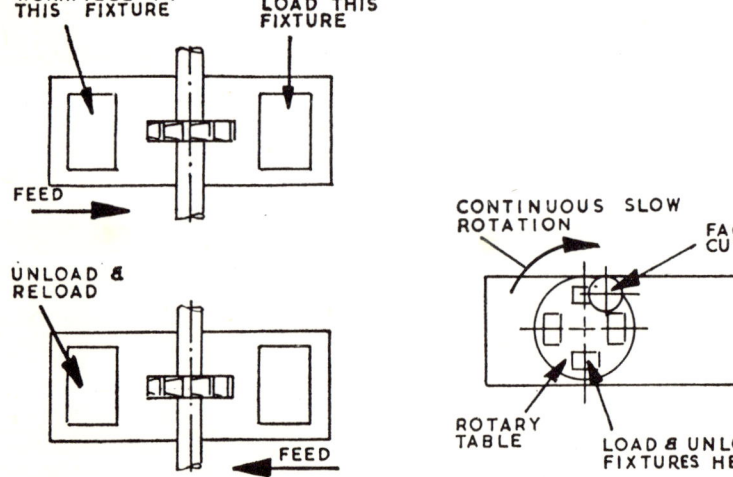

Fig. 9.6 Pendulum Milling (Two Fixtures Used)

Fig. 9.7 Rotary-Table Milling

9.2.5. Rotary-table milling. This is a method of continuous milling in which a rotary table is mounted on the machine table, and which rotates continuously past a facing cutter; a suitable number of fixtures is mounted on the rotary table, and loading and unloading done when each fixture in turn passes the operator.

MILLING FIXTURES

9.2.6. Profile milling. Complicated profiles can be milled by guiding the workpiece past the cutter. In fig. 9.8 a profile plate is located above the workpiece, and the machine table guided past the cutter by keeping the profile plate against the roller follower incorporated into the cutter arbor. In the method shown in fig. 9.9 a model is used, and a follower attached to the cutter head; the table is moved so that the model is kept against the follower.

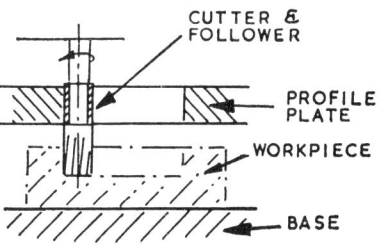

Fig. 9.8 Arrangement for Profile Milling

Fig. 9.10 shows a typical cutter and arbor arrangement as used with a profile plate, and fig. 9.11 shows how the workpiece profile and the roller diameter must be taken into account when the profile plate is designed.

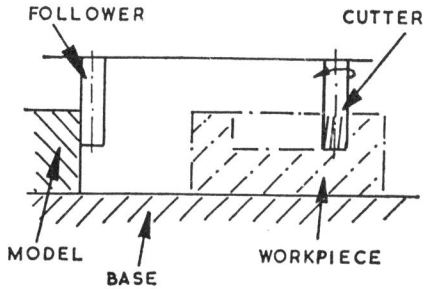

Fig. 9.9 Arrangement for Profile Milling

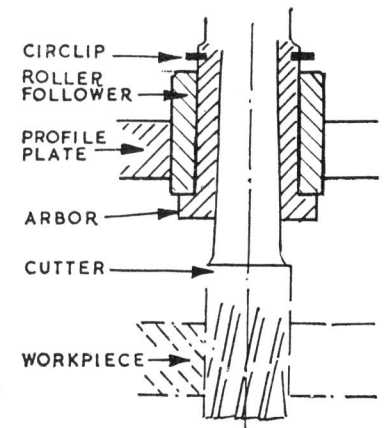

Fig. 9.10 Cutter Arrangement for Profile Milling

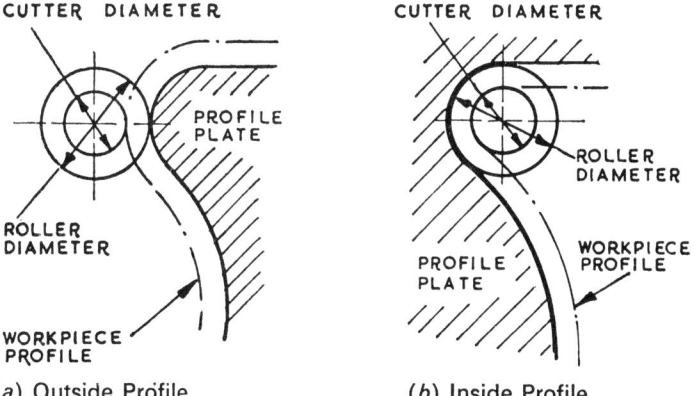

(a) Outside Profile   (b) Inside Profile

Fig. 9.11 Cutter, Roller, Workpiece and Profile Plate Relationship

9.3. The Location and Clamping on the Fixture on the Machine Table

The fixture is seated on the machine table and, if necessary, positioned relative to the table movement before it is clamped in position. It is the duty of the machine setter or operator to ensure that the machine table is clean before the fixture is placed upon it; but it is the duty of the designer to ensure that the seating area is adequate. Large fixtures should have a relieved seating face because there is a tendency for 'over-located' fixtures to spring.

Excepting when the fixture is used when milling a plane surface, it is necessary to position it relative to the machine-table movement. Location

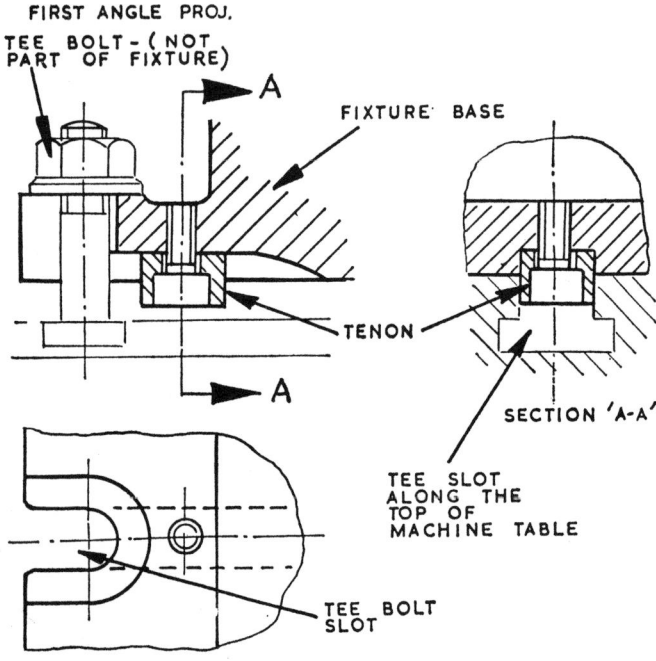

Fig. 9.12 Tenon and Bolt Slot

is obtained from the 'tee' slots that run the length of the table. Fig. 9.12 illustrates this form of location, and it will be seen that a rectangular tenon, secured to the underside of the fixture, locates in the narrow portion of the 'tee' slot. Two tenons, positioned as far away as possible to minimise errors, are used for this location; the location must be from **only one** slot, otherwise there will be redundant location (because of the effect of slot spacing). The workpiece location system is relative to the location slots for these tenons. The two tenons are part of the fixture assembly, and are of case-hardened steel.

The fixture is bolted to the table using 'tee bolts' in the tee slots; these bolts are not part of the fixture assembly. Two bolts may be adequate, but

MILLING FIXTURES 141

it may be necessary to use four bolts; when more than two bolts are used, bolting will be from more than one tee-bolt slot—this is permitted because the bolts do not locate the fixture. Fig. 9.12 also shows a tee bolt in position. When it is necessary to provide lugs from which to secure the fixture (as in fig. 9.17), care must be taken to ensure that the clamping lugs are not weak points.

9.4. Workpiece Location

The workpiece must be located on the fixture base—the shape and position of the location features being designed in accordance with the principles of location. As already stated, the location system must be positioned relative to the tenons. Milling is often the first operation to be done upon a casting; when this is the case, some of the seating location points must be adjustable for height.

9.5. Workpiece Clamping

The cutting forces during milling are very high and are interrupted, and unlike the forces produced by drilling, tend to move the workpiece away from its seating on the fixture. The clamps and bolts must be sturdy, and usually hexagonal nuts are used instead of hand nuts (as in drilling jigs). When many clamps are used they should be arranged so that the operator can 'move round' the clamps in a convenient sequence when clamping and unclamping.

9.6. Tool Setting

The fixture is positioned relative to the cutter with the aid of a case-hardened steel setting block as shown in fig. 9.13. The setting block is located on the fixture relative to the location system, and secured in position. As shown in fig. 9.13, the setting is done from the periphery and **one side** of the cutter (location from both sides would cause redundant loca-

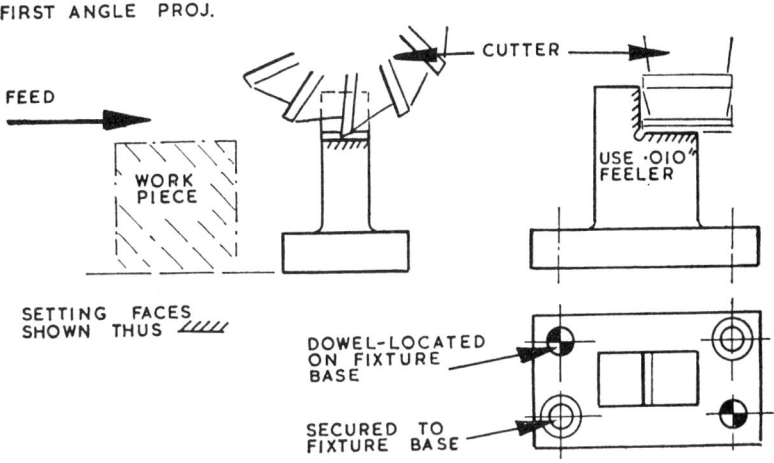

Fig. 9.13 Setting Block

tion); a feeler is introduced between the cutter and the setting faces—this ensures that the setting block will not be damaged by the cutter during the machining that follows. As explained in para. 9.2, when more than one cutter is involved, only **one** cutter is involved in the setting—the cutter-gang cutters having been previously spaced relative to each other. The vertical setting face is at the 'rear side' of the table so that it does not obstruct the setter's view when setting in the vertical direction.

9.7. General Features of Milling Fixtures

A milling fixture must be strong and very rigid. Unlike drilling jigs, which are usually moved about during a machining operation, a milling fixture is secured to the machine table and not moved until the completion of the batch; and so lightness is relatively unimportant. Milling is an operation that removes a large quantity of swarf, and so it is necessary to provide adequate swarf-clearance ports that enable swarf to be removed without having to invert the fixture. It has already been stated that milling is often done during the early stages of machining castings—adequate clearance must therefore be provided so that the operator can easily load and unload the workpiece even when it is oversized. It will also be appreciated that owing to the nature of the milling operation, the workpiece must be given adequate support during cutting, but because of the uneven surfaces of a rough casting these support points must be adjustable (see page 83 for details of adjustable support points). The body (or base) of a milling fixture is usually of grey cast iron—grey cast iron is used because it has good vibration-damping properties.

9.8. Special Vice Jaws

The machine vice should be regarded as a simple but effective milling fixture; by using special jaws it can be adapted to accommodate workpieces

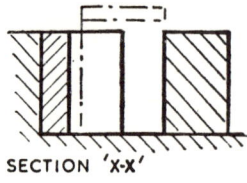

SECTION 'X-X'

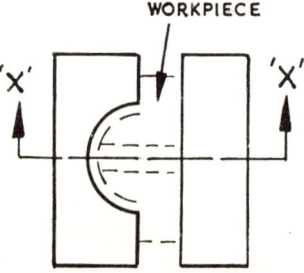

Fig. 9.14 Vice Jaws Shaped to Accommodate the Workpiece

MILLING FIXTURES 143

of awkward shape, or to carry a location system. Figs. 9.14, 9.15 and 9.16 show some typical vice jaws as used to adapt a standard machine vice.

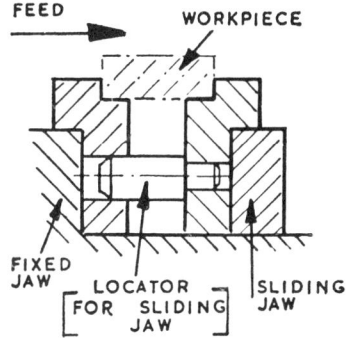

Fig. 9.15 Vice Jaws with Location for the Jaws

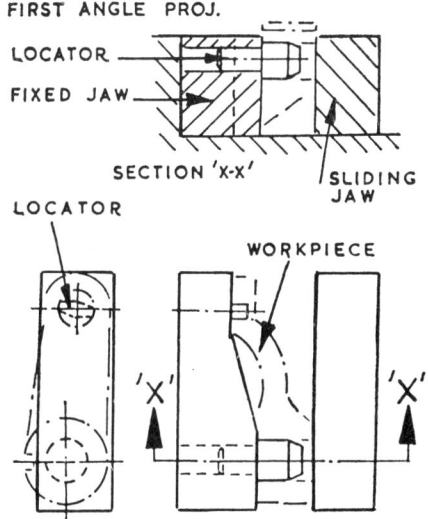

Fig. 9.16 Vice Jaws Shaped to Accommodate the Workpiece and with Workpiece Location

9.9. Milling Fixtures

Fig. 9.17 shows a simple milling fixture that includes the details already described—the base plate and clamp details used in this example are examples of the 'unit tooling' details that are available from certain manufacturers (see Chapter 13 and Chapter 17).

Fig. 9.18 shows a line (or string) milling fixture, in which six workpieces are arranged in a line, and a slot milled in the head of each; they are located and clamped by a series of vee pieces that are located in a guide system and operated by one screw. As an added refinement the movement

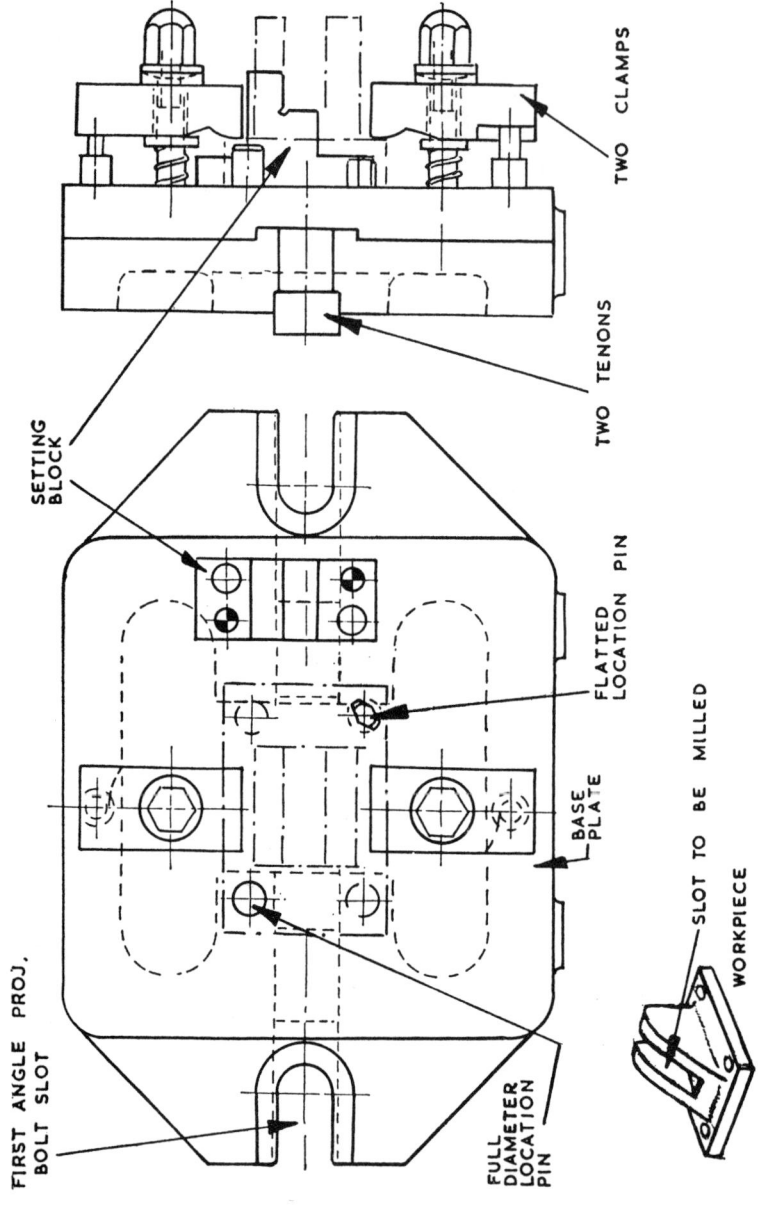

Fig. 9.17 Milling Fixture

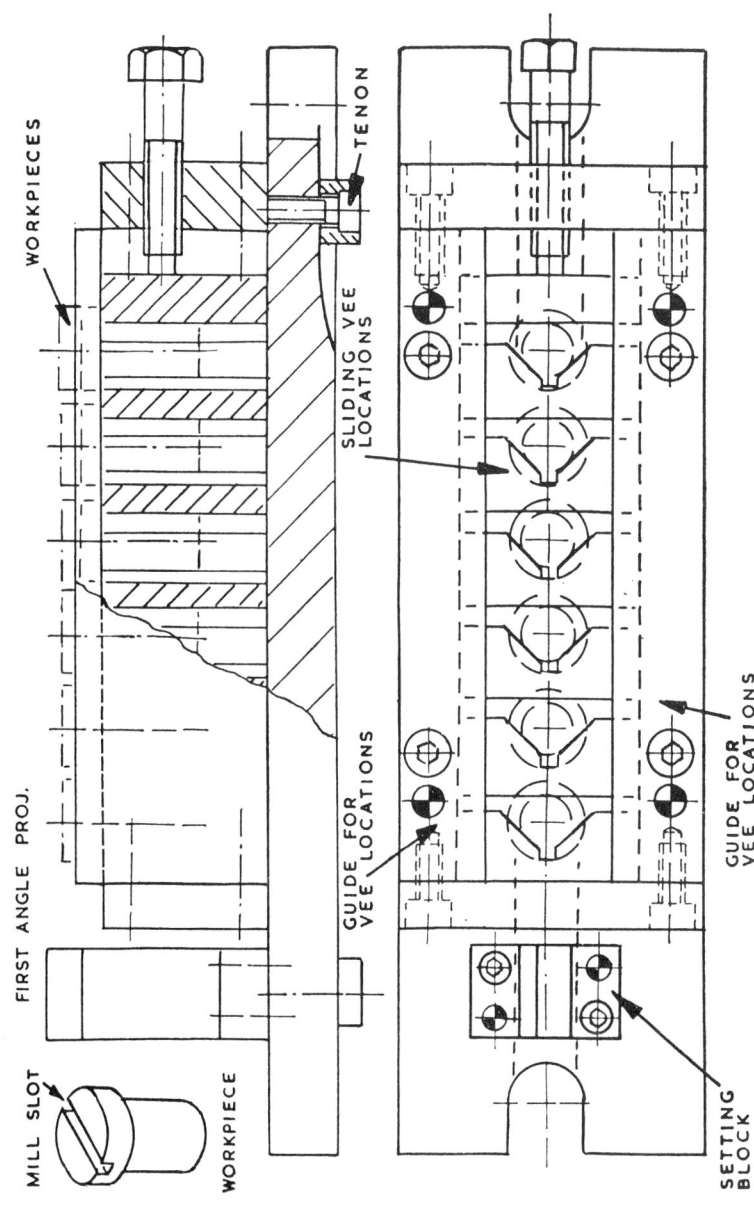

Fig. 9.18 Line or String Milling Fixture

of each of the vee pieces could be restricted by a pin so that it is less tiresome to load the workpieces.

Indexing milling fixtures are described in Chapter 11.

References
B.S. 122:Part 1:1953, Milling Cutters.

Chapter 10

MISCELLANEOUS WORKHOLDING FIXTURES

10.1. Workholding Devices For Turning

10.1.1. Collet chucks. Collet chucks are used in capstan-lathe work to hold bar or tube stock; they are usually classified according to the method of closing the collet. Fig. 10.1 shows a push-out collet; the collet being closed by pushing it against a taper in the hood. This type of collet is

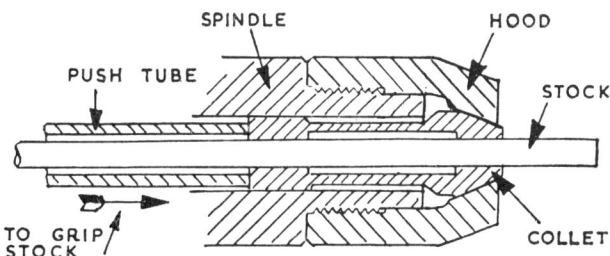

Fig. 10.1 Push-Out Collet Chuck

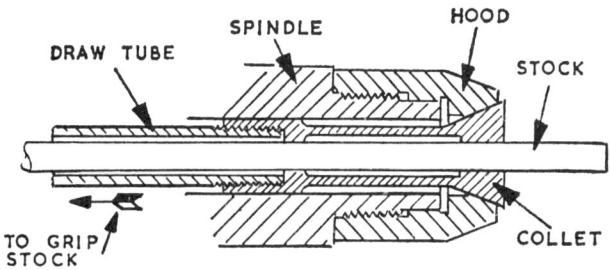

Fig. 10.2 Pull-In Collet Chuck

recommended for first-operation work, and gives a good control of length, because the stock is pushed out during clamping, and the protrusion length can be controlled by an end stop on the turret. In this arrangement the hood must be removed when the collet is changed.

Fig. 10.2 shows a pull-in collet; in this arrangement the collet is closed by pulling it into a taper in the hood. The disadvantages associated with this type of collet are that there is a poor control of length, and the capacity is reduced because the collet and operating tube must be threaded. The hood does not need to be removed when changing the collet.

Fig. 10.3 shows the collet end of a dead-length collet chuck. This collet is designed so that the workpiece is not moved longitudinally during its closing. This produces a good control of length, but the wear of the many components of this collet produces concentricity errors. Very often a master collet is used, and variations of shape allowed for by fitting a liner.

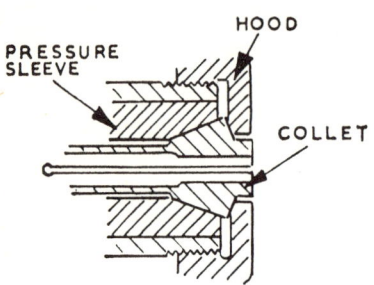

Fig. 10.3 Dead-Length Collet Chuck   Fig. 10.4 Soft Jaw

10.1.2. Jaw chucks. Jaw chucks are used for the early operations upon billets, castings and forgings. Soft jaws (see fig. 10.4) are attached to the radial jaw slide by collar screws, for second-operation work and for irregular-shaped workpieces. Soft jaws are made from case-hardened mild steel. Fig. 10.5 shows some applications of jaw chucks with specially-shaped jaws.

10.1.3. Expanding posts. Expanding posts are used to hold workpieces from a bore. Fig. 10.6 shows a typical expanding post.

10.1.4. Turning fixtures. Turning fixtures are located and clamped to the machine spindle, and carry location and clamping systems. The clamping system must secure the workpiece against the rotational forces as well as the cutting forces. When designing turning fixtures, care must be taken to ensure that there are no projecting details that can cause accidents, and also to provide means of balancing the fixture.

Fig. 10.7 on page 150 shows a simple fixture, and fig. 10.8 on page 151 shows a more complicated fixture of the 'faceplate' type. The fixture shown in fig. 10.8 is used to locate a workpiece from a surface that is not at right angles to that being machined. The supporting 'ledge' is part of the fixture casting, but can be an angle section attached to a circular fixture blank. This illustration also shows a setting face for use with tool setting pieces; the setting face is machined relative to the location fixture, and a set of setting pieces are supplied for the operation.

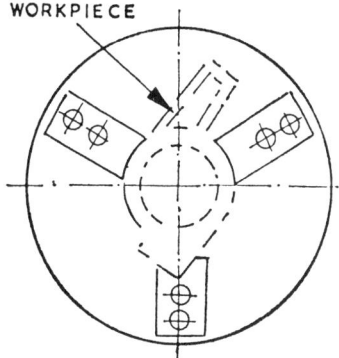

(a) Shaped Chuck Jaws

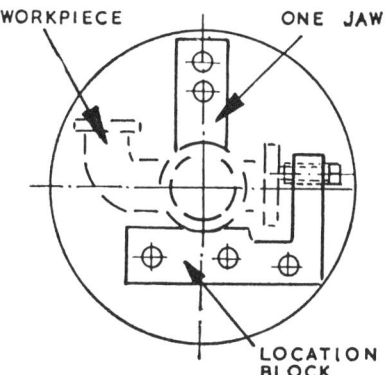

(b) Arrangement with One Chuck Jaw and Location Block

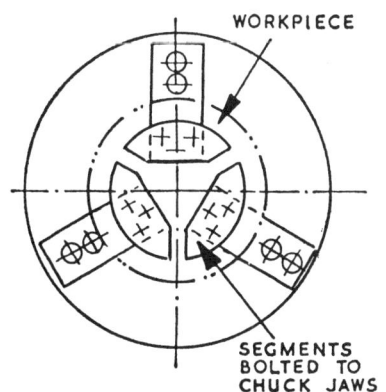

(c) Chuck Jaws for Holding From a Bore

Fig. 10.5

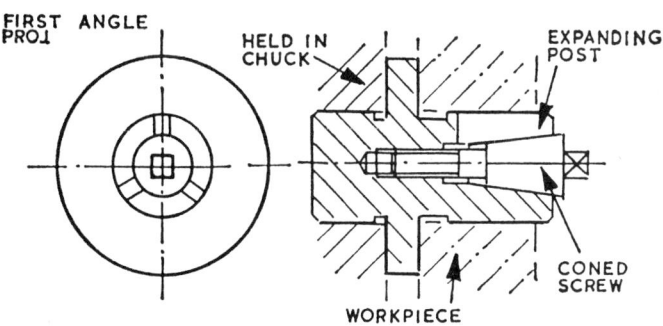

Fig. 10.6 Expanding Post

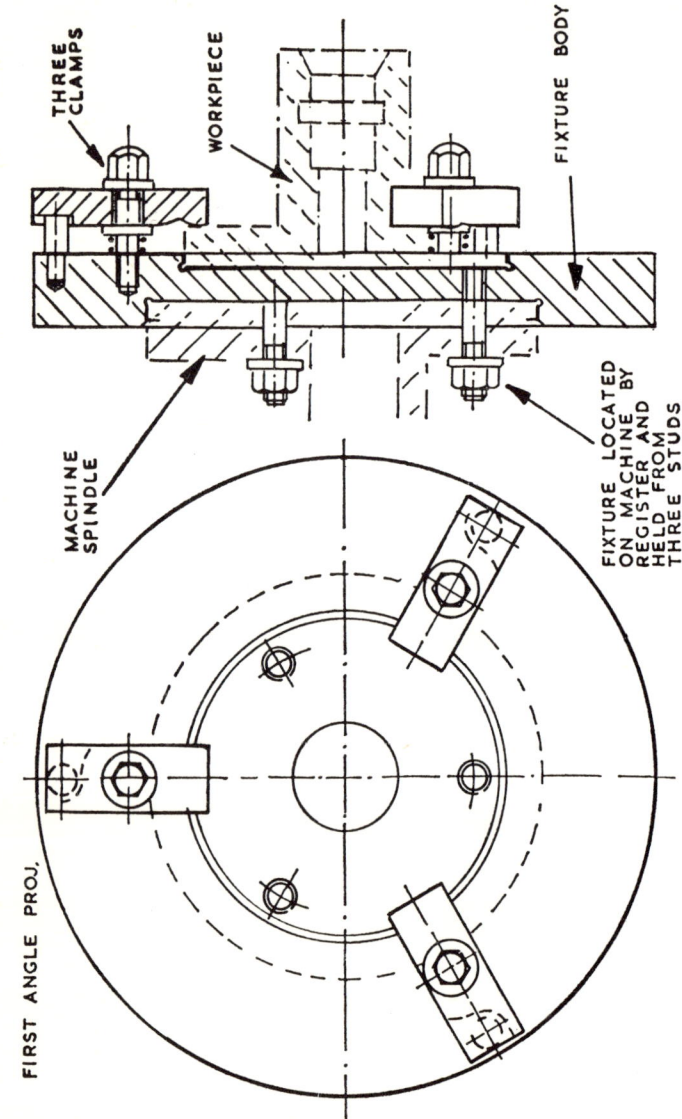

Fig. 10.7 Simple Turning Fixture

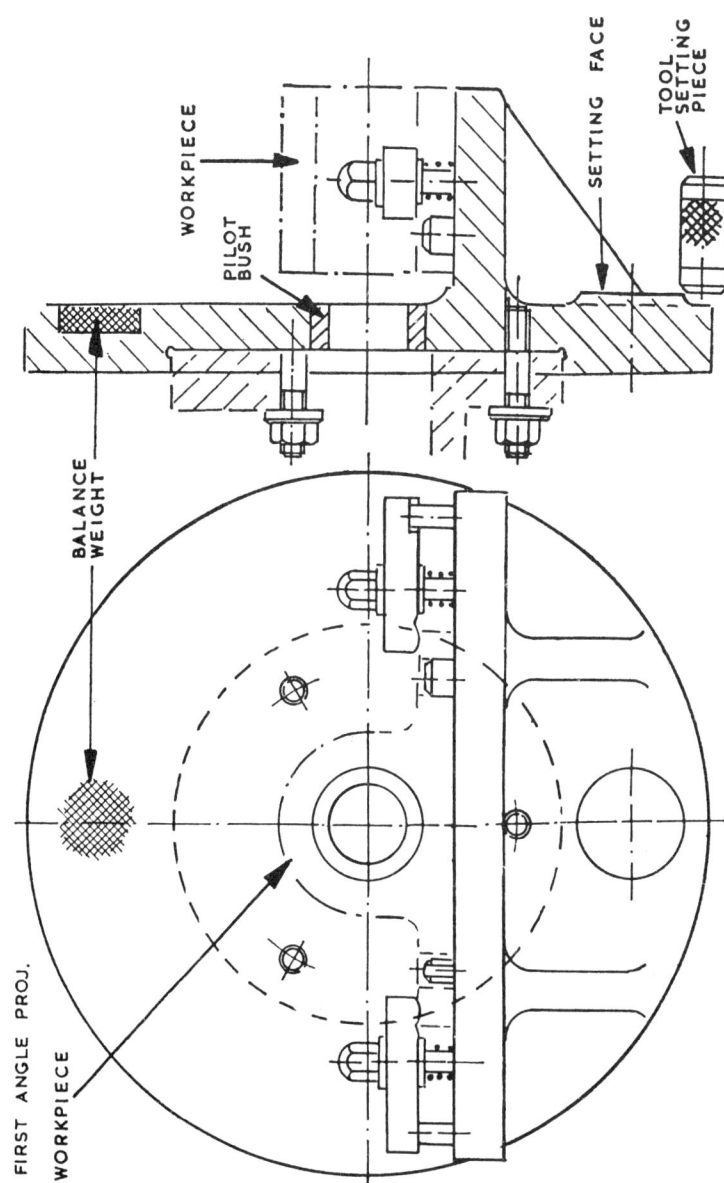

Fig. 10.8 Faceplate Turning Fixture

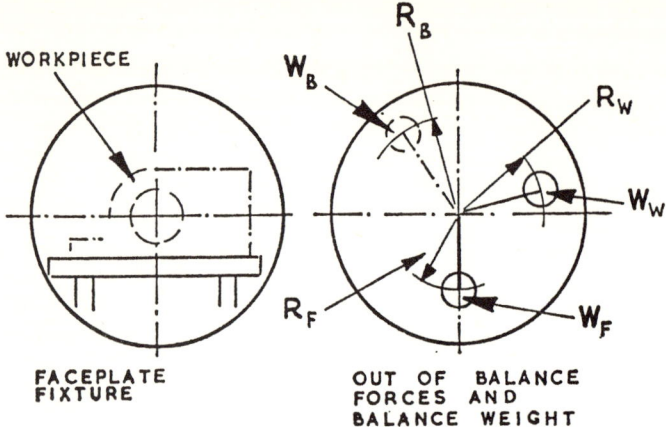

Fig. 10.9 Fixture with Out-of-Balance Forces

Fig. 10.9 shows an 'out-of-balance' faceplate fixture. The 'out-of-balance fixture' force is shown as W_F and acts at distance R_F from the axis; the out-of-balance workpiece force is shown as W_W acting at distance R_W from the axis. The position and magnitude of the balancing force can be found as shown in fig. 10.9 (*a*); in this scale-diagram the length of the lines

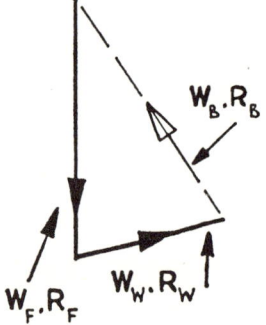

Fig. 10.9 (*a*) To Determine the Balancing Force

represents the moments, and their position indicates the moment direction. The closing line gives the balancing moment and its direction. Although the fixture and workpiece will be balanced in the tool room, this initial work will enable the designer to allow for the balance weight at design stage.

10.2. Workholding Devices for Grinding

10.2.1. Surface grinding. These fixtures are similar to milling fixtures but are more accurate.

10.2.2. Cylindrical grinding. A workpiece with a bore can be held on a mandrel as shown in fig. 10.10; several thin workpieces can be 'stacked' for

MISCELLANEOUS WORKHOLDING FIXTURES

grinding—this will speed up production, and also give the workpiece better support.

External and internal grinding can be done using a fixture that can be regarded as a more accurate version of a turning fixture. The fixture shown

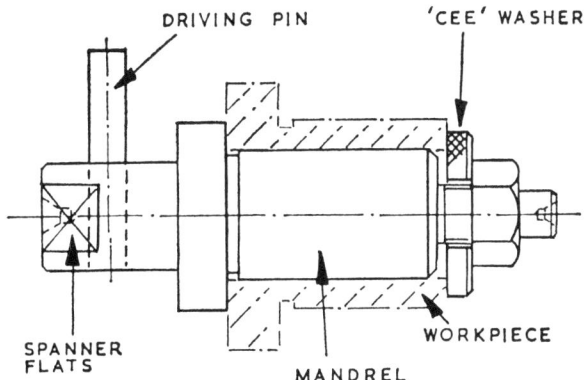

Fig. 10.10 Grinding Mandrel

in fig. 10.11 on page 154 is for internal grinding; it is held from the spigot diameter, but unlike a turning fixture, it is finally positioned up using a dial indicator from the 'clocking diameter', which is in the form of a groove to prevent it from being damaged. The fixture will be very carefully balanced.

10.3. Workholding Devices for Assembly

A workholding device will be used for an assembly operation if it is awkward to do. For example, a clock mechanism involves many gear and spindle units that are sandwiched between two plates containing the pivot bearings; if the reader has attempted to assemble a clock mechanism without a fixture he will appreciate the difficulty of this operation.

An assembly fixture must be designed so that all the parts are located and held during the assembly, but the fixture must be removed easily when the operation is completed. The fixture may contain movable parts; for example, an assembly that contains springs may require a fixture that enables the assembly to be done when the springs are not bearing upon the other components, and allows the springs to be held back until the final stages of the assembly operation.

10.4. Welding Fixtures

A welding fixture can be regarded as a special assembly fixture. It must incorporate means of locating and clamping the parts to be welded, and the usual requirements of simplicity of operation and of foolproofing must be fulfilled.

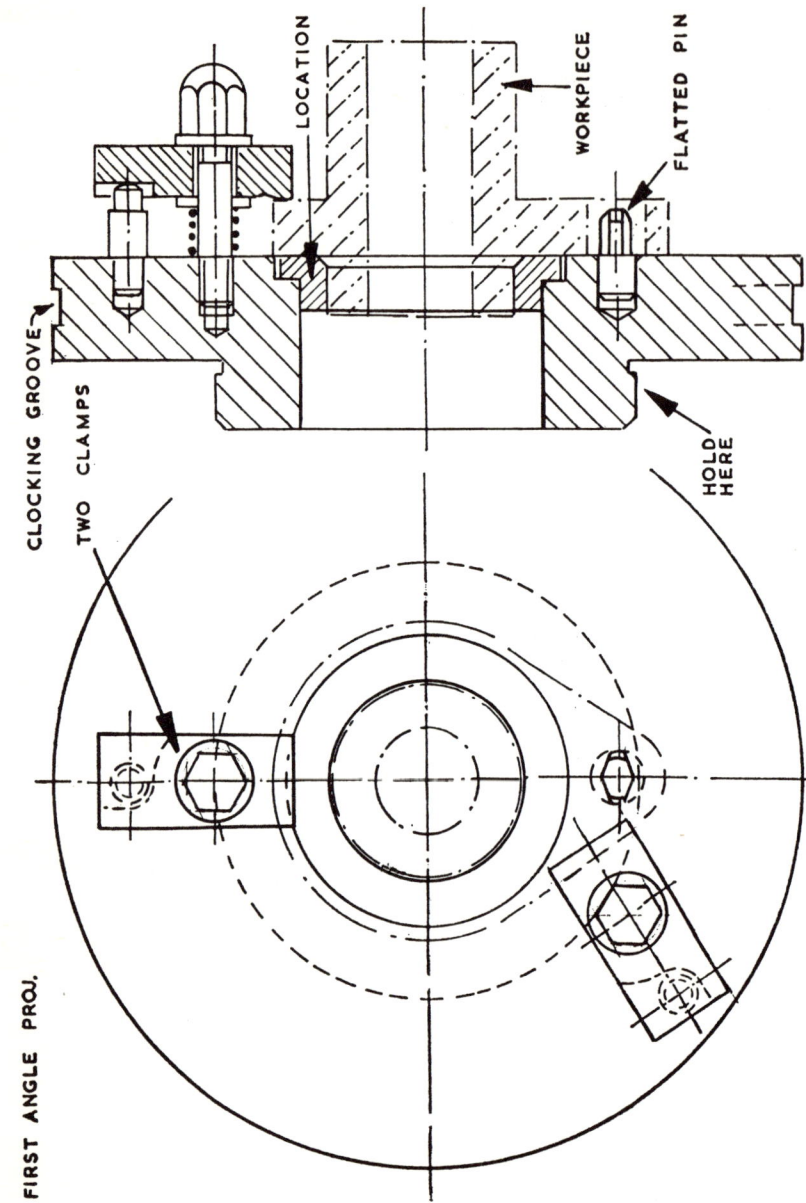

Fig. 10.11 Grinding Fixture

In addition to these requirements, the following special requirements must also be considered:

1. The welder must have access to all the joints, so that he does not need to tack-weld, and complete the welding when the assembly has been removed from the fixture.

2. The fixture must be designed so that it can be moved about during the welding operation to allow the welder to use the most convenient welding technique. The fixture must therefore have some form of trunnion mounting, be such that it can be mounted on a trunnion system or be light enough to allow the welder to move it about as required.

3. The design must allow for the high, concentrated heat associated with welding. The clamps must therefore hold the parts against the distortional forces, and the arrangement must be such that in the event of distortion taking place, the workpiece can be easily removed from the fixture. The position of the clamps and locators must be such that the heat does not damage them or cause errors.

4. Screw threads and similar fixture parts must be protected against weld splatter.

Chapter 11

INDEXING JIGS AND FIXTURES

11.1. Introduction

Indexing jigs and fixtures are used when it is necessary to move the workpiece relative to the machine table or machine cutter spindle between machining various features during an operation, but when the jig or fixture base must remain located relative to the machine. The dividing head used in milling practice is a universal indexing fixture, and indexing milling fixtures are, in fact, a simplified form of dividing head designed to perform direct indexing.

11.1.1. Applications of indexing. The following applications of indexing will show the scope of this technique.

Fig. 11.1 shows a long strip that is to be drilled in several places along its

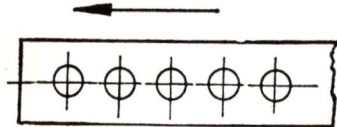

Fig. 11.1 Linear Indexing

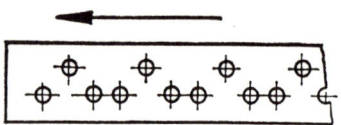

Fig. 11.2 Linear Indexing

length; a non-indexing drilling jig would require a machine with a large 'coverage'. In this example one drill bush is required, and the workpiece moved between drilling each hole so that it is located below the drill bush. More than one drill bush is required if the holes are not in line or when the spacing is irregular; fig. 11.2 shows a strip in which holes are in groups of three, and so an indexing drilling jig that incorporates three drill bushes is used. Figs. 11.1 and 11.2 are examples of linear indexing.

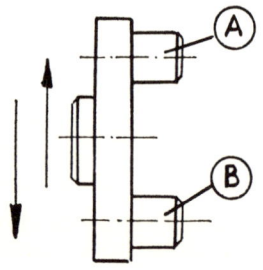

Fig. 11.3 Linear Indexing

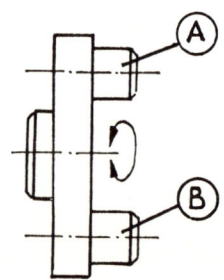

Fig. 11.4 Rotational Indexing

INDEXING JIGS AND FIXTURES

When a workpiece with two or more features on different axes is turned, it must be positioned so that the features are, in turn, positioned on the machine spindle axis. Fig. 11.3 shows how the problem is solved by linear indexing; the workpiece is moved so that the features *A* and *B* are in turn positioned on the machine axis. Fig. 11.4 shows rotational indexing applied to this problem; the workpiece being rotated about an axis between those of the two features so that, again, each is positioned. Rotational indexing is also used when holes have to be drilled on a large pitch circle. Fig. 11.5

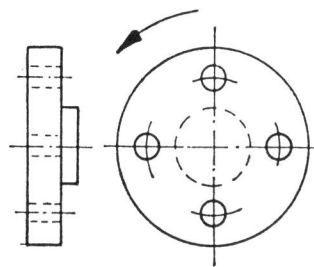

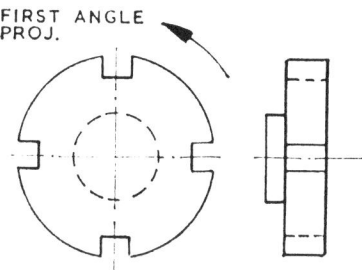

Fig. 11.5 Rotational Indexing   Fig. 11.6 Rotational Indexing

represents such a component; in this example one drilling station is used, and the component is rotated about the axis of the pitch circle to position it for drilling each hole in turn. Fig. 11.6 illustrates rotational indexing when producing radial features; in the example shown, four slots are to be milled in the periphery of a circular workpiece.

Indexing milling fixtures are also used in conjunction with pendulum milling (see para. 9.2.4 on page 137).

11.2. The Basic Features of an Indexing Jig or Fixture

The workpiece is located and clamped to a movable member that can be indexed to bring the workpiece to the required position relative to the cutter or drill bush, and then locked in that position whilst the feature is being machined. The jig or fixture must therefore include component location and clamps mounted on the movable member, a slide or a bearing system to control the movement of the movable member, an indexing device to position the member and a locking device to secure it in position during the machining. It must be emphasised that the locking device for the movable member must be **separate** from the workpiece clamp; it must be convenient to operate because it must be operated each time the workpiece is indexed—for example, in the example shown in fig. 11.6 the workpiece is clamped once during the milling operation, but the movable member must be clamped four times during the operation.

11.3. Indexing Devices

The indexing device is usually mounted in the fixed part of the jig or fixture, and it engages in slots or holes that are suitably spaced in the moving member. It is more convenient to arrange the indexing system in this way

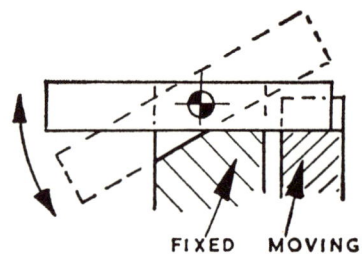

Fig. 11.7 Lever-Type Indexing Device

than to use the alternative, which is to mount the indexing device on the movable member and have the holes or slots in the fixed part, because the latter method would mean that the operator would take longer to perform the indexing movements, and because such an arrangement may be less

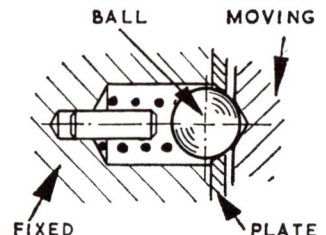

Fig. 11.8 Ball-Type Indexing Device

safe as the operator may have to lean across the machine table and near the cutters whilst performing the indexing movements.

Fig. 11.7 shows a simple locating device in which a lever engages in radial slots in the movable member; the lever can be made out of balance so that it drops into the locating slots under gravity, or as an alternative it can be spring-loaded.

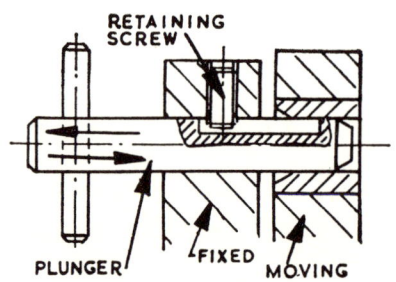

Fig. 11.9 Plunger-Type Indexing Device

INDEXING JIGS AND FIXTURES

Fig. 11.8 shows a spring-loaded ball that is retained by a plate, and located in recesses in the movable member; this device is useful for light work, but is less positive than the other methods described here.

The plunger shown in fig. 11.9 is a common device that gives positive location; it is retained by a screw as shown. As a refinement, the plunger

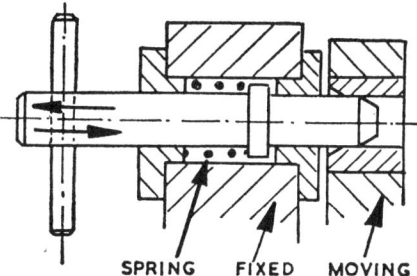

Fig. 11.10 Spring-Loaded Plunger

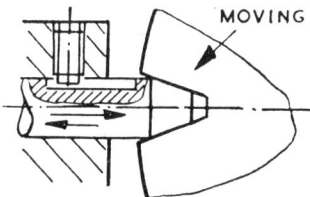

Fig. 11.11 Compensation for Wear

can be spring-loaded as shown in fig. 11.10. The end of the plunger can be coned and the locating holes in the movable member can be coned or vee shaped as shown in fig. 11.11 so that wear can be compensated for.

When the plunger is in an awkward position it may be operated by a rack and pinion system as shown in fig. 11.12.

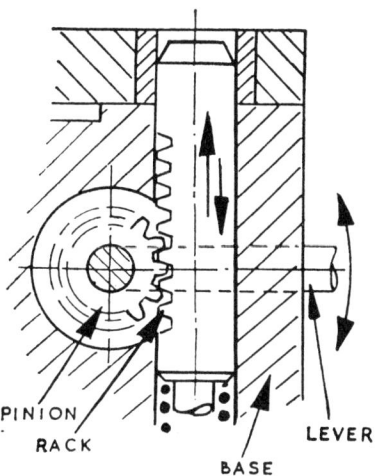

Fig. 11.12 Rack and Pinion Indexing Device

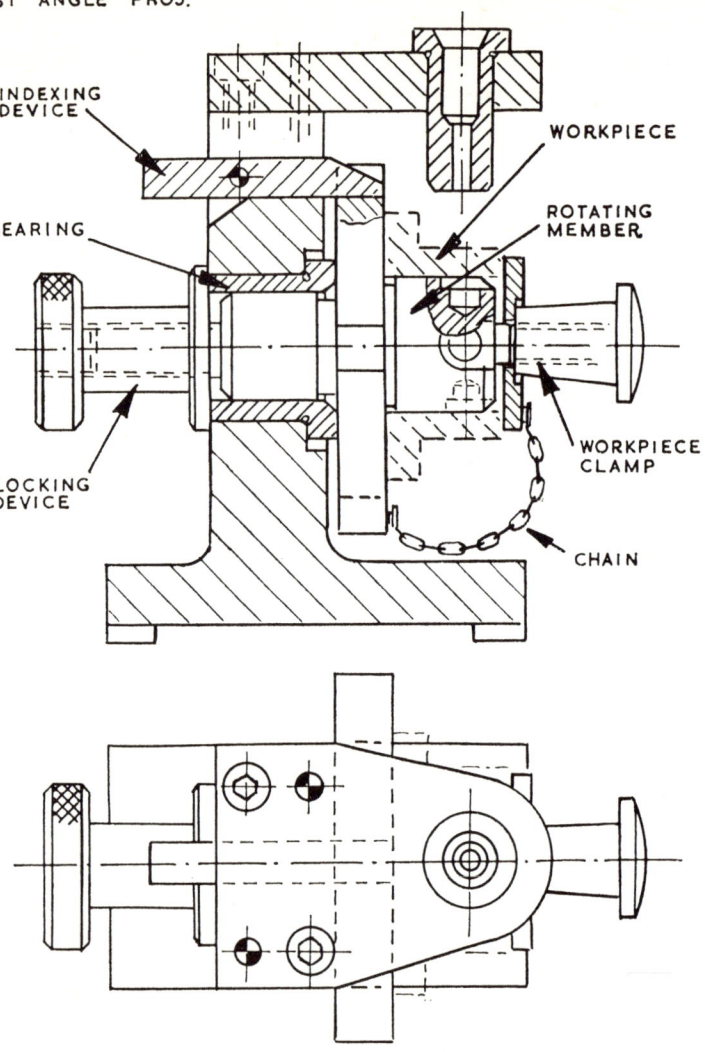

Fig. 11.13 Indexing Drilling Jig

11.4. Typical Indexing Jigs and Fixtures

Fig. 11.13 above shows a simple indexing drilling jig to produce four radial holes in the workpiece shown. The workpiece is located on the rotating member and secured in position. The indexing device consists of a lever that engages in turn in each of the four slots cut around the periphery of the rotating-member flange. The rotating member is locked in position during the drilling with a hand nut.

Figs. 11.14 and 11.14 (*a*) pages 161 and 162 show a typical milling fixture used to index heavy workpieces about a vertical axis. Before each indexing movement the movable fixture-table carrying the workpiece is raised from

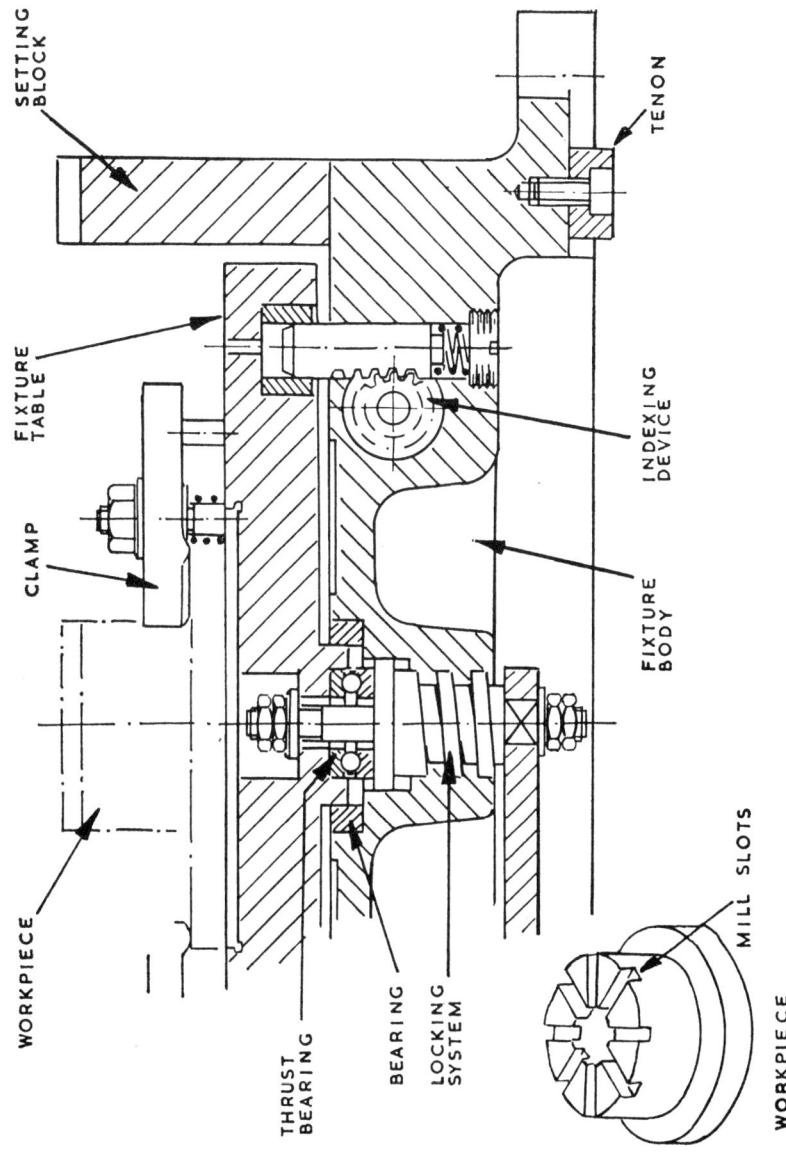

Fig. 11.14 Indexing Milling Fixture

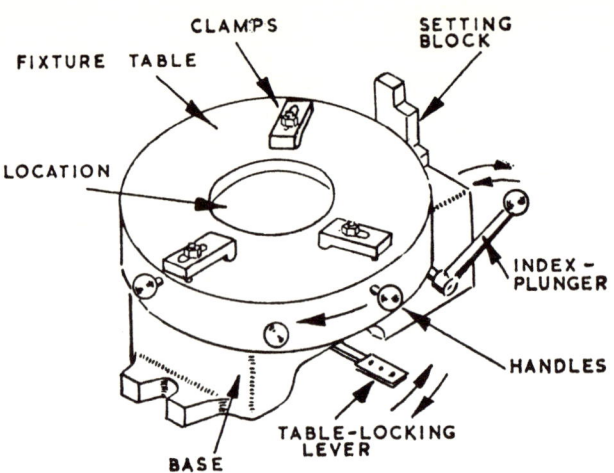

Fig. 11.14 (a) Indexing Milling Fixture

he fixture body by operating the table-locking lever so that the fixture-
able and workpiece are supported by the thrust bearing, and can be rotated
asily during indexing. The indexing device is of the rack and pinion type.
After indexing, the fixture-table is lowered and locked in position against
ie top face of the base by operating the locking lever.

Chapter 12
BROACHING

12.1. Introduction

Broaching is a metal-cutting forming process in which metal is removed by pushing or pulling a multi-toothed cutter across the surface to be machined. The cutter is called a **broach**, and it has a large number of teeth which engage the workpiece in turn; each tooth differs from its neighbour by a small amount, so that the required shape and size of the machined feature is gradually produced by the broach as it passes across the surface **once**. At the end of the stroke the broach is disengaged from the workpiece, and in the case of internal broaching, from the machine also. Usually one broach is sufficient, but if a large stock removal or a complicated shape is involved more teeth will be required, and so more than one broach may need to be used because the length of the broach is limited.

Broaching is usually used to produce complicated shapes that would be difficult or impossible to produce by any other method, but the process may be used to produce simple shapes, such as round holes and Keyways, on the grounds of economics or because the surface texture produced by broaching is more suitable for certain applications than that produced by a rotary cutter.

12.1.1. Advantages claimed for broaching. Broaching is a costly process and so its use must be justified. The following are the advantages claimed for broaching, and these may well influence one choice in favour of the process:

1. When applied to complicated internal shapes, such as serrations, the process permits greater accuracy because since the form on the broach will be external, it can be more easily produced.
2. Each tooth removes only a small amount of metal; this results in less tool wear than in other processes, and enables accuracy of product to be maintained over a longer time.
3. Each tooth is designed for a particular duty during the machining of the workpiece and is therefore more efficient than a cutter that is designed to cut during all stages of the metal-removing process.
4. Each tooth is in contact with the workpiece for part of the cutting time, and so overheating is less than in other machining processes.

12.1.2. Limitations of the broaching process. The following limitations must be taken into account when considering broaching:

1. The broaching machine is expensive and very specialised; the broaches are expensive to manufacture.
2. Broaching cannot be used to machine blind holes because the broach must pass through the hole; it cannot be used when the path of the broach or machine ram is obstructed by features of the workpiece that may be some way from the feature actually being broached.
3. The amount of stock that can be removed is small compared with similar metal-cutting processes.

The broaching process may be the only method of producing the required shape, and in order to overcome the problems associated with obstruction it may be necessary to redesign the workpiece; for example: when a blind hole can only be machined to the required form by broaching, it may be machined as a through-hole, and a blanking piece fitted at the end of the hole.

12.2. Types of Broaching

Broaching is usually classified as (*a*) internal and (*b*) surface. Internal broaching is done to form splines, serrations, etc., in holes that have already been drilled, or drilled and reamed; a characteristic of these operations is that the broach is cutting at many points within the hole, so that the workpiece itself takes the reaction that tends to push the broach away from the workpiece. Surface broaching is used to produce outside features; the broach does not cut all around the workpiece at once, and so it is necessary for the machine to take the reaction that tends to push the broach away from the workpiece. Keyway broaching is used to produce keyways in bores that have already been machined; this process is similar to surface broaching in that the reaction must be taken by the equipment, but it is usually classed as internal broaching because the problems of keyway broach design are more like those associated with internal broaching.

12.2.1. Internal broaching presents problems of broach design because the chips must be carried right through the hole, and the necessarily large chip clearances tend to weaken the broach, which must be sufficiently small so that it can pass through the hole being machined. Surface broaching enables stronger broaches to be used than when internal broaching because there is more space around the workpiece; broaches for surface broaching can be built up in small sections and secured in a holder that is, in turn, secured to the machine slide. This makes the manufacture of surface broaches relatively easy compared with that of internal broaches, and also means that if some teeth are damaged or become undersized as a result of regrinding, only the offending section need be replaced.

BROACHING

12.3. Broaching Machines

Broaching machines are classified broadly into: (1) internal broaching machines; (2) surface broaching machines. Within each group the machines are further classified according to the ram operation as follows:

(1) Internal broaching machines
- (*a*) Horizontal machines
    - (i) Pull type
    - (ii) Push type
- (*b*) Vertical machines
    - (i) Pull-up type
    - (ii) Pull-down type
    - (iii) Push-up type
    - (iv) Push-down type

(2) Surface broaching machines

12.3.1. Each type of machine is also classified according to the operation of the ram, i.e. screw, rack and pinion, or hydraulic operation, and into single-slide or ram, or double-slide or ram. The type of machine to be used and its size must usually be known before the broach design is started because this will control the design of the broach.

12.3.2. The design of broaches will now be considered under: (1) internal broaches; (2) surface broaches.

12.4. Internal Broach Design

Fig. 12.1 illustrates a pull-type broach for a round hole. The principal features are (i) the pull-end, (ii) the front and rear pilots, (iii) the follow-rest grip and (iv) the broach teeth.

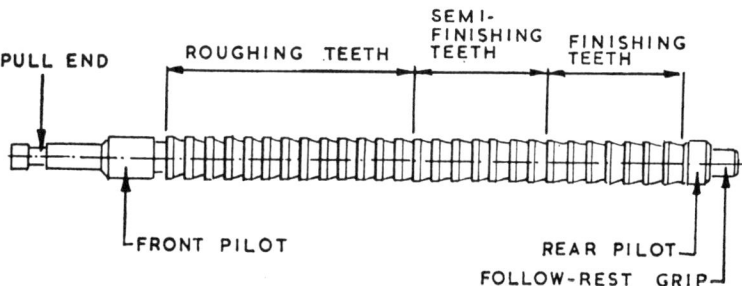

Fig. 12.1 Pull-Type broach

The pull-end is designed to suit the broach puller (see page 175). The front pilot locates in the hole in the workpiece that has already been produced, and the rear pilot is used to guide the broach whilst the last teeth are still passing through the workpiece. The follow-rest grip is introduced at the end of heavy broaches that are used on horizontal machines so that they can be gripped in a supporting device. The teeth are divided up into

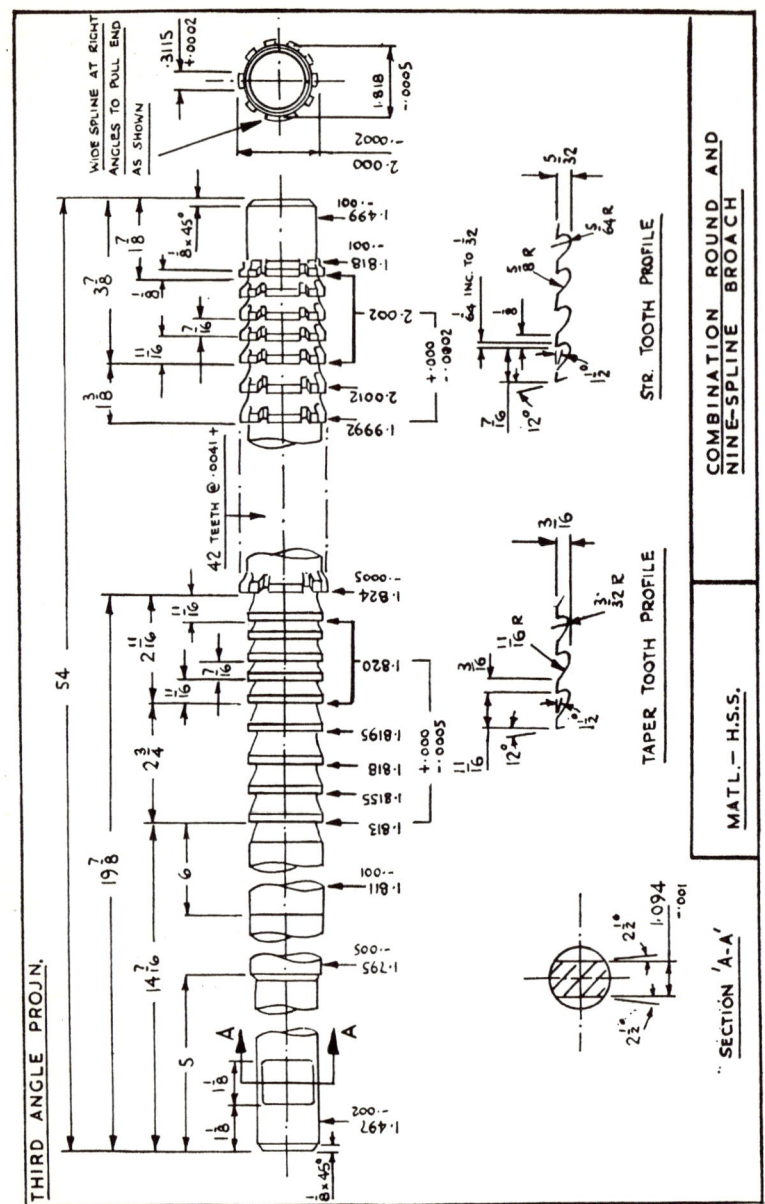

Fig. 12.2

roughing teeth, semi-finishing teeth and finishing teeth; there is often a gap between each set of teeth; the roughing and semi-finishing teeth are often called 'cutting teeth' and the finishing teeth 'sizing teeth'.

12.4.1. Fig. 12.2 on page 166 shows a drawing of a typical broach—certain dimensions and details of chip breakers (see para. 12.4.10 on page 170) have been omitted from the drawing for simplicity.

12.4.2. Fig. 12.3 shows the principal features of broach teeth; the factors that control these features will be considered in the sections that follow.

A broach must have adequate strength to prevent it from failing in tension if a pull broach, or buckling if a push broach. It must have adequate chip-carrying capacity because chip-packing can cause tooth failure.

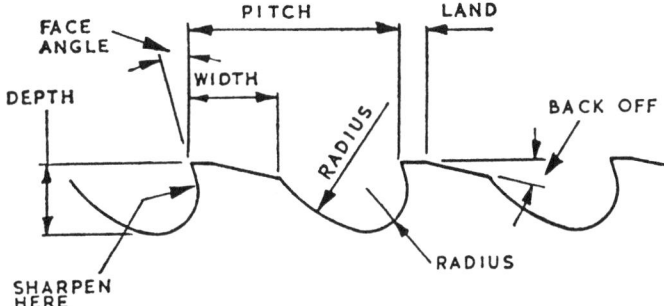

Fig. 12.3 Typical Broach Teeth

It must be emphasised that it is impossible to work to precise rules when designing broach teeth; such factors as workpiece material and workpiece shape will influence the design of broach teeth. In practice, broach manufacturers undertake to design broaches, and they are able to base their design upon carefully-recorded experiences.

12.4.3. Broach material. Case-hardened steel is suitable for small quantities, and high-speed steel is used for larger quantity production. A typical high-speed steel contains 0·7% carbon, 4% chromium, 18% tungsten and 1% vanadium.

12.4.4. Face angle (also called the hook angle). This is the rake angle, and is similar in magnitude to that used when turning; in general, brittle materials require a small face angle, and ductile materials require a large face angle.

12.4.5. Pitch of teeth. The tooth pitch influences the tooth space, the cutting force required and the length of the broach. The tooth space must

Workpiece material	Face angle
Aluminium alloy	10–12°
Brass	2°
Bronze	0°
Cast iron	5–8°
Copper	12°
Magnesium alloys	12°
Steel	8–15°
Zinc	6°

Table 12.1

be large enough to accommodate the chips; for safety, it is usual to make the tooth space large enough to accommodate the chips produced by three passes of the broach, so that the broach or the workpiece surface will not be endangered even if the operator fails to brush away the chips from the tooth spaces after the second pass of the broach. In order to promote a steady cut, four or more teeth should be cutting at a time, even if it means reducing the cut taken by each tooth to make this possible without broach failure.

The actual pitch is based upon an initial value, which is usually obtained from the following formula which takes into account the length being cut:

$$\text{Pitch} = 0.35\sqrt{L}$$

where L is the length to be cut.

In order to prevent the need for fine pitches and also to increase the rate of production, a number of workpieces can be 'stacked' during the broaching operation.

The initial pitch obtained from the above can be reduced when large bores are to be broached, and the consequent reduction in tooth space compensated for by increasing the tooth depth. The pitch is often varied from tooth to tooth to damp the vibration which would be produced if the teeth were to engage the workpiece at regular intervals of time, as would be the case if the teeth were given a uniform pitch.

Typical values of this **'stagger'** are:

Up to $1\frac{1}{2}$ in diameter—stagger $\frac{1}{64}$–$\frac{1}{32}$ in.
More than $1\frac{1}{2}$ in diameter—stagger up to $\frac{1}{16}$ in.

This stagger will increase the cost of broach manufacture and regrinding, and its use must be justified.

12.4.6. Tooth width. The tooth width, together with the pitch of the teeth and the shape of the tooth space, controls the strength of the tooth. The tooth width is usually between $\frac{1}{4}$ and $\frac{1}{3}$ that of the pitch.

BROACHING

12.4.7. Back off (also called the relief and the body clearance). The back-off angle (see figs. 12.4 and 12.5) should be sufficient to prevent the broach from rubbing, but must not be so large as to limit the life of the broach owing to loss of size following re-grinding. The roughing teeth are given a

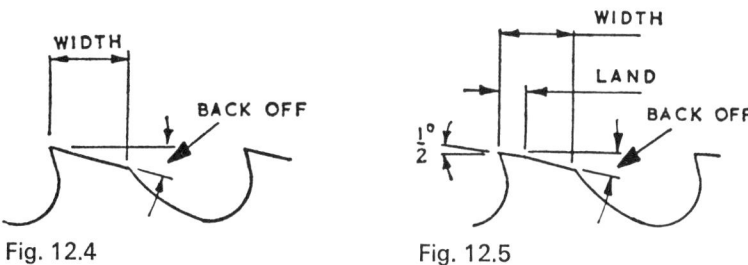

Fig. 12.4   Fig. 12.5

small ($1\frac{1}{2}°$) back-off angle, leaving a small land of up to 0·010 in. Finishing teeth (fig. 12.5) are given a land of about 0·030 in with a cutting clearance of about $\frac{1}{2}°$ to avoid loss of accuracy following re-grinding.

12.4.8. Side clearance. Broaches for cutting splines and gears, etc., may be given a side clearance to reduce friction. This is produced by grinding a taper of about 0·005 in over the length of the broach, on the sides of the teeth. This form of side clearance produces a small form error, so that, for example, a straight-sided spline would be slightly 'tapered' towards the major diameter; this error can usually be compensated for by designing the mating part accordingly.

12.4.9. Tooth space. The shape and size of the tooth space is very important because the shape of the tooth space controls the chip formation, and the strength of the teeth and size of the tooth space will control the performance and strength of the broach. The back angle (see fig. 12.6) is

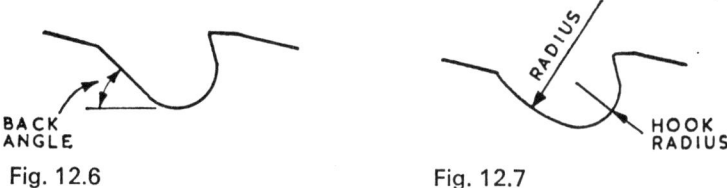

Fig. 12.6   Fig. 12.7

usually made no more than 45° and is blended into the hook radius. Fig. 12.7 shows an alternative shape.

The hook radius influences not only the tooth strength but also the chip formation. Fig. 12.8 shows the effect of a correctly-shaped tooth space, and fig. 12.9 shows how an incorrectly-shaped tooth space acts as a chip breaker, and produces malformed chips which pack the tooth space. The hook radius is usually 0·25 × pitch, but should not be less than 0·5 × tooth depth.

The tooth depth, together with the pitch of the teeth and the shape of the tooth space, controls the broach performance. The size of the tooth space (also called the 'gullet') must be adequate; the actual size required depends

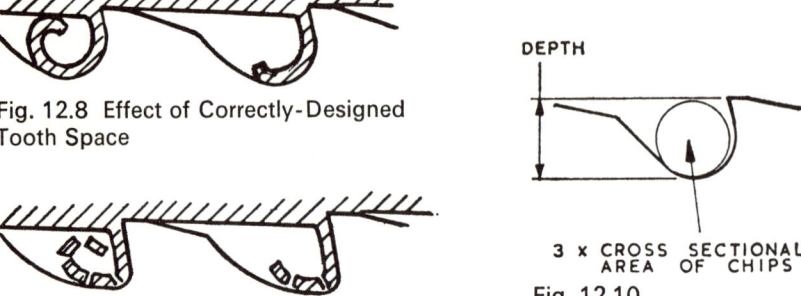

Fig. 12.8 Effect of Correctly-Designed Tooth Space

Fig. 12.9 Effect of Incorrectly-Designed Tooth Space

Fig. 12.10

upon the machining characteristics of the workpiece material, but as a guide, the tooth space should be large enough to contain a circle whose area is 3 × (cross-sectional area of chips produced by the tooth that follows); this is illustrated in fig. 12.10.

12.4.10. Chip breakers. Chip breakers are introduced around the roughing and semi-finishing teeth to enable the chips to be managed more easily. They take the form of small 'nicks' that are staggered from tooth to tooth so that the metal is completely cut (see fig. 12.11).

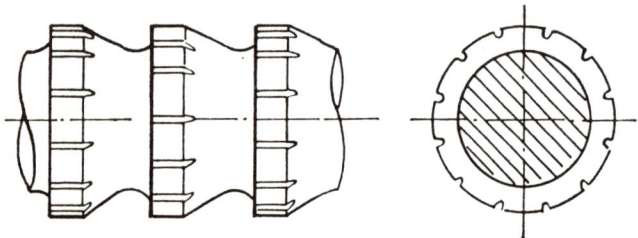

Fig. 12.11 Chip Breakers

12.4.11. Cut per tooth. The cut per tooth (i.e. the change in size from one tooth to the next) depends upon the material to be cut, length of cut, type of broach and the finish required. If too small, rubbing will occur, and if too large, tooth breakage will occur. The minimum cut per tooth is normally taken to be 0·000 5 in on diameter, and the maximum cut per tooth to be 0·006 in on diameter.

BROACHING 171

12.4.12. Burnishers. Some broaches include burnishers which operate behind the final teeth and tend to compress the surface of the workpiece to produce a smooth, hard surface. Figs. 12.12 and 12.13 illustrate typical burnishers.

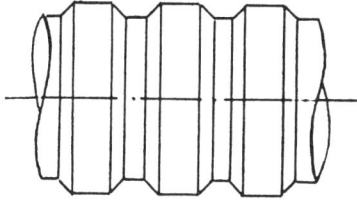

Fig. 12.12 Burnishers

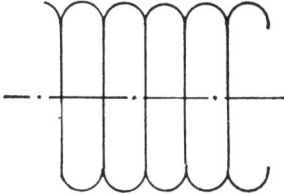

Fig. 12.13 Burnishers

12.4.13. Number of teeth. The number of teeth on the broach depends upon the amount of metal to be removed and the cut per tooth. In addition to these teeth a number of undersized teeth (called safety teeth) are introduced in front of the roughing teeth in case the hole is undersized, and three or four additional finishing teeth are introduced so that the life of the broach is not unduly reduced by re-grinding.

This can be simply expressed as:

$$N = \frac{T}{t} + S + F + B$$

where N is the total number of teeth;
T is the amount of metal to be removed;
t is the cut per tooth;
S is the number of safety teeth;
F is the number of additional finishing teeth;
B is the number of burnishers

$\dfrac{`T`}{t}$ in the above expression is representative since the cut per tooth will not be the same for all teeth.

12.4.14. Calculation to obtain a constant load per tooth. When a broach is used to cut serrations, splines and similar shapes, the width of the form is reduced as the required final shape is approached. Fig. 12.14 represents

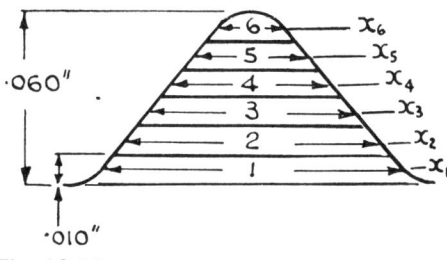

Fig. 12.14

one serration; if the hole has already been drilled and reamed, and the depth of serration is 0·060 in, the width of the cut taken will be reduced as shown in the fig. as the final size is approached. If the cut per tooth is kept constant the load per tooth will be greater at the roughing stages than at the finishing stages; the cut per tooth must therefore be increased from the roughing stages to the finishing stages if the load per tooth is to be kept constant.

A simple way to determine the cut per tooth for constant load is as follows:

1. Divide up a large-scale drawing of one serration space into a number of strips of equal depth—strips 1–6 in fig. 12.14. The area of each strip can be calculated, e.g. the area of strip 1 is 0·010 in × length x_1 (where x_1 is the mean length of strip 1 and found by scaling the drawing). The area of each strip is therefore proportional to its mean length.

2. Plot a graph as follows (see fig. 12.15). Draw the x and y axes, and divide the x axis into lengths that are proportional to lengths x_1, x_2, x_3 ... x_6, and the y axis into six equal divisions—each division representing 0·010 in (refer to the dimensions on fig. 12.14).

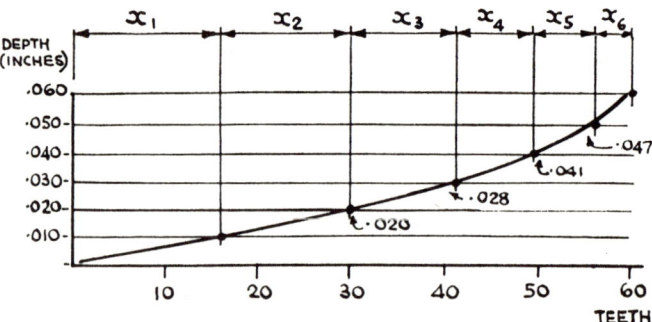

Fig. 12.15

Plot points as shown in fig. 12.15 to produce a curve. Assume that it has already been decided to have 60 teeth, divide up the x axis to represent 60 teeth. The curve will now give the cut that must be taken to ensure that the tooth loading is constant; in practice it is more economical to keep the depth of cut constant for a number of teeth, and so it is now necessary to satisfy both constant load and economy.

It will be seen that in this example the curve is almost a straight line up to the point that corresponds to '30 teeth', and the cut per tooth for the first 30 teeth is almost $\frac{0·020}{30}$ in. The curve is almost a straight line between '30 teeth' and '40 teeth', and the cut per tooth over this range is almost $\frac{0·008}{10}$ in.

BROACHING 173

Similarly the cut per tooth between tooth '40' and tooth '50' is almost $\frac{0 \cdot 013}{10}$ in, between tooth '50' and tooth '55' is almost $\frac{0 \cdot 006}{5}$ in, and for the last teeth $\frac{0 \cdot 013}{5}$ in.

It will be seen that the cut per tooth increases sharply towards the end; if the cut per tooth is excessive it will be necessary to redistribute the share of metal removal.

12.4.15. Section broaches. Broaches for large-diameter holes can be made up from short lengths of 'shell' broaches, located on a central stem, located radially relative to each other with suitable location and secured by nut. Fig. 12.16 shows part of a typical built-up broach. Section broaches are easier to heat-treat, and have an extended service life since undersized or damaged sections can be replaced.

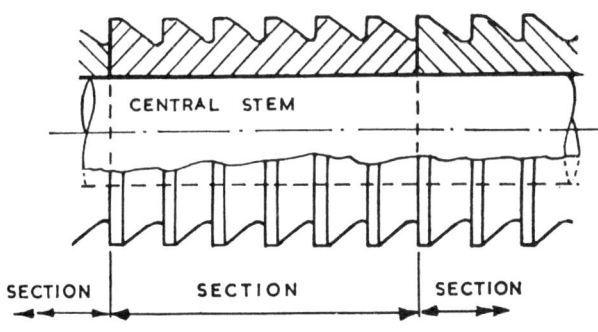

Fig. 12.16 Section Broach

12.4.16. Power required. Power required depends upon the number of teeth in contact with the workpiece at once, the chip size and the strength of the workpiece material. It has been found that the power required is greater when a small chip thickness is removed, and so a multiplier is used that takes into account the workpiece strength and the chip thickness. The power required can be obtained from:

$$L = N . C . W . K$$

where L = load in 100 000 lbf;
        N = number of teeth in contact at a time;
        W = chip width (in the case of a circular hole this will be the circumference) in inches;
        C = chip thickness (i.e. cut per tooth) in inches;
        K = multiplier that allows for workpiece strength and the chip thickness effect (see Tables 12.2 (*a*) and (*b*)).

Cut per tooth in inches	K value		
	Cast iron	Low-carbon steel	Alloy steel
0·000 5	6·6	7·6	8·0
0·001 0	5·1	5·6	7·25
0·001 5	4·0	4·6	6·0
0·002 0	3·4	3·85	5·2
0·002 5	3·0	3·4	4·8
0·003 0	2·7	3·0	4·5
0·003 5	2·5	2·8	4·25
0·004 0	2·32	2·7	4·0
0·004 5	2·3	2·6	3·95
0·005 0	2·25	2·56	3·8
0·005 5	2·21	2·55	3·65
0·006 0	2·2	2·45	3·58
0·007 0	2·2	2·4	3·4
0·010 0	2·1	2·4	3·3
0·015 0	2·0	2·2	3·18

Table 12.2 (a)

Cut per tooth in mm	K value		
	Cast iron	Low-carbon steel	Alloy steel
0,010	6,7	7,7	8,1
0,025	5,1	5,6	7,25
0,040	4,0	4,6	6,0
0,050	3,4	3,85	5,2
0,060	3,1	3,4	4,9
0,075	2,8	3,1	4,6
0,100	2,4	2,8	4,1
0,120	2,3	2,6	3,95
0,128	2,25	2,56	3,8
0,140	2,2	2,55	3,65
0,150	2,3	2,5	3,6
0,178	2,2	2,4	3,4
0,250	2,2	2,4	3,4
0,380	2,0	2,2	3,18

Table 12.2 (b)

12.4.17. Push broaching. Fig. 12.17 illustrates push broaching; it will be seen that there is a plain portion at the end of the broach so that the final broach teeth are pushed through the workpiece before the ram touches it. Push broaches cannot be as long as pull broaches because they tend to bend at the early stages of broaching.

BROACHING

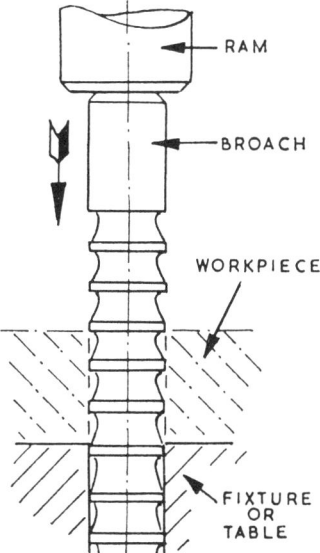

Fig. 12.17 Push Broaching

12.4.18. Pull broaching. Pull broaches can be made longer than push broaches; the end of these broaches is shaped to fit the puller which is attached to the machine ram. The pullers and associated broach pull-ends will now be considered in some detail.

12.4.19. Broach pullers. The broach is attached to a holder (called a 'puller') which in turn is attached to the machine pull-head. Fig. 12.18

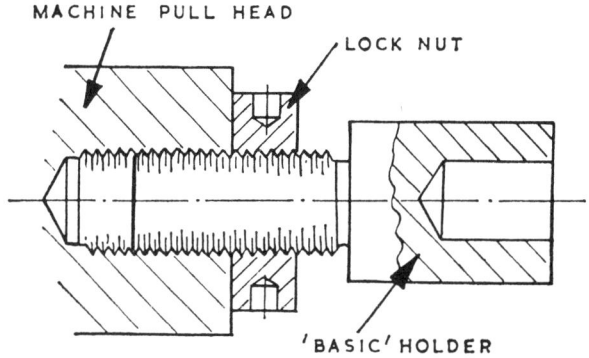

Fig. 12.18 Method of Attaching Holder to Pull-Head

shows a typical method of attaching the puller to the pull-head; a 'basic' shape is shown—the precise shape of the puller depends upon the method of attaching the broach to the puller.

Keyway broaches are usually attached to the holder by screw thread, and not removed from the holder between each pass. Instead, the finished part is removed and the next part is 'threaded' over the full length of the broach; this is possible because the keyway broach is very much smaller than the hole in which the keyway is to be cut.

Small broaches can also be attached to the holder by screw thread; they can be easily released if the threaded part is flatted and the threaded hole is elongated as shown in fig. 12.19.

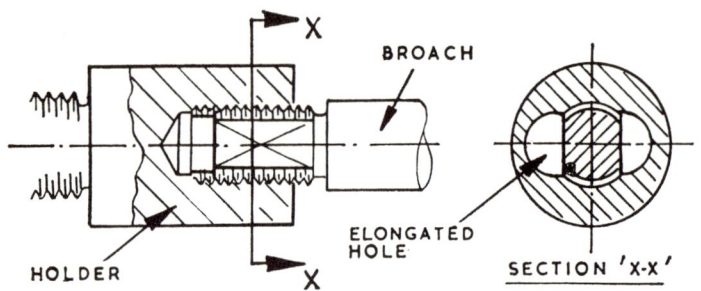

Fig. 12.19 Broach Puller for Screw-Thread Attachment

Pullers using a key or a cotter are inexpensive to make and easy to operate. Fig. 12.20 illustrates such a system; the broach must have a hole in the pull-end to take the taper cotter or key.

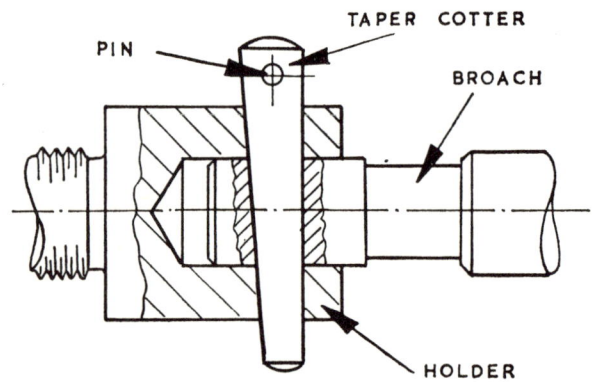

Fig. 12.20 Cotter-Type Puller

When the hole to be broached is small there may not be room to cut a slot in the broach, which must be small to pass through the broached hole; in these cases a pin-type puller is used as shown in fig. 12.21.

A special pin puller is shown in figs. 12.22 and 12.22 (*a*); the pin disengages the broach when moved backwards and engages the broach when moved forwards; an automatic version of this puller has a pin on the front face which operates the pin when pushed.

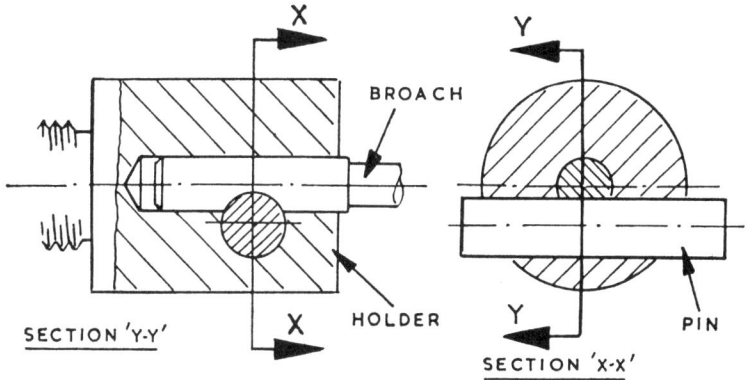

Fig. 12.21 Pin-Type Puller

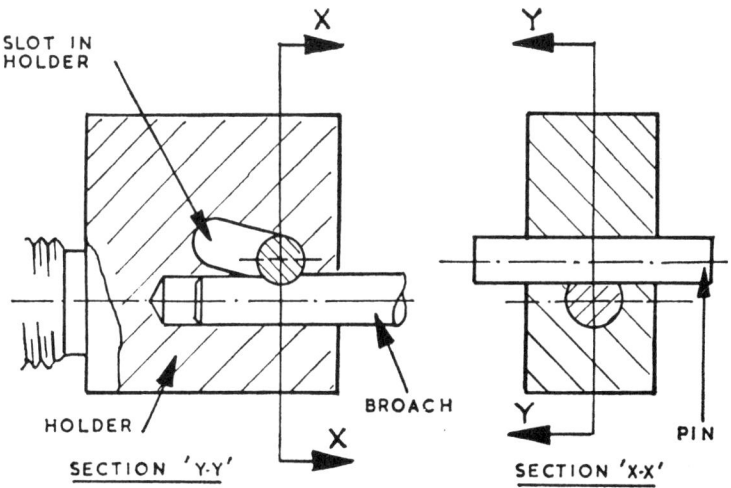

Fig. 12.22

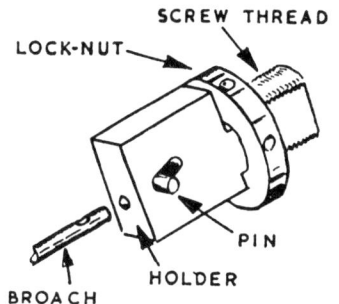

Fig. 12.22 (a)

Fig. 12.22 and Fig. 12.22 (a) Special Pin-Type Puller

Figs. 12.23 and 12.23 (*a*) show a pull-plate-type puller; the pull-plate locates in slots cut in the holder, and engages with slots cut in the pull-end of the broach. This system permits quick release of the broach.

The pullers shown in figs. 12.20–12.23 all enable the broach to be radially

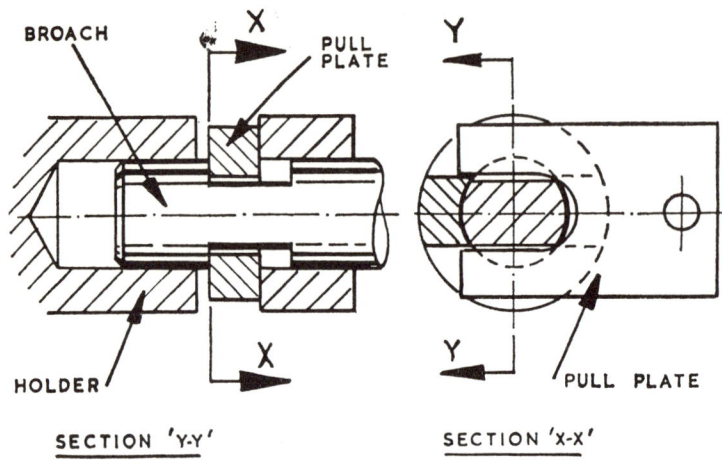

Fig. 12.23

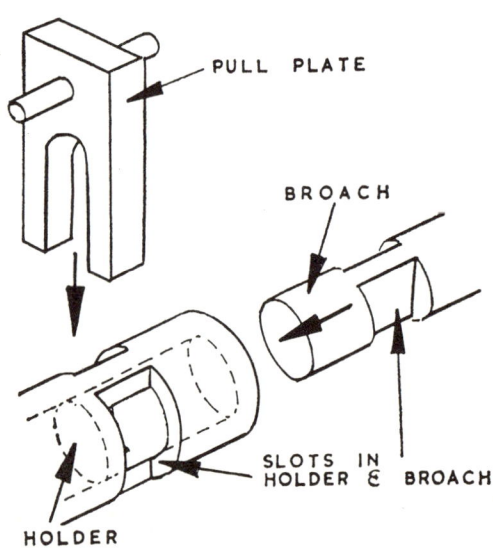

Fig 12.23 (*a*)

Fig. 12.23 and Fig 12.23(*a*) Pull-Plate Puller

located—for example, when broaching serrated holes with one serration left uncut so that it locates a mating part upon assembly.

Fig. 12.24 illustrates an automatic round puller. The sleeve is pulled either by hand or machine stop so that the jaw segments are free to open and to release the broach when the latter is pulled. After the broach has

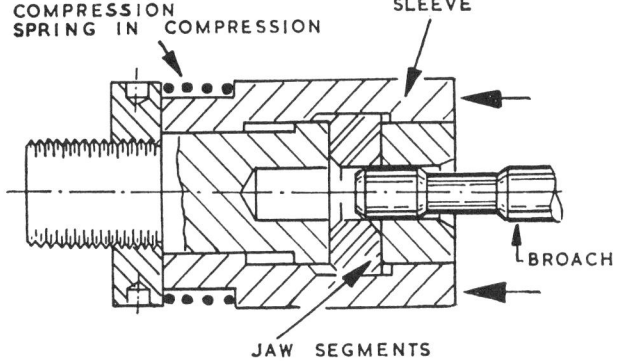

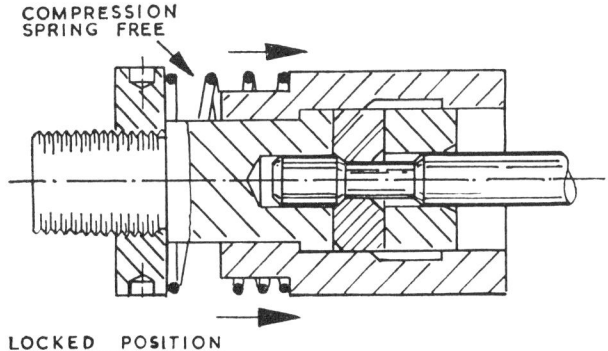

Fig. 12.24 Automatic Round Puller

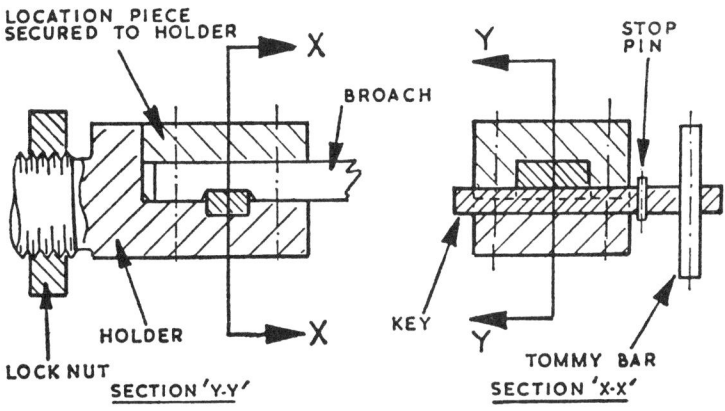

Fig. 12.25 Flat-Type Puller

been replaced in the holder for the next stroke, the sleeve is released, and causes the jaw segments to close in and lock the broach.

Fig. 12.25 shows a flat-type puller that is particularly suitable when it is desired to have a broach with a rectangular body to make re-grinding easier. The holder is made in two parts for easier production—the location piece is grooved to provide location for the broach, and is located and secured to the holder, which has a cross-slot to take the key which secures the broach to the holder unit.

12.4.20. Broach pull-end. The shape of the broach pull-end depends upon the type of puller to be used as shown in the following examples.

Fig. 12.26 shows the pull-end used for keyway broaches; the broach is

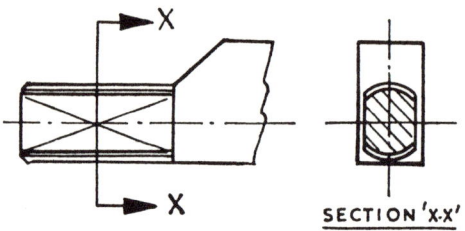

Fig. 12.26 Pull-End for Screwed Puller

secured to the holder (see fig. 12.19) and the workpiece threaded over the broach.

Fig. 12.27 shows a slotted pull-end used in conjunction with a key-type puller (fig. 12.20); the pull-end will have a round hole when the cotter-type puller is used.

A slot across the broach pull-end (fig. 12.28) is used in conjunction with the holders shown in figs. 12.21 and 12.22.

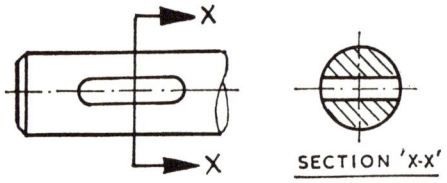

Fig. 12.27 Pull-End for Key-Type Puller

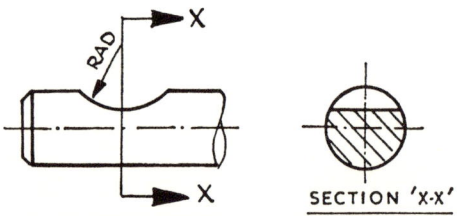

Fig. 12.28 Pull-End for Pin-Type Puller

A double-slotted pull-end (fig. 12.29) is used with the pull-plate puller as shown in fig. 12.23.

A pull-end with a groove (see fig. 12.30) is used with the automatic puller shown in fig. 12.24.

Fig. 12.31 shows the straight-grooved pull-end as used with the flat-type puller shown in fig. 12.25.

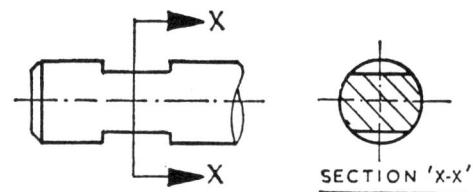

Fig. 12.29 Pull-End for Pull-Plate Puller

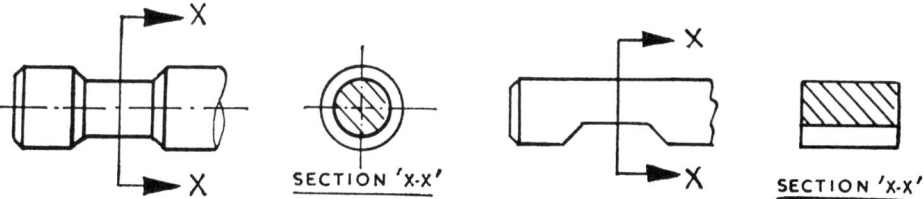

Fig. 12.30 Pull-End for Automatic Round Puller

Fig. 12.31 Pull-End for Flat-Type Puller

12.4.21. Spiral broaching (more correctly called 'helical broaching'). This technique is used to cut internal splines on a helical path, in a hole that has already been drilled and reamed. In order to generate a helix, the broach must rotate about its axis whilst cutting the splines. To avoid interference, the tooth form must lie along a helix of the same helix angle and 'hand' as that of the spline to be cut. Fig. 12.32 shows three teeth of a spiral spline broach.

If the helix angle is less than 8° the broach can be held from a ball-bearing holder, or alternatively, the workpiece can be held in a ball-bearing fixture, so that either the broach or the workpiece can rotate to generate the required helix. If the helix angle exceeds 8°, this 'self-generating' will not take place, and a lead screw, or similar system, must be used to produce a positive rotation.

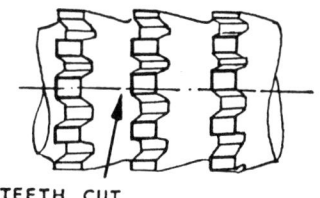

Fig. 12.32 Spiral Spline Broach Teeth

12.5. Surface Broaching

Fig. 12.33 illustrates the arrangement for surface broaching. Reference to the illustration will show that the workpiece is clamped to the fixture table in such a way that it is given maximum support. The broach is secured to

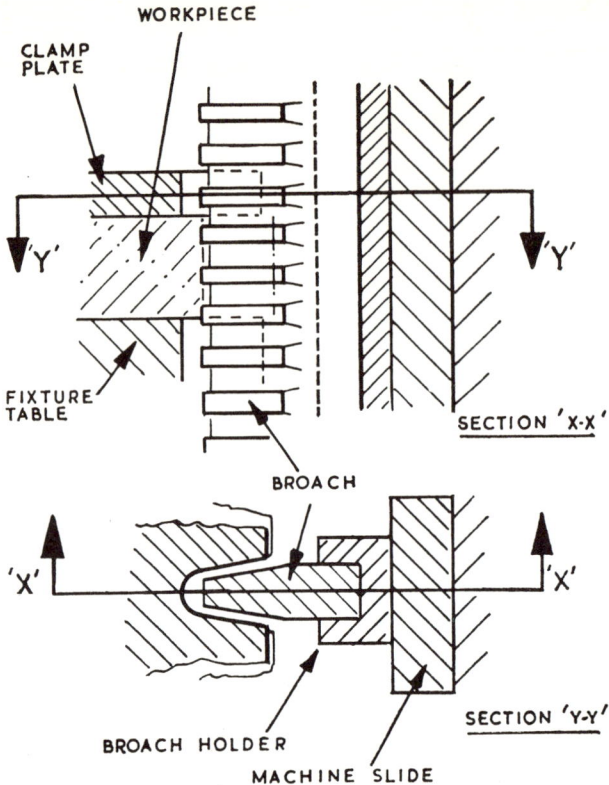

Fig. 12.33 Surface Broaching

the holder, which in turn is bolted to the machine slide. The method of locating the holder on the slide is not shown in the illustration, but it depends upon the shape of the slide. An indexing fixture is used when a number of slots or similar features is to be broached.

12.5.1. Surface broach design. Many of the principles associated with the design of internal broaches can be applied to surface broach design. The principal differences are as follows:

1. Unlike internal broaches, surface broaches are built up in sections (also called 'inserts') that are secured to a holder, which in turn is secured to the machine slide. Some methods of securing the broach to the holder are shown on page 184. The holder may be located on the slide by key and keyway, and then bolted in place, or it may be located and clamped in position by a dovetail system; the method used depends upon the shape of the machine slide. Short broaches are easier to heat-treat than long broaches because there is less tendency for distortion to occur. The life of surface broach sections is extended by 'downgrading' finishing broaches to become roughing broaches when they become undersized as a result of re-grinding.

BROACHING

FIRST ANGLE PROJ.

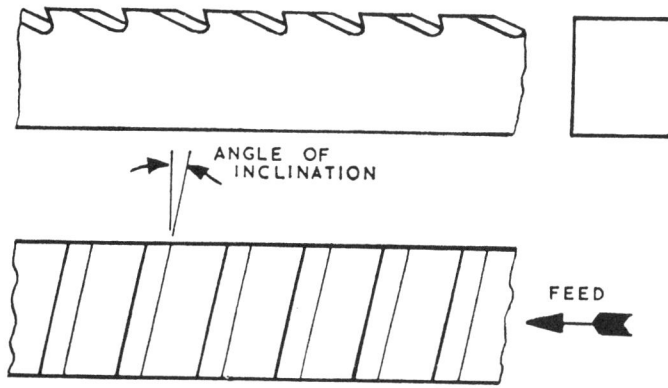

Fig. 12.34 Surface Broach Teeth

2. When an internal broach is, for example, cutting splines, the teeth cut to the same width and the depth is increased; but when surface broaching, the teeth may either:

(a) cut to constant width but increased depth; or
(b) cut to constant depth but increased width; or
(c) Gradually increase the depth and the width.

3. Surface broaches are often made with teeth at an angle (called the angle of inclination on fig. 12.34) to produce a smoother cutting action and reduce the tendency for vibration to occur. This angle is not recommended when deep slots are cut because there is a tendency for the chips to be forced against the side of the slot and so cause inaccuracies. The angle of inclination tends to cause a side thrust that can be minimised by staggering the teeth as shown in fig. 12.35, or by using two broaches to produce a 'herringbone' effect (see fig. 12.36).

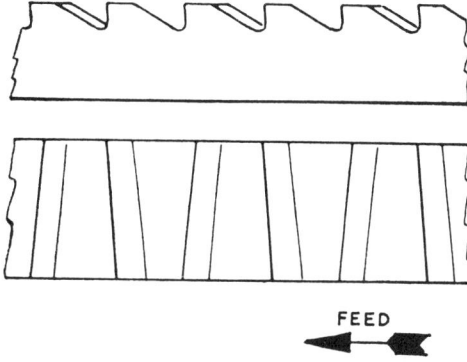

Fig. 12.35 Surface Broach with Staggered Teeth

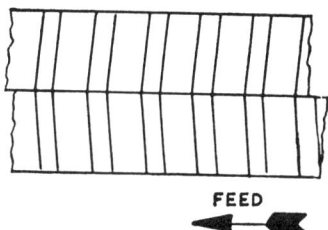

Fig. 12.36 Surface Broaches arranged in Herringbone Arrangement

4. The minimum pitch of teeth is given by:

$$\text{Pitch} = 3\sqrt{L \times w \times r}$$

where L is the length of surface to be broached;
  w is the chip thickness per tooth;
  r is a multiplier that takes tooth-space packing into account. For roughing r is between 3 and 5, and for finishing r is 8.

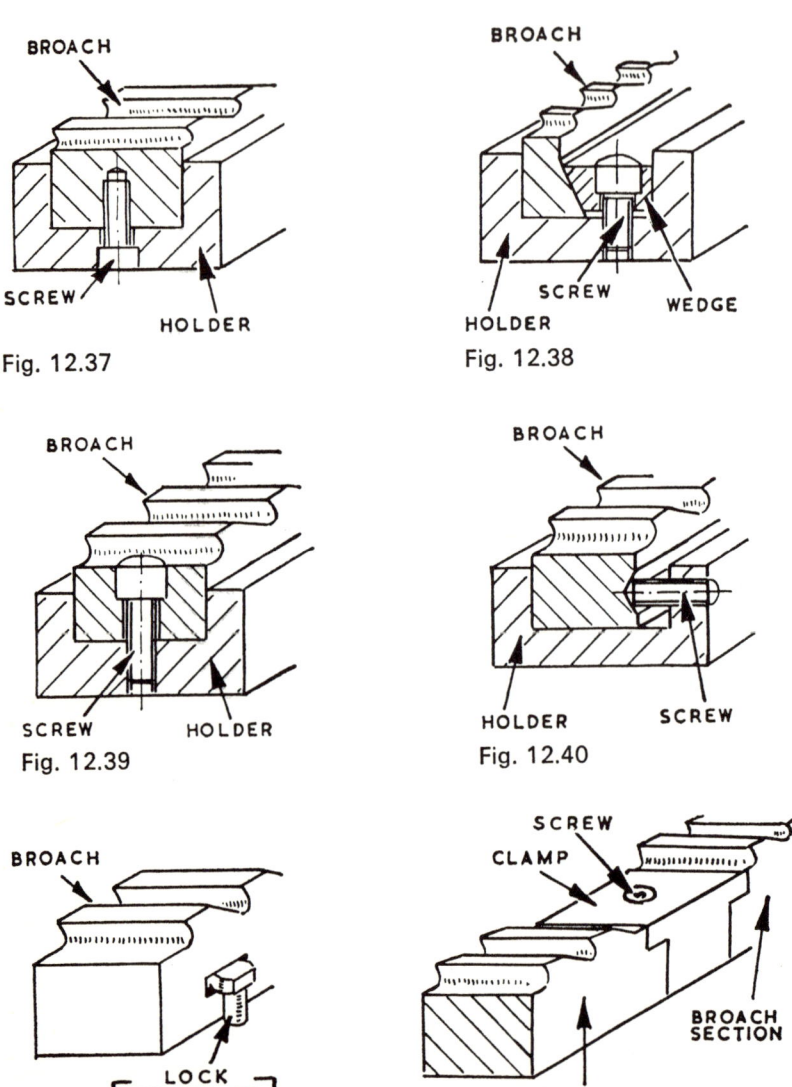

Figs. 12.37–12.42 Methods of Securing Surface Broaches to Holders

BROACHING 185

12.5.2. Surface broach holders. Figs. 12.37–12.42 show some methods of securing broaches to holders.

12.6. Broaching Fixtures

Fixtures for internal broaching. These fixtures are very simple, and consist of a simple locator which is located and clamped to the machine faceplate; it is unusual for clamps to be incorporated if the workpiece is seated on a

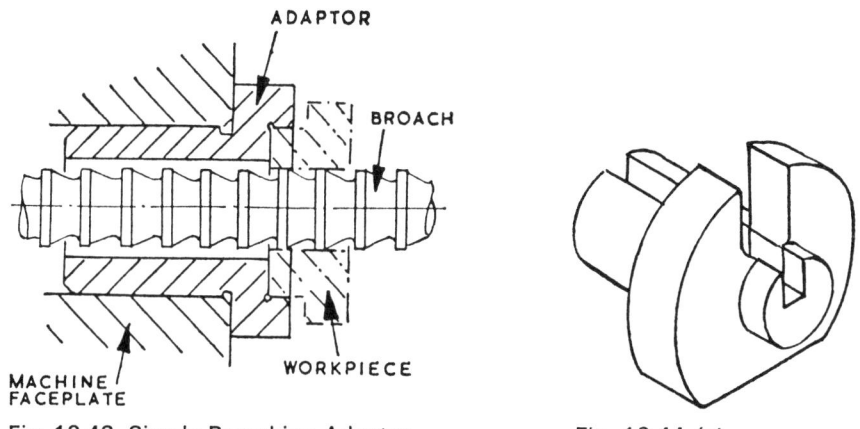

Fig. 12.43 Simple Broaching Adaptor

Fig. 12.44 (a)

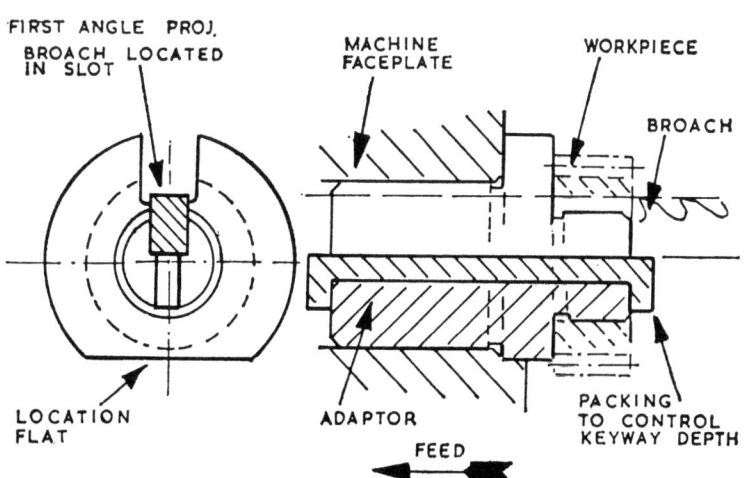

Fig. 12.44 Adaptor for Keyway Broaching

vertical face because the broach will force the workpiece against the fixture.

Fig. 12.43 shows a simple adaptor for use when broaching a plain round, serrated or splined hole, and fig. 12.44 shows an adaptor for keyway broaching.

G

When it is important that the radial position of the splines, etc., be controlled, the adaptor must be radially located on the machine, and the workpiece located similarly on the adaptor; the broach must also be located on the puller. Fig. 12.45 shows a fixture for broaching a keyway in a tapered bore. The adaptor is located relative to the machine faceplate, the locator is positioned relative to the adaptor and the workpiece is located radially

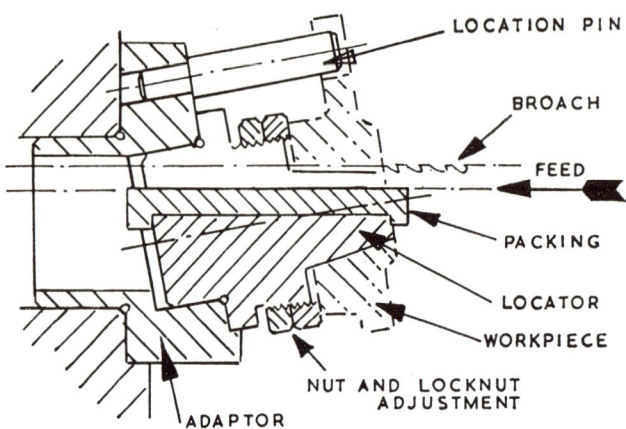

Fig. 12.45 Fixture for Broaching a Keyway in a Tapered Bore

as well as from its bore. The depth of keyway is controlled by the adaptor; a packing strip is used so that the depth of key can be varied. Variation in workpiece length is taken up by the nut and locknut arrangement.

Fixtures for surface broaching. The fixtures are slightly more complicated than those used for internal broaching, and are similar to milling fixtures. It must be noted that the broaching force is on one side of the workpiece, and also that care must be taken to ensure that the path of the broach is not obstructed. The fixture and clamping plate must therefore be designed to supply adequate support for the workpiece as close as possible to it without obstructing the passage of the broach. Fig. 12.33 on page 182 shows the principle of surface broaching, and the 'local' clearance can be seen in this illustration.

Chapter 13

PRODUCING THE DESIGN AND THE WORKING DRAWINGS OF PRODUCTION EQUIPMENT

13.1. Design Study

As a result of studying the examples of typical jigs and fixtures that have been presented in the earlier chapters, the reader may have formed the impression that the design to be used for a specific operation will be obvious from a short examination of the component and the operation to be performed. Usually, the opposite is the case; most successful equipment designs, like successful product designs, are evolved by continual self-criticism. The reader is warned against the common error of making an over-elaborate drawing of the workpiece and then spending so long drawing part of the jig or fixture that a poor design is produced, either because of lack of time or lack of self-discipline when it means rubbing out a beautiful drawing of an incorrect design. The reader is also warned against treating design work like a machine-drawing exercise, and producing a complete elevation, without reference to the plan view. Design is a three-dimensional problem that is solved using two-dimensional conventions; this can only be done by working upon several views at the same time.

Before starting the actual shape design, a design study should be made. The object of such a study is to examine the location and clamping systems, and then to determine any problems that are likely to occur; a more detailed examination is made when the assembly drawing is produced.

13.1.1. When making a design study an outline drawing of the relevant parts of the workpiece is made, showing it in the machining position; at this stage the loading and location problems can be considered. The machining forces are then studied, and the clamping points indicated. Figs. 13.1 and 13.2 show two typical design studies. It must be emphasised that at the design-study stage the designer must keep a perfectly open mind as regards the final design because if he pre-judges the issue the object of the study will be defeated.

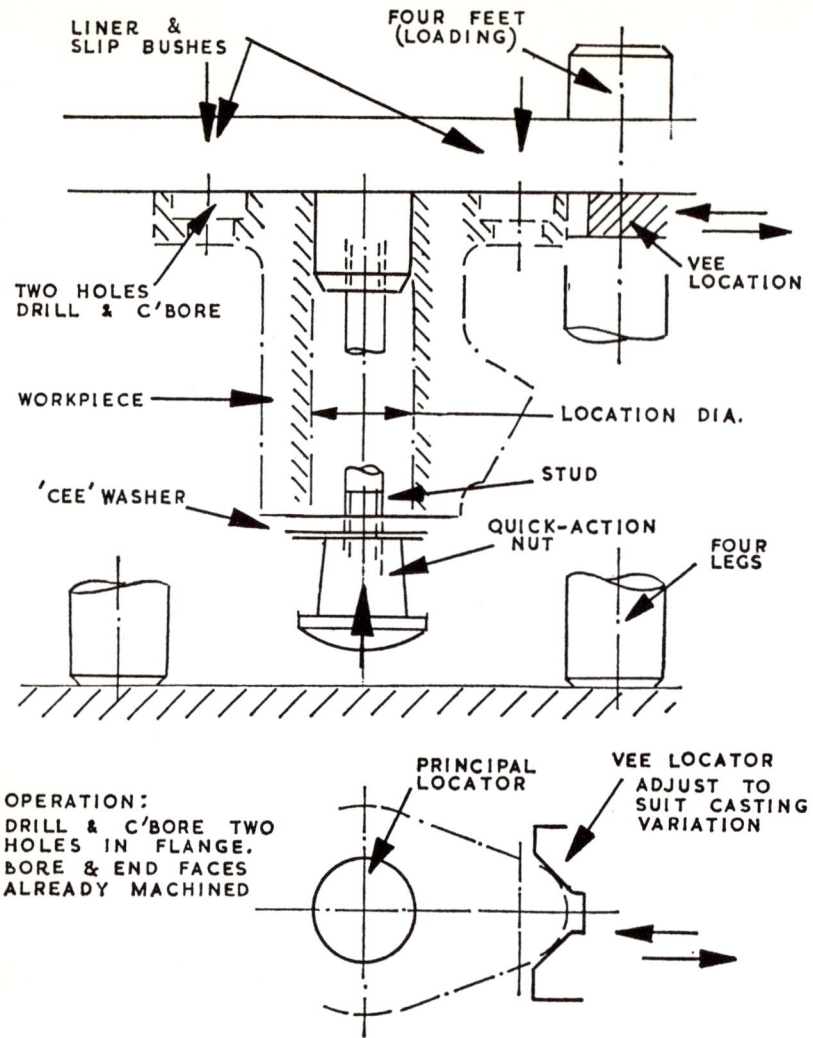

Fig. 13.1 Design Study for Drill Jig for Machining Holes in Zinc-alloy Gravity Die-casting

13.1.2. During the design study the wisdom of the choice of location, clamping, etc., can be examined by 'self-interrogation' on the following lines:

1. Is the workpiece position suitable for the operation and the machine?
2. From where will the workpiece be located?
3. Does this location system follow the 'location programme' laid down in the operation plan?
4. Does the location system constrain the workpiece just adequately?

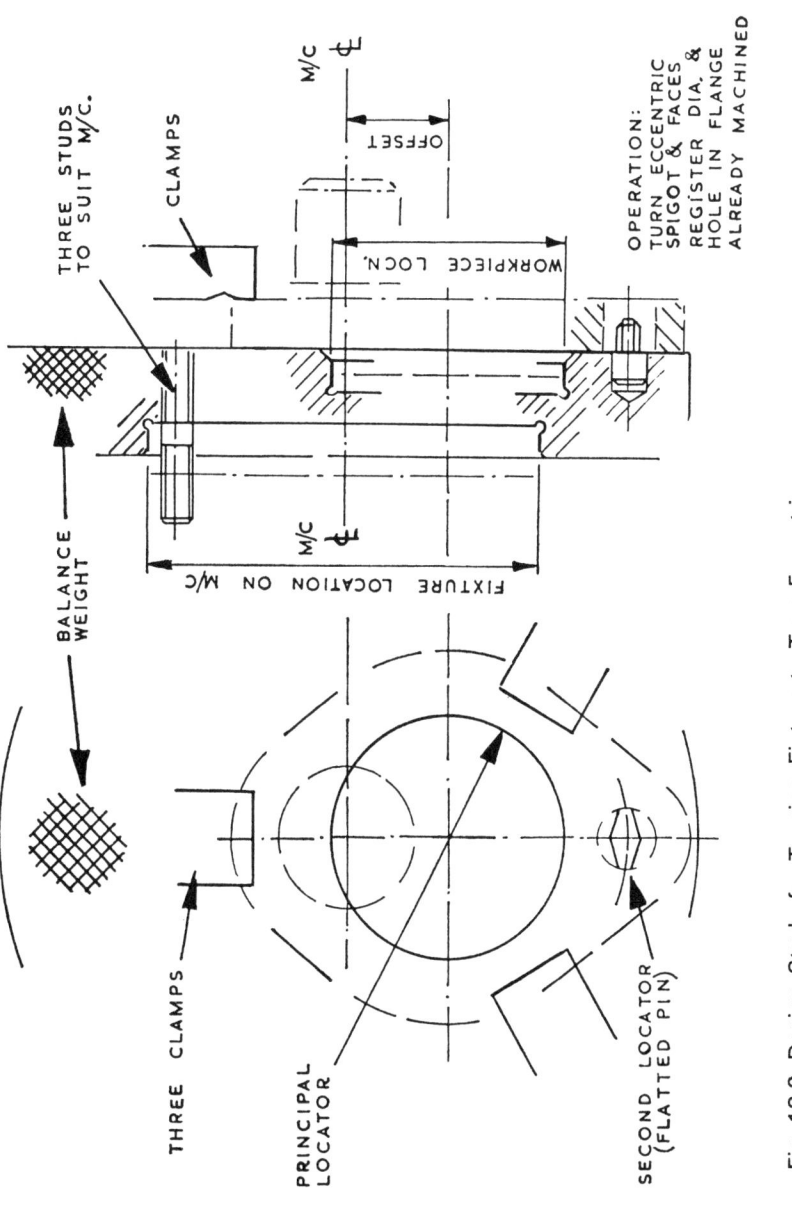

Fig. 13.2 Design Study for Turning Fixture to Turn Eccentric

5. Has redundant location been avoided?
6. Is the location system convenient for the operator? Can he see the locators when in his working position? Is there a 'follow on' system when more than one locator is used?
7. Has location adjustment been provided where required?
8. Does the clamping system secure the workpiece adequately?
9. Is it easy for the operator to operate the clamps?
10. Is the clamping system safe?
11. Has variation in workpiece size been allowed for?
12. Is there plenty of room for the operator's hands when he is loading and unloading the workpiece?
13. Is there likely to be swarf-clearance trouble?
14. Has the jig or fixture been correctly located on the machine tool?
15. Has the bolting of the jig or fixture to the machine tool been considered?
16. Has provision been made for tool setting or tool guiding?

It will be realised that certain of the above will demand more detailed treatment, but notes regarding clearances, etc., can be made on the design-study drawing.

13.2. Consideration of Method of Construction

When the design study has been completed the basic features of the design will have been established. Before the design can be examined in greater detail, the method of construction will need to be considered. The main body of the equipment will be constructed by one of the following methods: (1) casting; (2) welding; (3) fabricating from several parts by screw and dowel, etc. The main characteristics of these methods will now be considered.

13.2.1. Casting. A cast body or base is rigid and has good damping properties, can be of complicated shape and, if dropped, will either be obviously broken or not suffer any damage (a base constructed by welding or fabricating may not be obviously damaged when dropped, and cause inaccurate work for some time).

The disadvantages associated with casting are that a single casting is expensive to produce because a pattern is required, that casting is a slower process than welding and fabricating, and that a cast body is heavier than one that is produced by welding or fabricating.

When designing with a view to casting, care must be taken to ensure that the pattern can be removed from the mould without having to use loose pieces and that the parting line (at the greatest section) is straight. For best results, the wall thickness should be uniform so that stresses caused by inequality in the rate of cooling are eliminated, and so that sponginess at local heavy sections is also eliminated; where it is essential that the section thickness changes, the change must be made gradually. Stiffening ribs should be staggered as shown in fig. 13.3 (*a*) to reduce a local heavy section, and when a boss is stiffened by several ribs the boss should be cored as shown in fig. 13.3 (*b*) to reduce the section. Local heavy sections are also

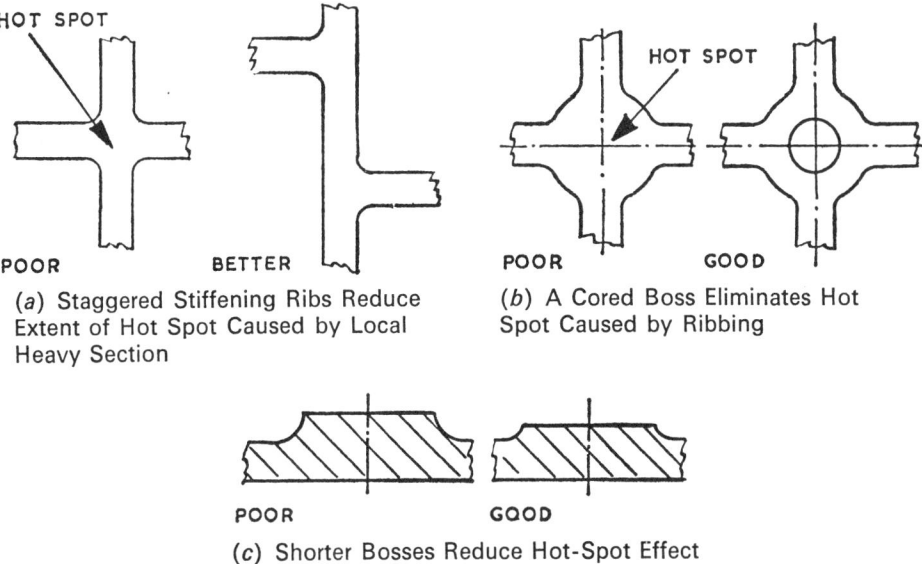

(*a*) Staggered Stiffening Ribs Reduce Extent of Hot Spot Caused by Local Heavy Section

(*b*) A Cored Boss Eliminates Hot Spot Caused by Ribbing

(*c*) Shorter Bosses Reduce Hot-Spot Effect

Fig. 13.3

caused by making bosses too tall (see fig. 13.3 (*c*)); bosses are expensive to produce and, if possible, spotfacing should be used instead.

Cast iron is the usual material for making cast bodies because it is reasonably inexpensive and because it absorbs vibrations; it is poor in tension but good in compression, and so the more highly-stressed parts of the casting should be designed to be in compression as shown in fig. 13.4.

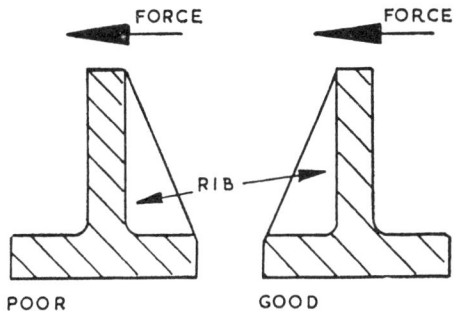

Fig. 13.4 Stiffening Ribs Should Be Positioned so That They Are in Compression

13.2.2. Welding. Equipment can be produced quickly by welding, but is less strong than cast or fabricated assemblies; welded equipment is, however, light and is more easily handled when in use. When designing with a view to welding, it must be arranged so that the minimum number of pieces are used, and so that the joints are at regions of minimum stress. Fig. 13.5 shows some typical welded joints.

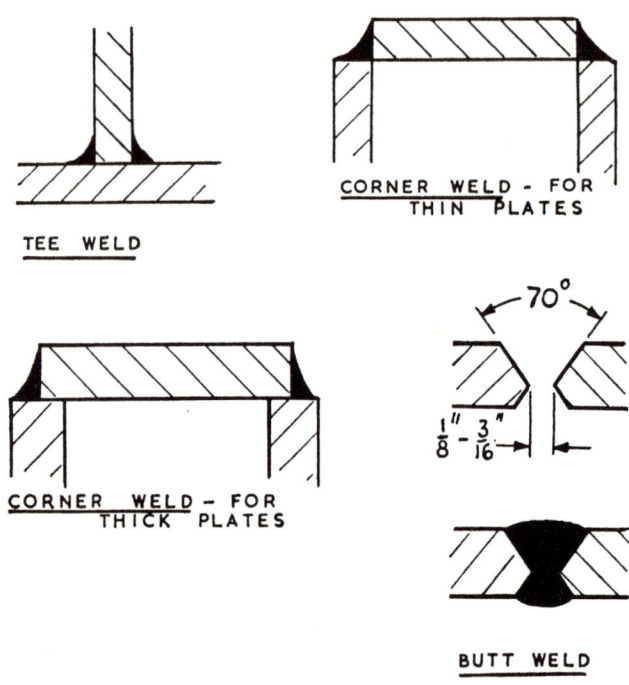

Fig. 13.5 Some Typical Welded Joints

13.2.3. Fabricating from several parts. Fabricated (or build-up) assemblies are lighter than cast bodies, and the parts can often be used again for other equipment. The disadvantage associated with this method of construction is that the parts must be machined to locate and secure them together. Most jigs and fixtures are produced by having some parts cast and some parts machined from stock, and then joining them by using dowels and set-screws. Very extensive use is made of 'unit tooling parts' marketed by the Purefoy Unit Tooling Company Limited, and by the Woodside Die Sinking Company Limited; these companies produce a range of standard items, such as bases, blanks, clamps, jig feet, etc., and a number of other companies produce drill bushes.

Figs. 13.6 and 13.7 show typical drilling jigs produced from standard parts.

The manufacturers of 'unit tooling parts' produce tracing templates so that the designer does not have to waste his valuable time constructing views of these items; fig. 13.8 illustrates a typical tracing template. It is

DESIGN OF PRODUCTION EQUIPMENT

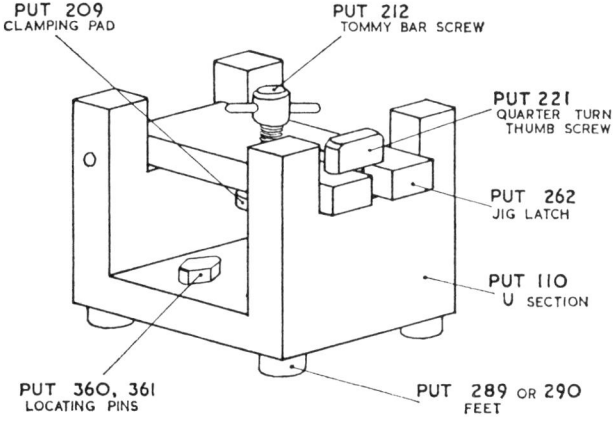

TYPICAL APPLICATION OF PUREFOY STANDARD PARTS
Fig. 13.6 (Courtesy of Purefoy Unit Tooling Ltd.)

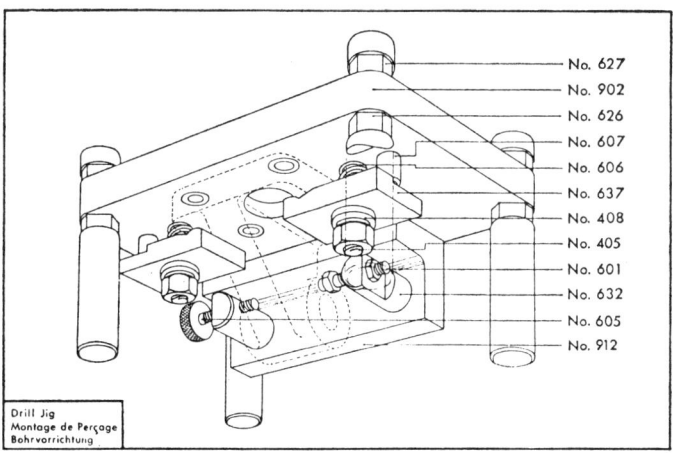

Fig. 13.7 Typical Application of W.D.S. Tooling Aids (Courtesy of W.D.S. Tooling Aids Ltd.)

considered that at the early stages of his study the student should be able to design his own clamps and similar details, but that at the final stages of his study he should be encouraged to make use of all the drawing aids used in the tool design office.

13.2.4. Use of plastics materials. Knobs and handles are often made from plastics materials and have the special advantage of being self-coloured so that identification is easy. Certain assembly fixtures are often made from glass-fibre using the wet lay-up method, and are inexpensive to make and are very light. Drill plates may be made from fairly thin plate when the

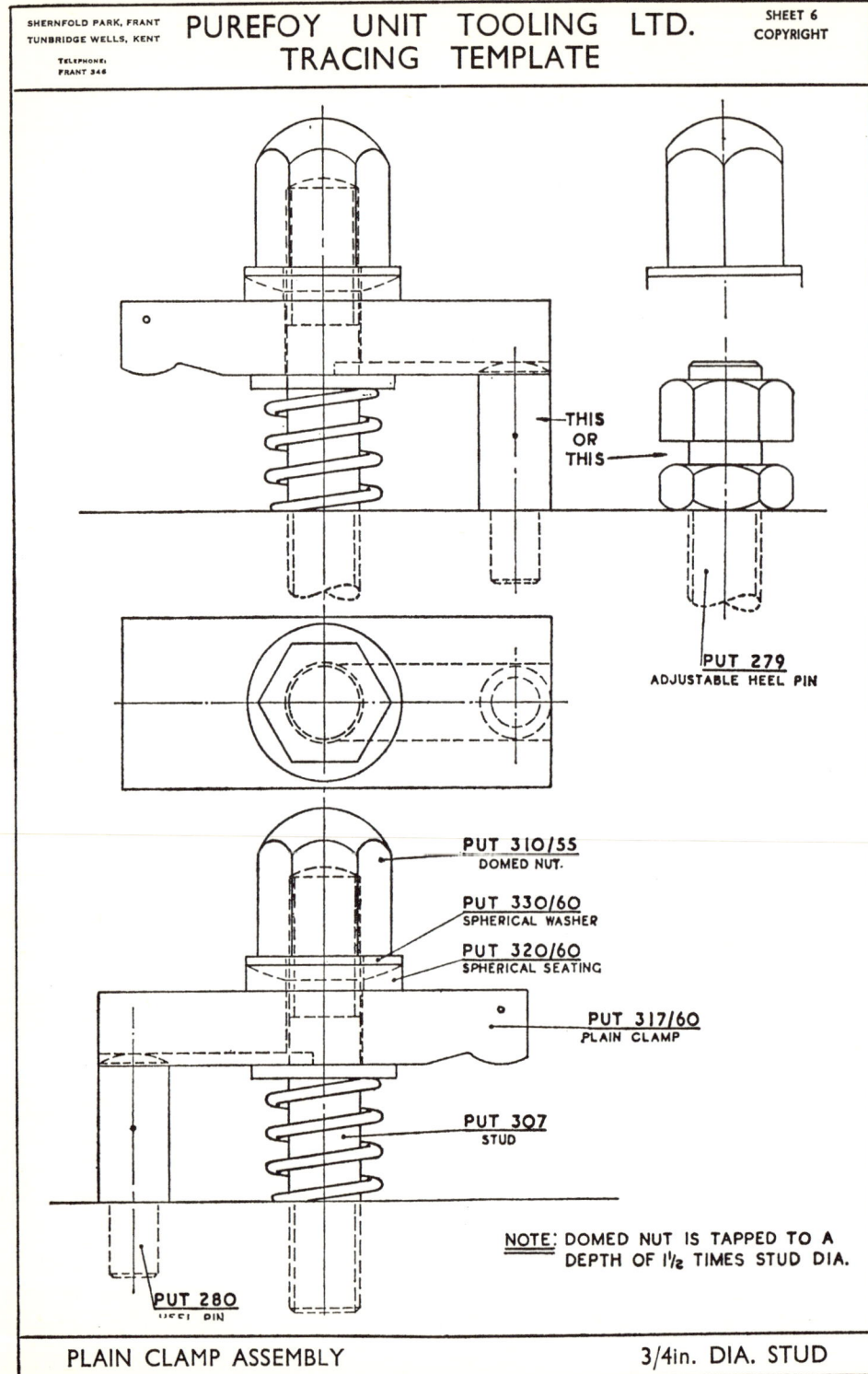

Fig. 13.8 (Courtesy of Purefoy Unit Tooling Ltd.)

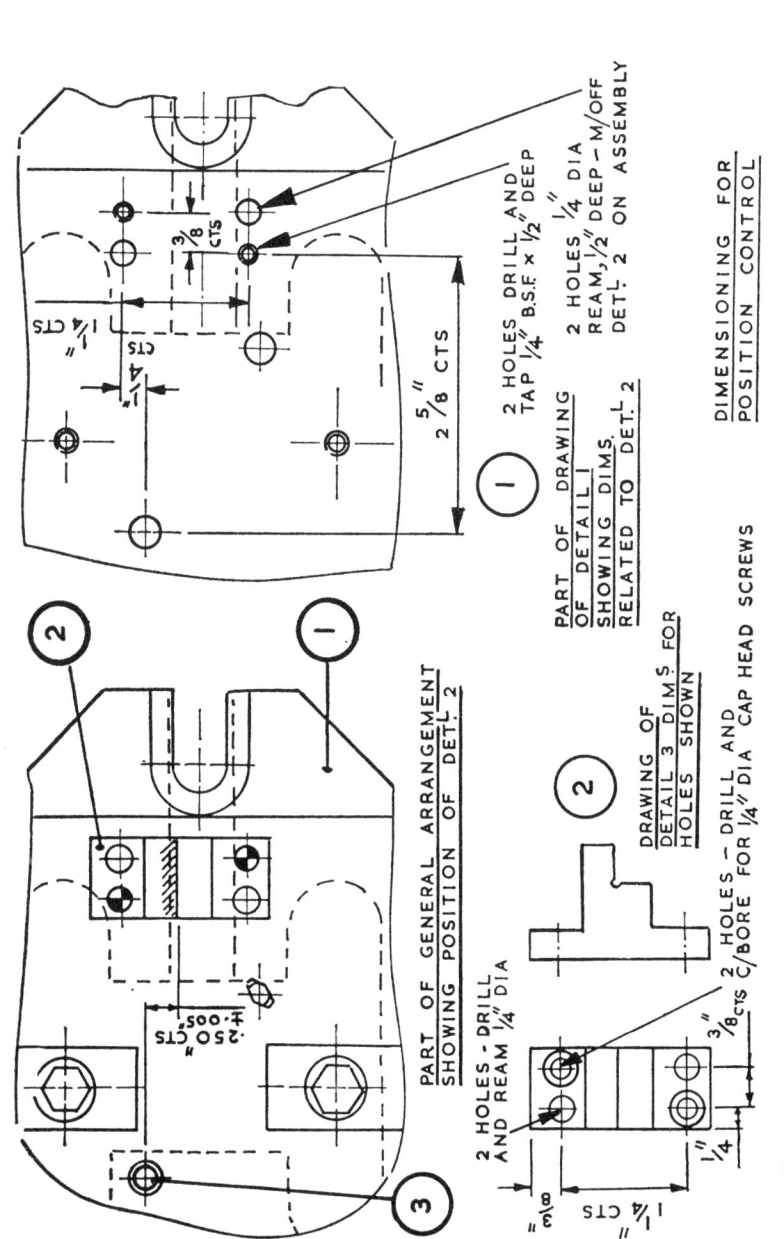

Fig. 13.9

work is not required to be extremely accurate, and special bushes are marketed that can be knocked into holes that are drilled in the plate.

13.3. Working Drawings of Production Equipment

The working drawing should include an assembly drawing with suitable 'balloon references' and parts list to assist in the ordering of the parts and the assembly of the equipment; some balloon references will be seen on fig. 13.9 on page 195, and reference is made to the use of a parts list on page 257. The drawing must also give all the required information about the 'details', and if the drawing needs to be modified after it is issued, details of modifications must be given on the drawing (see page 260).

Some parts will need to be fully dimensioned, but British Standard parts, 'unit tooling parts' and other 'bought out' parts are only drawn and dimensioned when they are machined after they are purchased (and then only the modified dimensions are specified).

13.3.1. When several details are to be positioned relative to each other it is less expensive to part-machine them and set, or finally machine them upon assembly, than to attempt to machine them so that they can be located directly in position. For example, fig. 13.9 illustrates part of the general arrangement and part of the detail drawings associated with the position of the setting block (Detail 2). The drawing of the setting block shows the position of the two dowel holes and the two holes for the retaining screws; the drawing of the base (Detail 1) shows the position of the tapped holes for the retaining screws, but only illustrates the dowel holes. When the associated parts are manufactured, all the holes will be produced in the setting block, but only the tapped holes will be machined in the base (the dowel holes in the setting block are usually drilled ready for reaming). Upon assembly the setting block will be lightly secured on the base, and then positioned relative to the location system for the workpiece; when the position of the setting block is correct it will be lightly secured, and the two dowel holes drilled in the base, using the holes in the setting block as a guide. The dowel holes in the setting block and the base will then be reamed and the dowels fitted.

Chapter 14

LIMIT GAUGES

14.1. Introduction

When the working drawing of a part is dimensioned it is necessary to stipulate the permitted variation in its size because errors will occur owing to inaccuracies in the machine tools, jig and fixtures, and measurement equipment used in its production. The extremes of size that are permitted are called the **limits of size** (or more simply the 'limits'), and the difference between these limits is called the **variation tolerated** (or more simply the 'tolerance'). The tolerance should be as large as possible to minimise the cost of the component, but be sufficiently small to ensure that the required fit or degree of interchangeability between parts is ensured.

14.1.1. The limit that is associated with the greatest amount of metal is often called the **maximum metal limit**, and that associated with the least amount of metal is often called the **minimum metal limit**. The largest shaft size permitted is the maximum metal limit, and the smallest shaft size permitted is the minimum metal limit; similarly, the largest hole size permitted is the minimum metal limit, and the smallest hole size permitted is the maximum metal limit.

14.1.2. One method of inspecting parts is to use limit gauges; these gauges are designed to accept the workpiece if its size and shape lies within the specified limits. A limit gauge (or pair of limit gauges) consists of a 'GO' member that will pass over or through a correct feature, and a 'NOT GO' member that will not pass over or through a correct feature. The 'GO' member is used to check the maximum metal limit and the 'NOT GO' member is used to check the minimum metal limit. The main disadvantages associated with the use of limit gauges are that the extent of error is not indicated when a workpiece is rejected, and that the system imposes smaller tolerances than stipulated on the working drawing, to allow for gauge tolerances and allowance for gauge wear.

14.2. The Taylor Principle

(Stated in 1905 by William Taylor, of Messrs Taylor, Taylor and Hobson.) As stated above, the 'GO' member is used to check the maximum metal limit; the 'GO' member should be full form, because, as shown in fig. 14.1, a

'GO' member that is not full form will accept an incorrectly-shaped workpiece. Taylor stated that **the 'GO' gauge should incorporate the maximum metal limit of as many dimensions as it is convenient and suitable to check in one operation.**

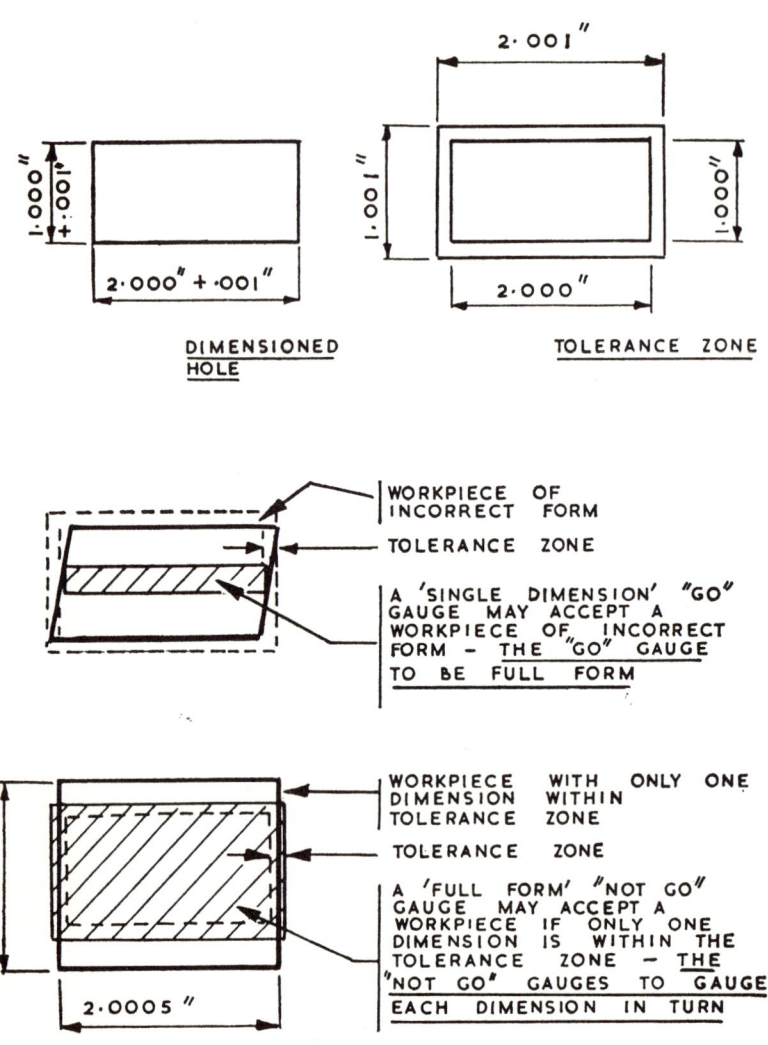

Fig. 14.1 THE TAYLOR PRINCIPLE

Taylor also stated that **the 'NOT GO' gauges should be separate, and check the minimum metal limit of each dimension in turn**; this is because, as shown in fig. 14.1, if more than one dimension is checked at a time by the 'NOT GO' member, that gauge will accept an incorrect workpiece as long as one dimension is within limits.

It will be appreciated that a 'GO' gauge cannot always be full form, and a 'NOT GO' gauge cannot always check only one dimension. For example, a plain plug gauge would be heavy and awkward to use if it was long enough to gauge a very long hole for its full length, and would wear quickly and be awkward to use if made to check the diameter of a hole from point to point.

14.3. Design of Limit Gauges

B.S. 1044: Part 1:1964, Specification for Gauge Blanks, gives general notes regarding the design of gauge blanks, standardises handles and gauging members for screw and plain plug gauges, blanks for screw and plain ring gauges, and lists recommendations for the design of adjustable ring and caliper gauges. The specification also states that all non-gauging sharp edges should be removed, and that hardened-steel blanks, particularly for larger sizes, should be stabilised before they are completed (stabilising treatment recommended consists of heating to 150°C, soaking at that temperature for approximately 10 hours followed by slow cooling).

14.4. Limit Gauge Tolerances

It is necessary to allow manufacturing tolerances when designing a gauge, but it is also necessary that these tolerances do not permit inaccurate parts to be passed by the gauge, or excessively reduce the workpiece tolerances. B.S. 969, Plain Limit Gauges—Limits and Tolerances, recommends that the limits on the 'GO' gauge member are such that its size is **within** the limits of size for the workpiece, and that the limits on the 'NOT GO' are such that its size is **outside** the limits of size for the workpiece. When made to these limits the 'NOT GO' will not accept faulty parts, but the 'GO' will reject parts that are close to the maximum metal limits; but the latter condition will rarely occur in practice. The gauge tolerance is usually one-tenth of the workpiece tolerance being gauged.

14.5. Allowance for Wear

B.S. 969 recommends that when the workpiece tolerance exceeds 0·003 5 in additional metal be left on the 'GO' gauge surface to allow for wear (this has the effect of placing the manufacturing limits for the 'GO' member still further within the workpiece limits). The standard also recommends that if the workpiece tolerance is too small to permit this, the gauge should be made from a specially-hard-wearing material.

14.6. Materials for Limit Gauges

Gauges are usually made from case-hardened steel that is heat-treated during the manufacture of the gauge, or from cast steel (with between about 0·7 and 1·2% carbon) that is usually hardened, but may be hard enough 'as received'. Larger gauges are sometimes made from grey cast iron, or steel that is chromium-plated to increase its wear resistance.

14.7. Typical Limit Gauges

The following represent the principal types of limit gauge; many of them are covered by British Standards, and where applicable the relevant standard and size range for which the gauge is recommended is quoted.

14.7.1. Plain plug gauges. The gauging portion may be integral with the handle (a 'solid' gauge), or the gauging portion and handle may be separate and engaged together to form an assembly (a 'renewable' gauge). The 'GO' and 'NOT GO' members may be separate, at each end of one handle, or combined in one member to form a 'progressive' gauge. Fig. 14.2 illustrates a double-ended solid plug gauge, and fig. 14.3 illustrates a progressive sc

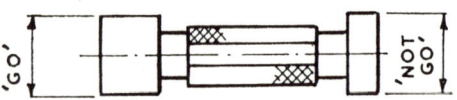

Fig. 14.2 Double-Ended Solid Plug Gauge

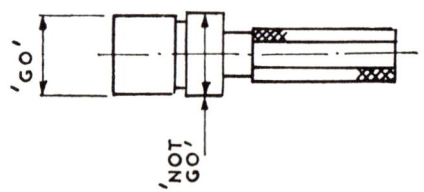

Fig. 14.3 Progressive Solid Plug Gauge

plug gauge. B.S. 1044 does not include solid gauges because this type of gauge is becoming obsolete, but they have been included in this book because solid gauges are often called up when they are to be produced within a manufacturing company's own tool room.

B.S. 1044 lists the following renewable-end-type gauges: collet type (see figs. 14.4, 14.5 and 14.6), taper-lock type (see fig. 14.7) and trilock type see fig. 14.8).

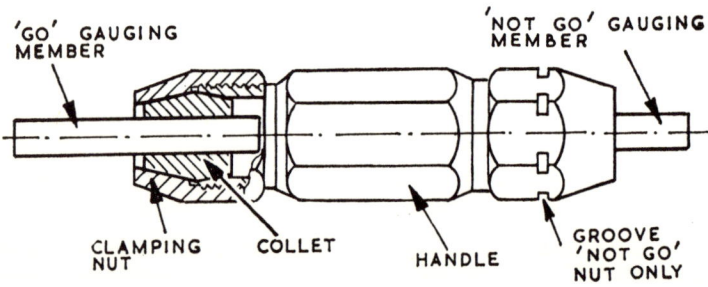

Fig. 14.4 Collet-Type Plug Gauge (between 0·015 and 0·760 in)

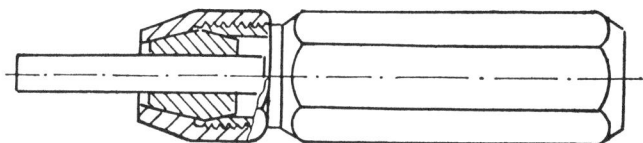

Fig. 14.5 Collet-Type Plug Gauge

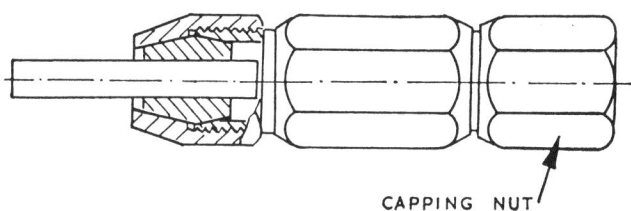

Fig. 14.6 Collett-Type Plug Gauge

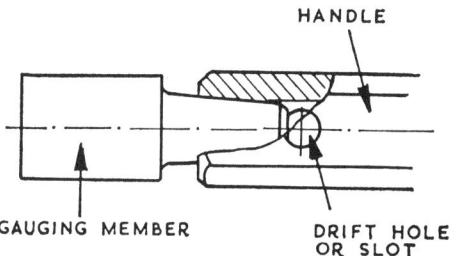

Fig. 14.7 Taper-Lock Plug Gauge (between 0·059 and 2·510 in)

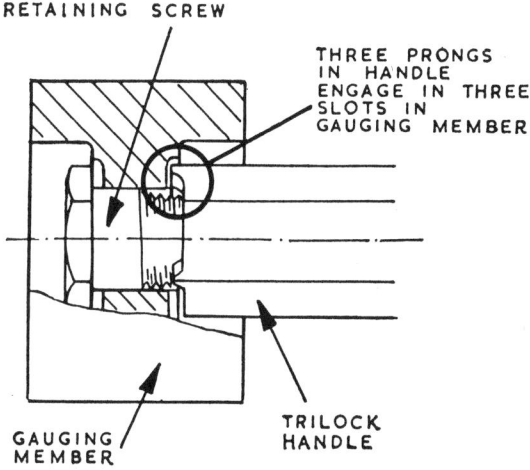

Fig. 14.8 Trilock Gauge (between 1·510 and 2·510 in)

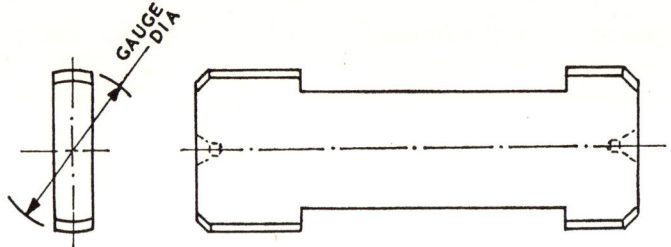

Fig. 14.9 Segmental Cylindrical Gauge (up to 2·510 in)

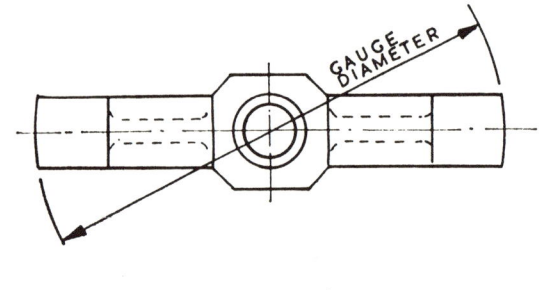

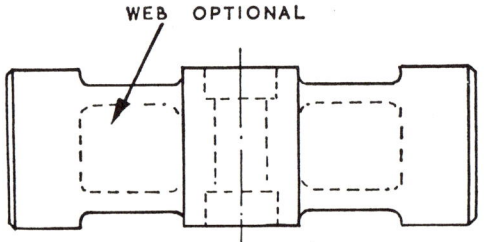

Fig. 14.10 Segemental Cylindrical Gauge (between 2·510 and 8·010 in)

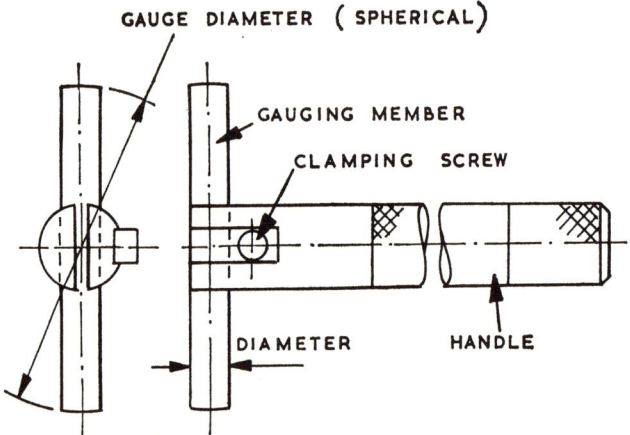

Fig. 14.11 Spherical-Ended Rod Gauge (between 0·510 and 6.010 in)

LIMIT GAUGES

B.S. 1044 also lists the following gauges for gauging large holes in rigid parts: segmental cylindrical gauges (figs. 14.9 and 14.10; the latter for use with a handle), and spherical-ended rod gauges (figs. 14.11 and 14.12).

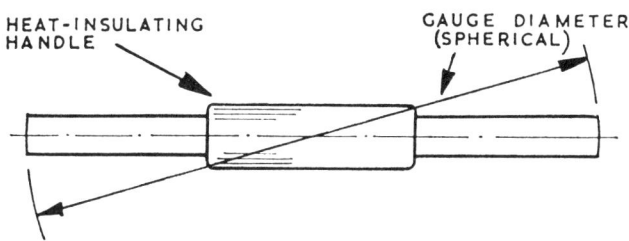

Fig. 14.12 Spherical-Ended Rod Gauge (between 6·010 and 12·010 in)

The ends of large plug gauges should be protected from becoming burred when being placed on the machine table by providing a 'guard extension' as shown in fig. 14.13. This should be applied to gauges of more than about $2\frac{1}{2}$ in diameter, excepting those for testing holes to their full depth.

The centres should be good quality; they should not be large, and the length of the cone should be kept short. The mouth of the centre should be protected by a small recess $\frac{1}{32}$ or $\frac{1}{16}$ in deep (see fig. 14.14).

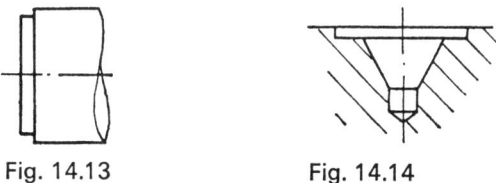

Fig. 14.13   Fig. 14.14

Large-diameter holes are gauged by annular-type gauges as shown in figs. 14.15 and 14.16.

Adequate air venting should be provided when small gauges are used for gauging blind holes; when a gauge is more than 4 in diameter lightening

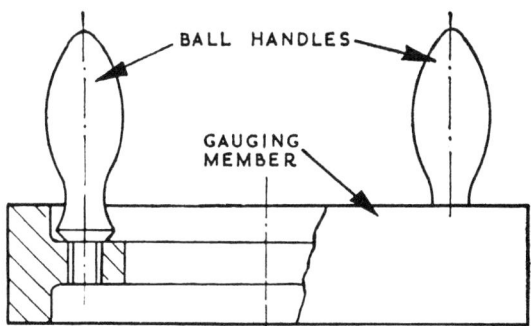

Fig. 14.15 Annular Plug Gauge with Ball Handles (between 8·010 and 12·010 in)

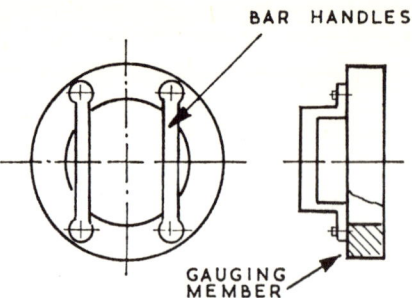

Fig. 14.16 Annular Plug Gauge with Bar Handles (between 8·010 and 10·010 in)

holes are incorporated, which also give the required air venting. The marking of gauges should be kept to a minimum; the marking should include the limiting dimension controlled by the gauge, 'GO' or 'NOT GO', and the gauge serial number.

14.7.2. Gauges for shafts. In order to satisfy Taylor's principle, a shaft should be gauged by a full form ring 'GO' gauge, and a gap-type 'NOT GO' gauge. When recess diameters are to be gauged it is necessary to use gap-type gauges for both 'GO' and 'NOT GO' gauging. Fig. 14.17 shows a typical ring gauge, fig. 14.18 shows a plain gap gauge, fig. 14.19 shows an adjust-

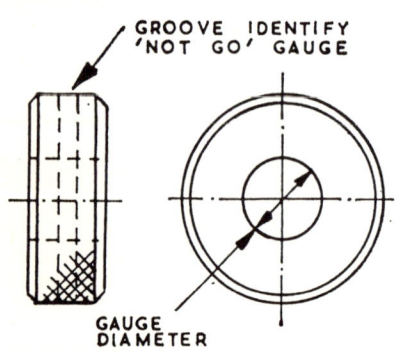

Fig. 14.17 Ring Gauge

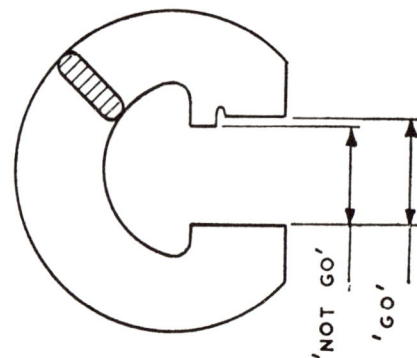

Fig. 14.18 Plain Gap Gauge

able caliper gauge, and two variations of anvil. B.S. 1044 lists recommendations to ensure satisfactory performance of these gauges; these recommendations include the following: the frame should be rigid and strong, the metal distribution should produce a nice balance and feel, supports should be provided for clamping during manufacture and grinding, the adjustment of the anvils should be by sliding and not rotational movement, the distance between the 'GO' and 'NOT GO' anvils should permit the work to be free when past the 'GO' anvils and before meeting the 'NOT

LIMIT GAUGES

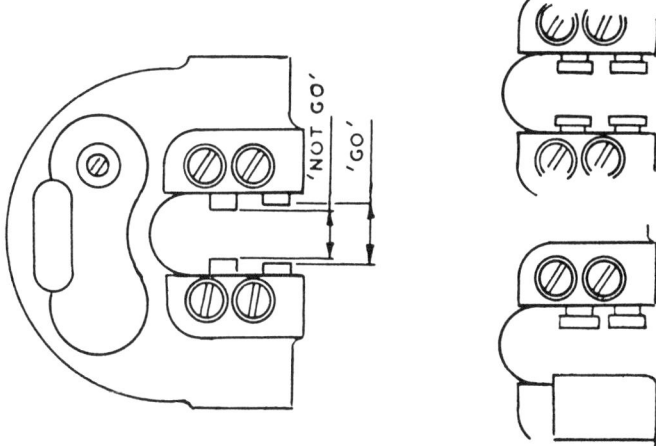

Fig. 14.19 Adjustable Caliper Gauge

GO' anvils, and provision should be made to seal the gauge when adjusted for size.

14.7.3. The gauging of screw threads. External screw threads are usually gauged with a plain gap gauge for the major (outside) diameter, and a thread gauge for the effective diameter. The effective diameter is the diameter of an imaginary cylinder whose generator cuts the thread such that the distance between the points where it cuts the flanks of the thread groove is equal to one-half the pitch of the thread.

The 'matrix' thread gauge is illustrated in fig. 14.20. This gauge has two

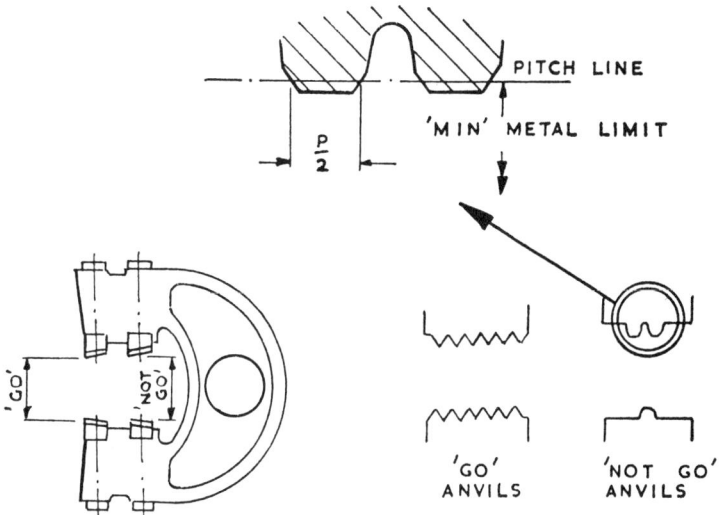

Fig. 14.20 'Matrix' Thread Gauge

sets of adjustable anvils (these may be as illustrated, be threaded rollers or be portions of threaded cylinders); the front anvils form a full form 'GO' gauge, and the rear anvils form a 'NOT GO' effective diameter gauge. The rear anvils gauge only two threads so that pitch error will not interfere with their function, and are truncated and cut so that they do not contact the crest and root of the thread. Both the front and the rear anvils are shaped so that the helix angle of the thread being gauged will not cause interference. The 'GO' and 'NOT GO' anvils are illustrated in fig. 14.20.

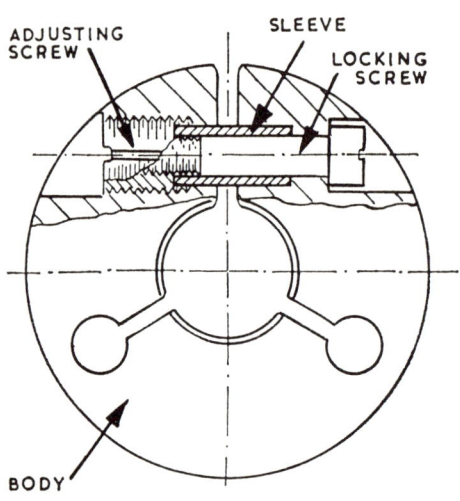

Fig. 14.21 Adjustable Screw Ring Gauge

Adjustable screw ring gauges, widely used in the U.S.A., have been developed for gauging of threads having clearance between the crests and roots (they are not recommended for use on Whitworth and B.A. threads); this type of gauge is illustrated in fig. 14.21. The adjusting screw is threaded internally and externally and is split longitudinally—when this screw is tightened it presses against the sleeve and expands the ring; the gauge is locked when set by tightening the locking screw, which causes expansion of the adjusting screw; the sleeve is accurately fitted, and serves as a large dowel to maintain the alignment of the gauge.

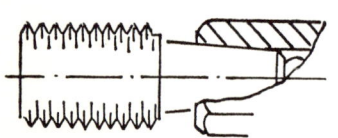

Fig. 14.22 Screw Plug Gauge Member

Internal screw threads are usually gauged with a plain plug gauge for the minor (inside) diameter, and a double-ended screw plug gauge for the effective diameter (see fig. 14.22). The 'GO' member is full form, and the 'NOT GO' member gauges the effective diameter, and so is shaped so that it does not contact the crest and root of the thread being gauged. It is common practice for plug

LIMIT GAUGES

screw gauges to have a dirt-clearance groove cut axial to the thread to a depth slightly below the root of the thread. The general notes regarding plain plug gauges also apply to screw plug gauges.

14.7.4. Thickness and length gauges. Gauges of thickness are shown in figs. 14.23, 14.24 and 14.25, and a length gauge is shown in fig. 14.26. These gauges are usually made from gauge plate, but may be built up as shown in fig. 14.25. They may be double-ended or single-ended.

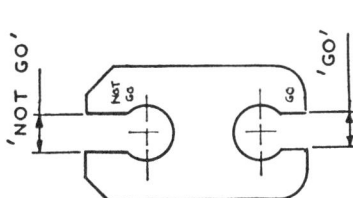

Fig. 14.23 Thickness Gauge

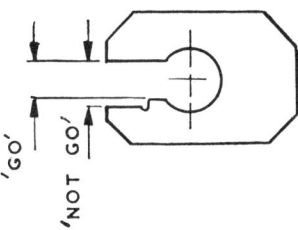

Fig. 14.24 Thickness Gauge

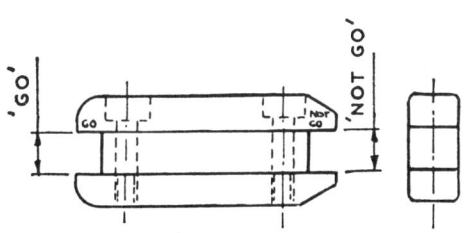

Fig. 14.25 Thickness Gauge

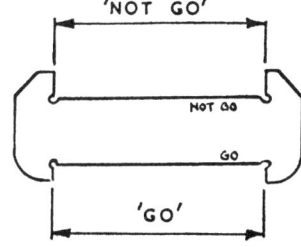

Fig. 14.26 Length Gauge

14.7.5. Recess gauges. Fig. 14.27 shows a plate gauge for checking the position of the recess; care must be taken when this type of gauge is designed, to ensure that the leading end of the gauge will enter the hole, and that the gauge will be seated on the workpiece face, when in both of the extreme positions of gauging. The recess width can be gauged with a simple plate gauge as shown in fig. 14.28.

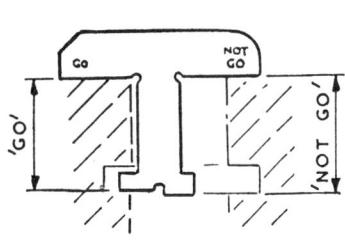

Fig. 14.27 Recess Position Gauge

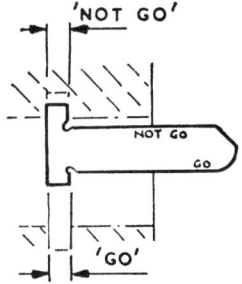

Fig. 14.28 Recess Width Gauge

The diameter of the recess is more difficult to gauge because the gauge must enter the smaller-diameter hole before it gauges the diameter of the recess. Fig. 14.29 shows a typical gauge that locates in the smaller-diameter

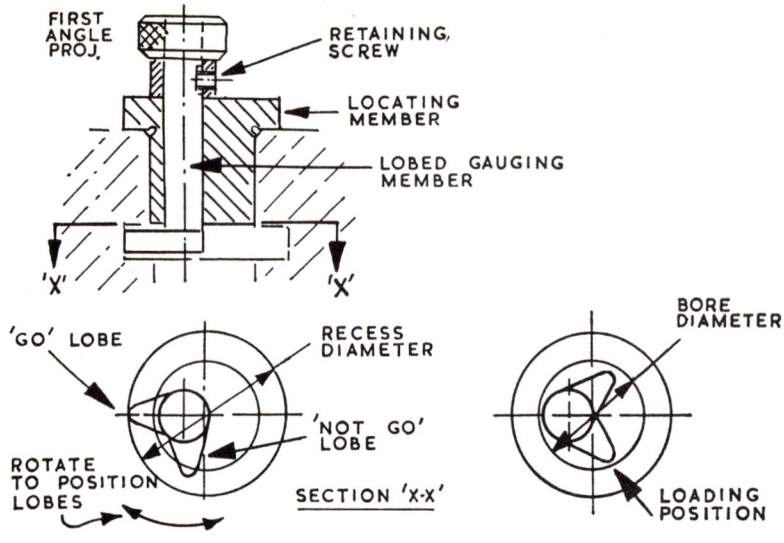

Fig. 14.29 Recess Diameter Gauge

hole, and the recess diameter is gauged by rotating the lobed members. The positions for 'GO' and 'NOT GO' must be indicated on the locating member and the gauging member.

14.7.6. Step gauges. These gauges are designed so that the workpiece is accepted if one step is below the datum face, and the other step is above the

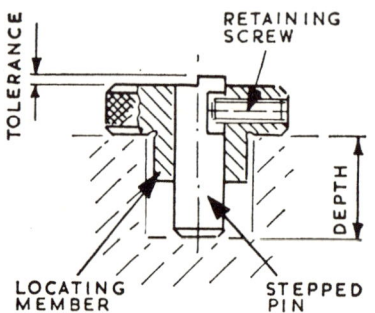

Fig. 14.30 Stepped-Pin Depth Gauge

datum face; they are convenient to use if the step is at least 0·010 in. Fig. 14.30 shows a simple stepped-pin depth gauge. Fig. 14.31 shows a stepped taper-plug gauge, and fig. 14.32 shows a stepped taper-ring gauge; in both these gauging systems the datum is the workpiece face, and the

LIMIT GAUGES 209

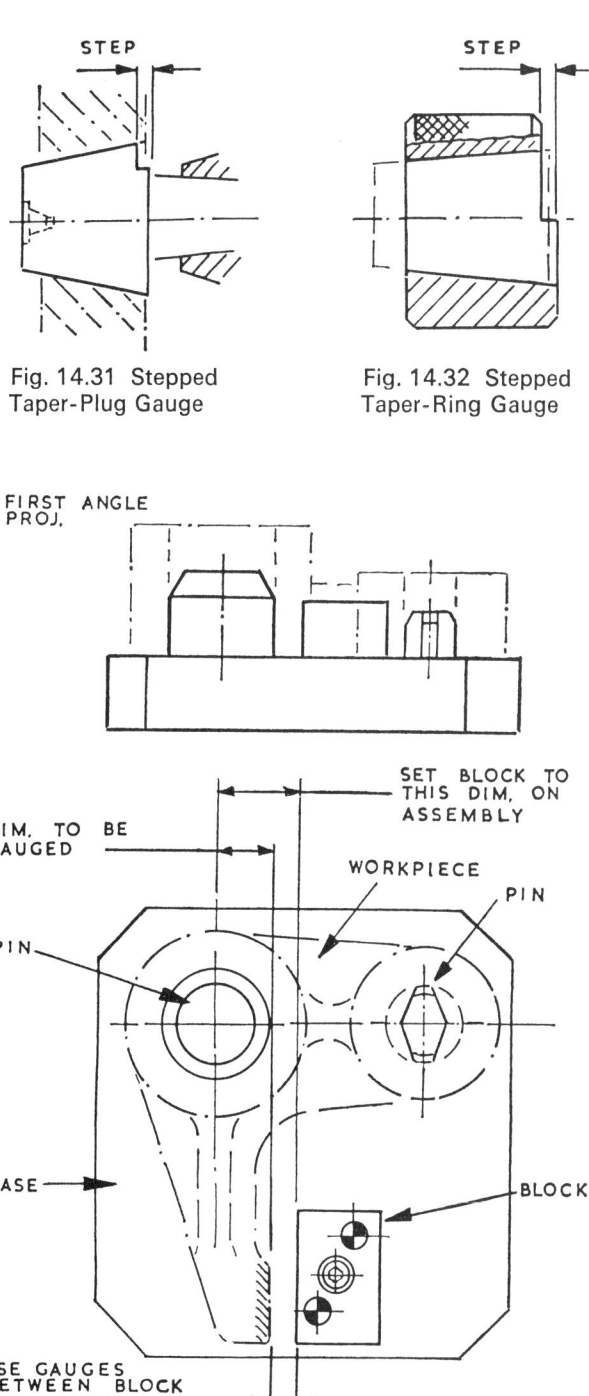

Fig. 14.31 Stepped Taper-Plug Gauge

Fig. 14.32 Stepped Taper-Ring Gauge

Fig. 14.33 Position Gauge

tolerance on the large end is gauged. The step is $X \times Y$, where Y is the tolerance on the diameter, and the taper on diameter is X in one.

14.7.7. Position gauge. This type of gauge is used to check the relative position of several features. Fig. 14.33 shows a position gauge to check the relative position of two holes and a face; the workpiece is located from two pins, and the position of the face is then checked by inserting gauges between the block and the face.

14.7.8. Receiver gauges. Receiver gauges are used to check that a part will assemble correctly with the mating part. Fig. 14.34 shows a receiver gauge

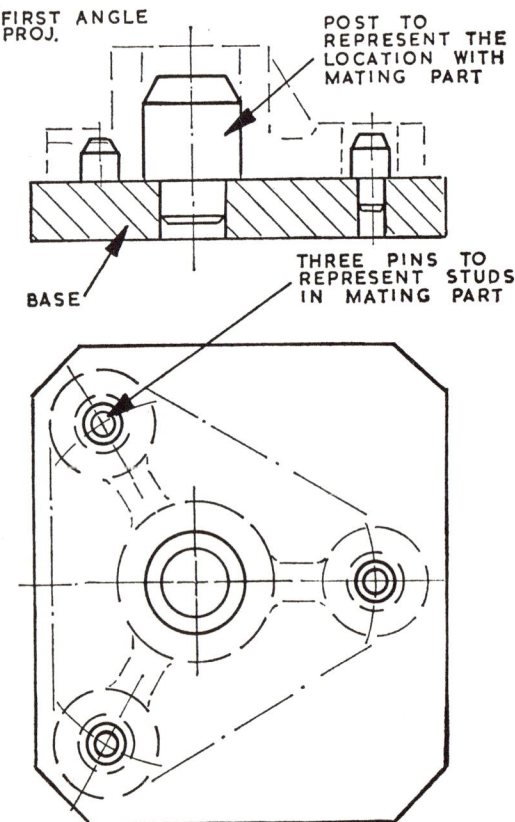

Fig. 14.34 Receiver Gauge

for a workpiece that is to be located on a spigot that is part of the mating part, and secured by three studs. The receiver gauge includes a post to represent the spigot, and three pins to represent the three studs; if the workpiece is accepted by the gauge it will assemble correctly with its mating parts.

References

B.S. 1916:1953, Limits and Fits for Engineering.
B.S. 1044:Part 1:1964, Specification for Gauge Blanks.
B.S. 969:1953, Plain Limit Gauges—Limits and Tolerances.
B.S. 919:1960, Screw Gauge Limits and Tolerances.

Chapter 15

PRESSWORK

15.1. Introduction

Press tools are used to cold-form and to cut thin metal that is usually supplied in the form of sheet or strip. Presswork operations can be simplified to a few basic operations involving a punch and die, or a punch and form block. Fig. 15.1 illustrates some typical parts produced by presswork

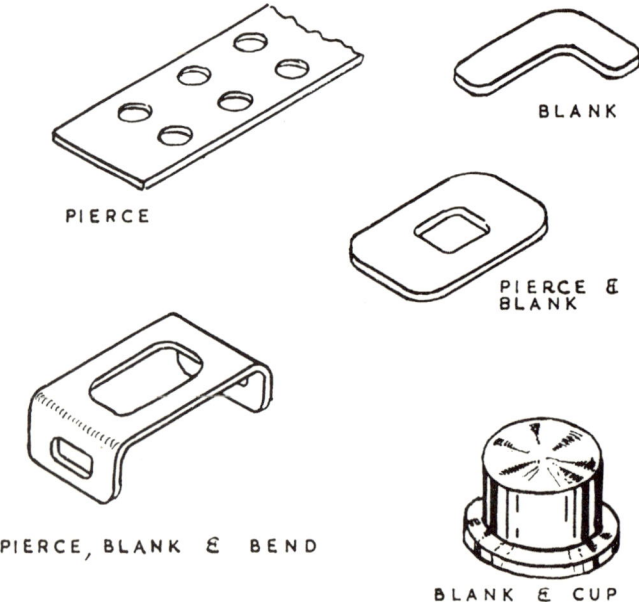

Fig. 15.1 Typical Examples of Presswork

operations, and also names the operations involved in their production. Pressings are more usually made from low-carbon steel, 70/30 brass, aluminium alloys and copper.

15.2. Presswork Operations

The following illustrate the main features of the more common presswork operations.

15.2.1. Piercing. Piercing implies the cutting of holes within the outside contour of the finished workpiece; the 'punching' (the name given to the

PRESSWORK 213

metal that is removed) is the scrap, and the metal that is left is the workpiece. Fig. 15.2 illustrates the piercing operation, and fig. 15.3 illustrates the production of perforated strip by employing several punches that cut simultaneously (note the 'feed' in this example).

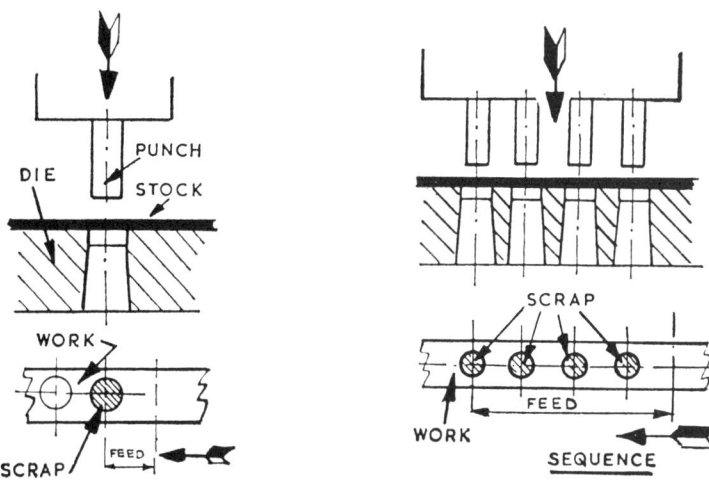

Fig. 15.2 Piercing   Fig. 15.3 Multiple Piercing

15.2.2. Blanking. Blanking implies that the outside contour of the workpiece is produced by removing metal from the stock; the 'punching' is therefore the workpiece. Fig. 15.4 illustrates simple blanking. Multiple blanking, in which several tools cut simultaneously, can be used to speed up production.

15.2.3. Piercing-and-blanking. A 'follow-on' die set can be used to produce a pierced-and-blanked workpiece at each stroke of the machine. Fig. 15.5

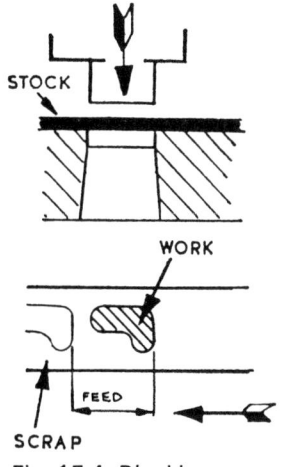
Fig. 15.4 Blanking

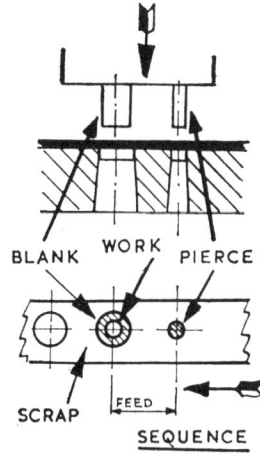
Fig. 15.5 Piercing-and-Blanking

illustrates such a die set; in this example a washer is produced at each stroke. The piercing tool produces the hole, the stock is fed to bring the hole so produced under the blanking tool and the washer is blanked at the next stroke; from then on one completed washer is produced at every stroke. If the circumstances are suitable, two blanking tools and two piercing tools can be used, so that two workpieces are produced at each stroke of the press.

15.2.4. Bending and forming. In bending and forming no cutting is involved, but the material may have been previously pierced and blanked. The basic features are the forming punch and the forming block. The included angle of the punch and the block is usually smaller than that to be produced, to allow for spring-back. In the example shown in fig. 15.6 the material is bent to an angle of 87° to allow for spring-back to produce the required 90° angle.

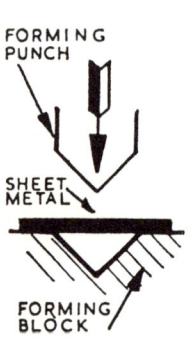

Fig. 15.6 Bending

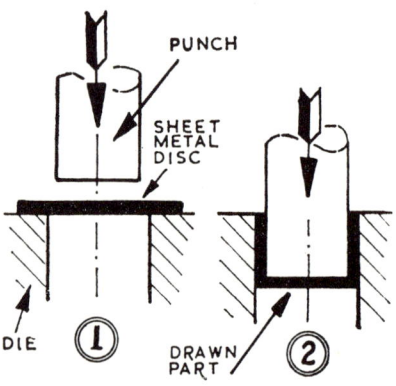

Fig. 15.7 Drawing

15.2.5. Drawing. In this operation a blank is pushed completely through, or partly through, the die to produce a cup or cup-like shape (see fig. 15.7). Deep cups are produced by 'deep drawing' in which the final shape is produced in stages, with annealing between each stage to restore the ductility that is lost following work-hardening. The final stage is often an 'ironing' operation in which the space between the punch and the die is less than the thickness of the stock before that stage.

15.2.6. Embossing. Embossing involves pressing sheet metal between dies; the shape being produced by local bending and stretching with very little change in the thickness of the material. This operation produces a component of which one side is the 'negative' of the other (see fig. 15.8).

15.2.7. Coining. This operation (shown in fig. 15.9) differs from embossing and other presswork operations because slightly thicker workpiece material

PRESSWORK 215

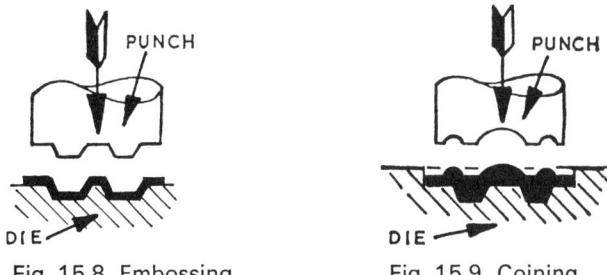

Fig. 15.8 Embossing   Fig. 15.9 Coining

is used, and because the material is caused to flow and to fill the die cavity by local indentation and extrusion. The part so produced has sides that are not 'related'.

15.2.8. Crimping. A second part can be held in a cup-shaped part by crimping, in which the sides of the cup are pinched or squeezed to hold the second part.

15.3. Presses

The press set is located in a press which supplies the required movement and pressure. Presses can be classified as: (1) hand- or foot-operated presses; (2) power-operated presses.

15.3.1. Hand- or foot-operated. The fly press shown in fig. 15.10 is a typical example of this class of press. A foot-operated press is useful for light assembly work because both of the operator's hands are free.

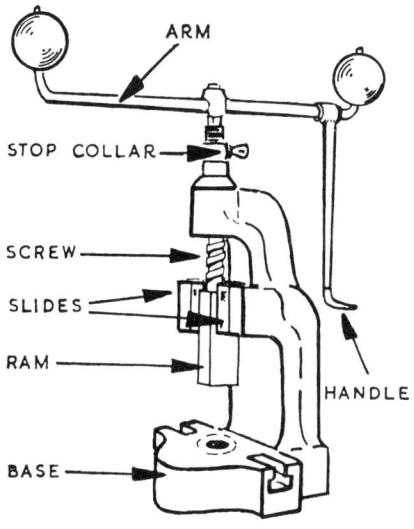

Fig. 15.10 Fly Press

15.3.2. Power presses. Power presses are operated from line shaft or by electric motor. Care must be taken to ensure that the operator's hands are away from the space between the ram and the table when the press is operating, and so power presses are usually equipped with a guard that pushes the operator's hands from the danger zone before the machine can start its pressing cycle. Power presses are classified according to the ram action.

15.3.3. Single-acting presses. These presses have one ram or slide. This slide may be operated by crank, eccentric, toggle or screw; the latter type is a mechanised version of the fly press, and operated by friction drive. The other machines can be set either to run continuously or to disengage at the top of the stroke. Continuous operation must, in the interests of safety, only be used with an automatic feed; intermittent operation is used for 'second-operation' work where the workpiece must be positioned by hand. Single-acting presses can be open-fronted (with a 'C'-shaped frame) or be double-sided, with a box-like frame; the former type is more convenient, but the latter type is more rigid. The open-fronted type can be stiffened up by the use of tie bars. Single-acting presses are often made inclinable—the frame being tipped back so that the workpiece can escape from the machine by gravity.

15.3.4. Double-acting presses. These presses have two rams or slides, and are used when it is necessary to use a separate blank holder and punch

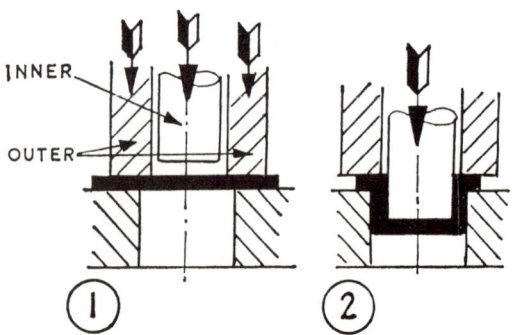

Fig. 15.11 Movements of a Double-Acting Press

arrangement. When a large pressing is produced the outer slide carries the blank holder, and is set in advance of the inner slide, which carries the punch (see fig. 15.11). The holder presses the sheet against the face of the die before the former pushes the sheet into the die; the outer slide is stationary when the punch is operating upon the material.

The inner slide is usually operated by crank and connecting rod linkage, but the inner slide can be operated either by cam or by toggle. The mechanism of the outer slide describes the press: e.g. 'double-acting cam-

PRESSWORK 217

operated'. The toggle-operated type is better for larger work because the machine frame takes the pressure, but the cam-operated type is easier to set. Double-acting presses are not usually continuously-operating.

15.3.5. Triple-acting press. This type of press is similar to the double-acting press but has a third slide, so that the material can be pushed up into a recess in the bottom of the punch. Fig. 15.12 illustrates the operation of this type of press.

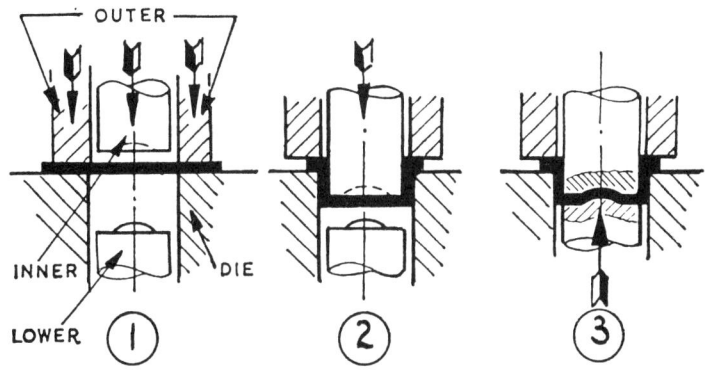

Fig. 15.12 Movements of a Triple-Acting Press

15.3.6. Cut-and-cupping press. This type of press is similar to the double-acting press except that both slides are moving during the cycle; the outer slide carries the blanking punch and returns when the blanking is completed. The inner slide carries the cupping tool which follows through, and pushes the blank into a cup shape.

15.3.7. Press capacity. The presses in each of the main groups already described are classified according to their capacity. The press capacity includes the size of the press (stroke, minimum distance between ram and press, and the bed size), the press speed and the pressure that it can exert. The latter requirement can often be checked by making a simple calculation, but if this is difficult it may be necessary to establish the force by trial and error.

15.3.8. Feeding. The feed of strip material for blanking, piercing, etc., may be by hand- or roll-feed (see fig. 15.32 on page 227). Second-operation work, such as the cupping of blanked material, may be fed by hand or automatically by hopper, etc.

15.3.9. The ejection of finished workpieces. When the workpiece is produced by pressing into blind dies or into former blocks, or when the workpiece cannot be knocked through the die, it is necessary to incorporate an ejector. This may be part of the press and be either a pneumatic cushion or

H

powerful springs, the machine may be horizontal so that the workpiece falls from the die when it is opened or, alternatively, the ejector may be part of the die set (as shown on page 238).

15.4. Standard Die-Sets

Standard die-sets are marketed by the makers of 'unit tooling parts'. The use of these sets produces a saving of tool-room time. Standard die-sets consist of two main parts: (1) the top plate; and (2) the bolster and guide pillars—as shown in fig. 15.13. The top plate and the bolster can be made either of cast iron or of steel; the guide pillars are made of steel. The top plate is located relative to the bolster by the two pillars, so that alignment can be obtained easily. Die-sets are available in a range of 'configurations' as shown in fig. 15.14. The principal dimensions are indicated in fig. 15.15.

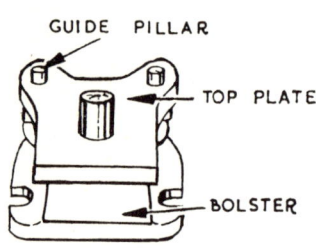

Fig. 15.13 Standard Die-Set

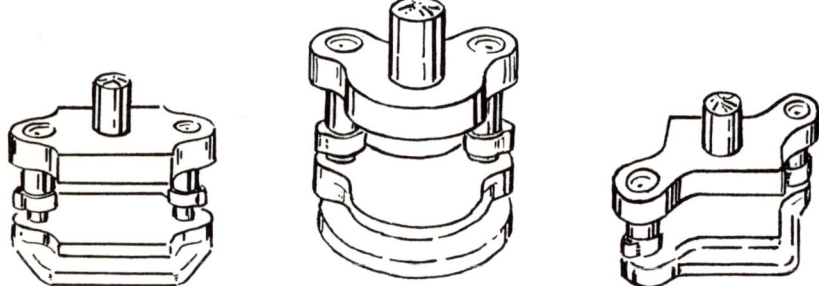

Fig. 15.14 Some Other Configurations of Standard Die-Set

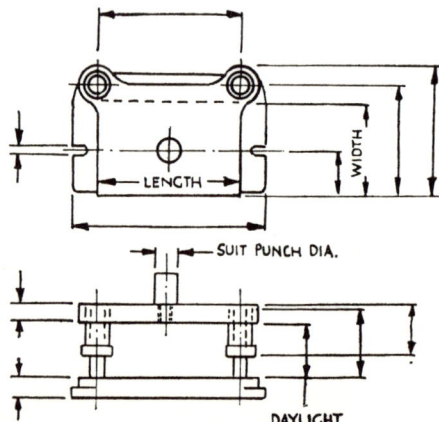

Fig. 15.15 Principal Dimensional Features of a Standard Die-Set

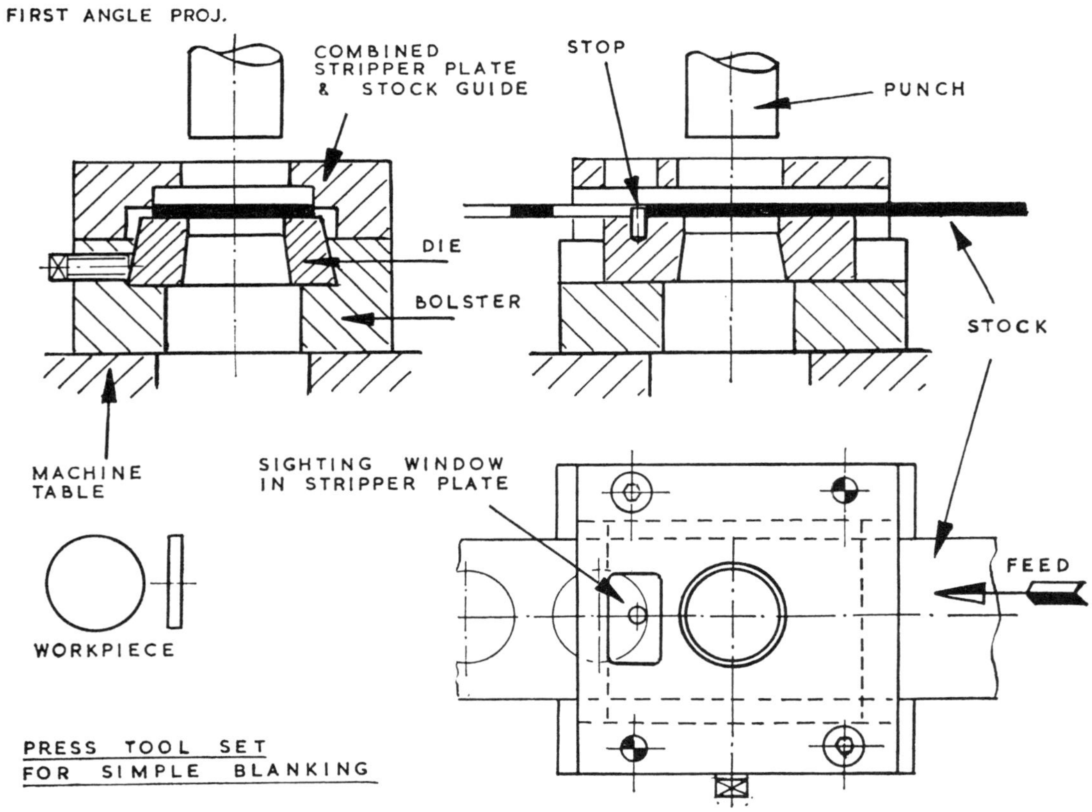
FIRST ANGLE PROJ.
COMBINED STRIPPER PLATE & STOCK GUIDE
STOP
PUNCH
DIE
BOLSTER
STOCK
MACHINE TABLE
SIGHTING WINDOW IN STRIPPER PLATE
FEED
WORKPIECE
PRESS TOOL SET
FOR SIMPLE BLANKING

Fig. 15.16

15.5. Piercing and Blanking

The following examples will illustrate the main features of press-tool sets for piercing and blanking operations.

Fig. 15.16 Simple blanking (see page 219)

The set shown includes the following basic features:

Punch—to shear the workpiece
Die—to act with the punch to shear the workpiece
Bolster—to hold the die
Guide—to ensure that the stock is located
Stop—to position the stock, using the hole already blanked, between each stroke
Stripper—to remove the stock from the punch during the return stroke. In this example the stop can be seen through a 'window' in the stripper.

Fig. 15.17 Pierce-and-blanking (see page 221)

The set shown pierces the stock at station 1; the stock is then moved to position 2 where it is blanked. One complete workpiece is produced at each stroke of the press. When the stock is fed, it is initially positioned using the stop, and then finally located by the pilot on the end of the blanking punch so that the outside and inside shapes are correctly positioned relative to each other.

The design of equipment for piercing and blanking will now be considered in detail.

15.5.1. Material for punches and dies

Application	*Type of steel*
Light stock	High-carbon, high-chromium steel
Light stock (extra long runs)	Tungsten–Chromium–molybdenum steel
Medium stock	High-carbon steel Carbon–manganese steel
Heavy stock	Chromium–vanadium steel Molybdenum–vanadium steel

Table 15.1

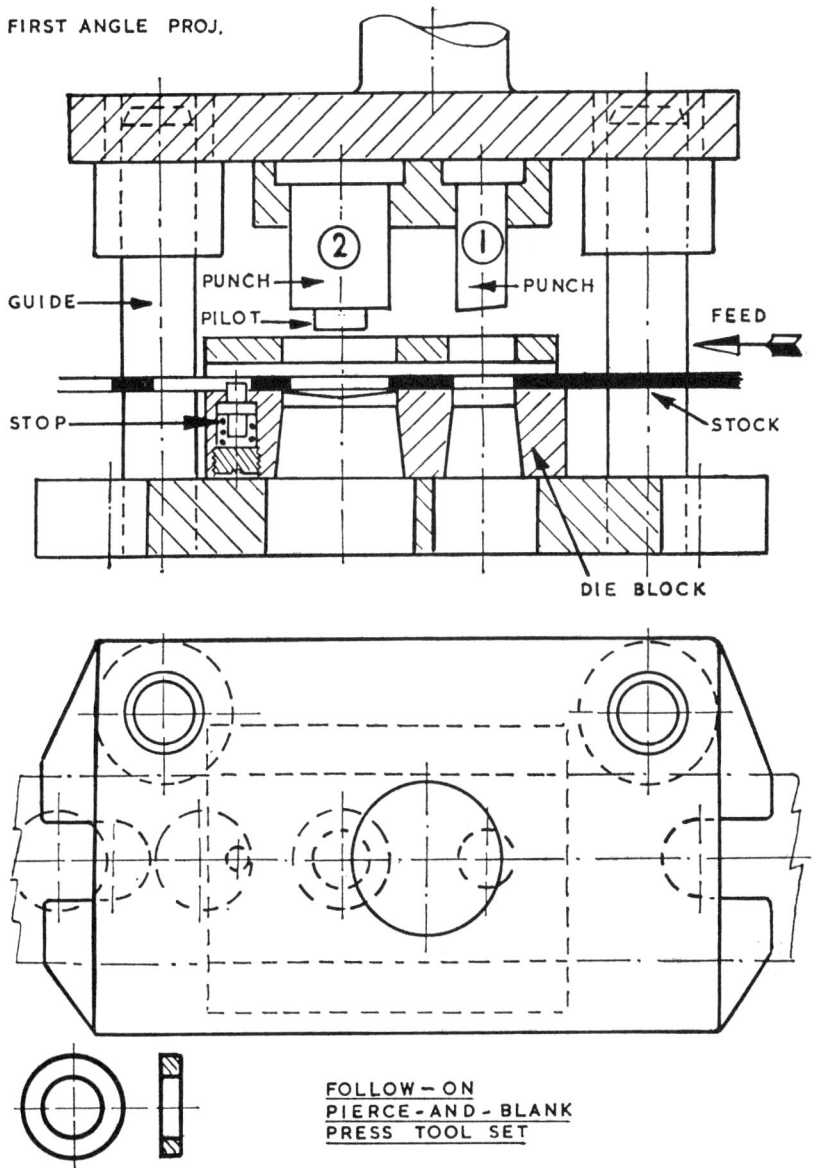

Fig. 15.17 FOLLOW-ON PIERCE-AND-BLANK PRESS TOOL SET

15.5.2. Punch retention. Fig. 15.18 shows the punch embedded in a wedge-shaped punch pad (tight fit to prevent shock from causing slackness) and retained by set-screw. The pressure is taken by the hardened pressure plate.

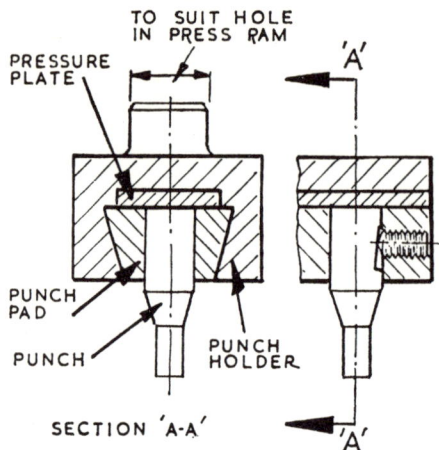

Fig. 15.18 Punch Retention

In the method shown in fig. 15.19 the punch is flanged, and this flange takes the pressure; the punch is retained by set-screw.

Fig. 15.20 illustrates a method used in large presses. The punch is flanged and fits into a recess in the punch pad; the pressure is taken by the pressure plate. This method produces good punch location in the pad,

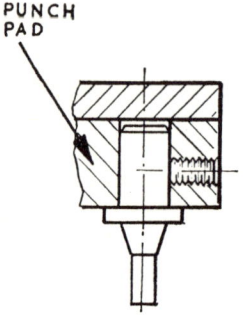

Fig. 15.19 Punch Retention

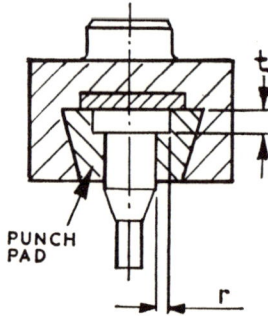

Fig. 15.20 Punch Retention

which in turn can be dowel-located in the holder. The flange thickness t should be $2r$ (see fig. 15.20).

In the arrangement shown in fig. 15.21 the flange takes the pressure and the punch is retained by set-screw.

In general, piercing punches should be at least of the same diameter as the stock is thick. Slender punches may be located in the stripper plate as

PRESSWORK

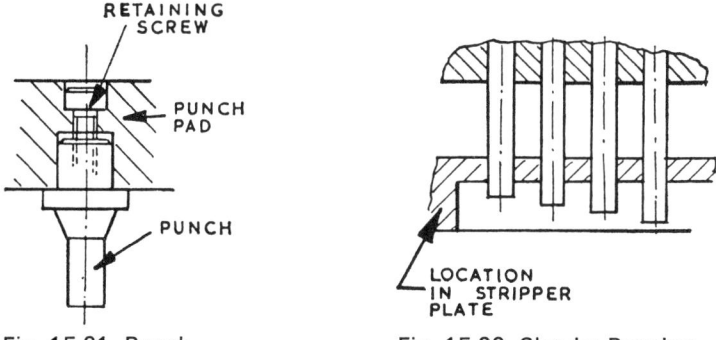

Fig. 15.21 Punch Retention

Fig. 15.22 Slender Punches

shown in fig. 15.22; for good location the stripper plate should be bushed or locally hardened.

It is often necessary to prevent the punching from sticking to the punch. Fig. 15.23 shows a spring-loaded plug (the spring being 'pinched' to the

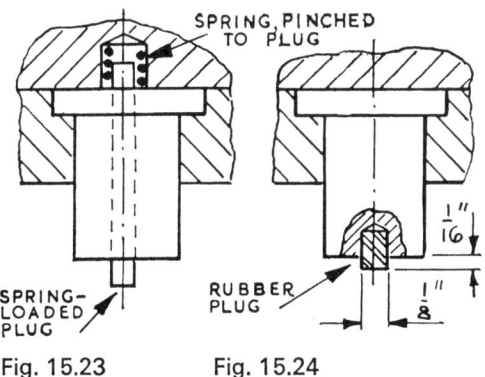

Fig. 15.23   Fig. 15.24

plug); the plug is pushed into the punch by the resistance of the workpiece during cutting, and allowed to spring out during the return stroke to ease off the punching. Fig. 15.24 shows a simple rubber plug that produces a similar effect.

15.5.3. Blanking and piercing dies. The die is secured to the bolster; some typical methods are shown in fig. 15.25.

Fig. 15.25 (a) shows a simple system in which the die has tapered sides and is pressed against the bolster by screws; this arrangement allows rapid die-change. Fig. 15.25 (b) shows a more permanent method of retaining the die; this method is more expensive and it takes longer to change the die. Fig. 15.25 (c) shows a very effective method in which the die is retained by a case-hardened steel retaining ring; the bore of the ring and the outside of the die are tapered (about 10°) to produce the locking effect.

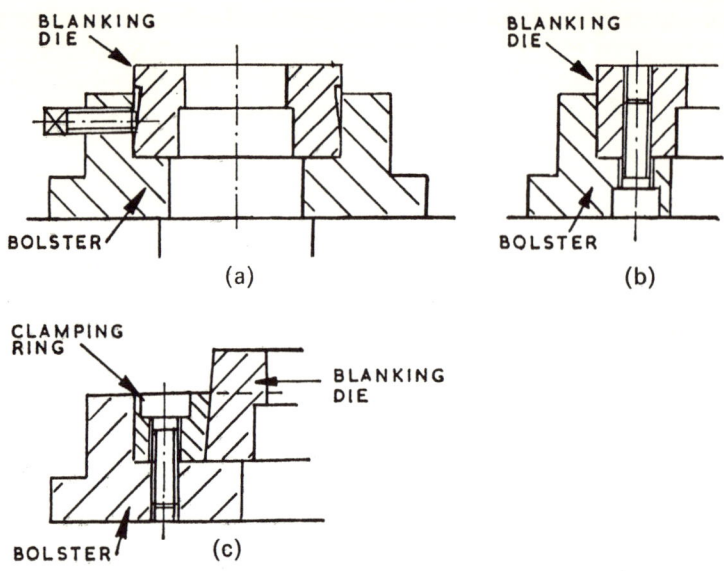

Fig. 15.25 Retention of Die

15.5.4. Clearance between punch and die. The clearance between punch and die depends upon the workpiece material and the stock thickness—i.e.

$$\text{Overall clearance} = Kt$$

where t is the stock thickness.

Values for K	
Material	*K*
Steel	0·07
Brass	0·05
Aluminium	0·10
Magnesium	0·03

Table 15.2

When piercing, the punch is nominal size, and the die is made oversize to produce the clearance. When blanking, the die is nominal size, and the punch is made oversize to produce the clearance. The clearance is very important—a correct clearance produces a clean, smooth edge.

15.5.5. Angular die clearance. Angular die clearance is necessary to prevent the punchings from jamming in the die. Fig. 15.26 shows the angular clearance—angle θ is about $\frac{1}{2}°$ leaving $\frac{3}{16}$ in land (about 8 mm) so that the clearance between punch and die is not increased by re-grinding.

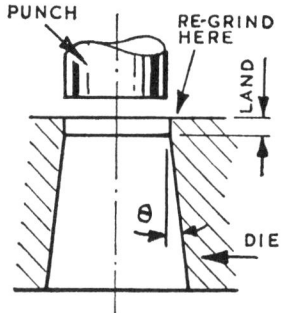

Fig. 15.26 Angular Die Clearance

15.5.6. Stripper and guides. The stripper is used to remove the stock from the punch during the return stroke. The guides, used to locate the stock, are usually incorporated into the stripper assembly.

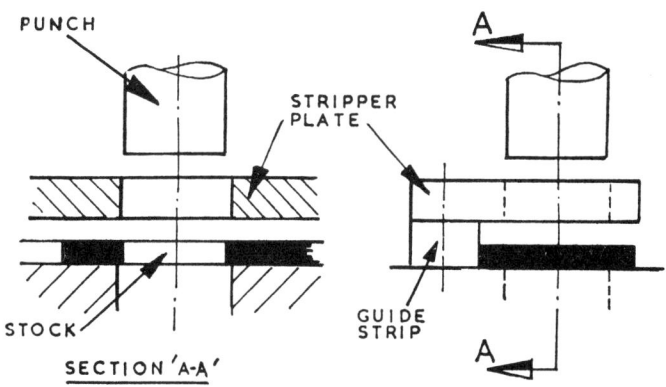

Fig. 15.27 Guide and Stripper Unit

The arrangement shown in fig. 15.27 incorporates a single guide strip, and the stock must be pushed back against the guide for location.

Fig. 15.28 shows a channel-type stripper that produces a more positive stock location.

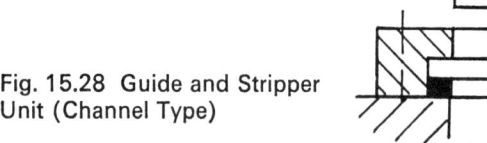

Fig. 15.28 Guide and Stripper Unit (Channel Type)

15.5.7. Pressure pads. Fig. 15.29 shows a combined stripper and pressure pad which is spring-loaded, and presses the stock against the die face during the punching operation to control the flow of the metal. During the return stroke the pad acts as a stripper to remove the stock from the punch.

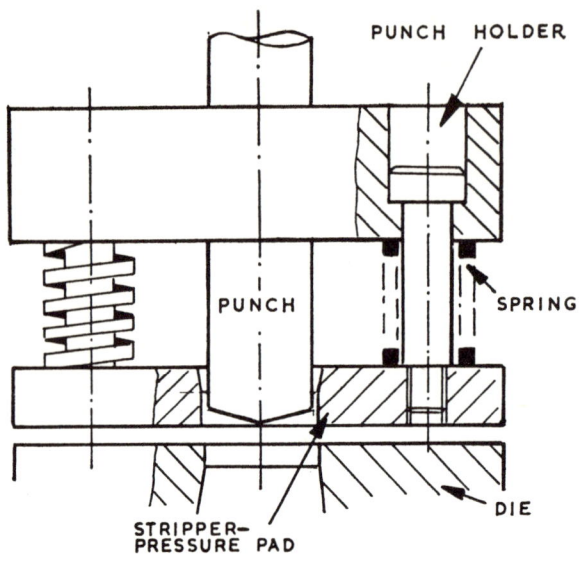

Fig. 15.29 Pressure Pad

The springs must be carefully selected to ensure that the pressure is adequate to hold the stock during punching, and to provide the stripping action.

15.5.8. Pilots. Pilots are often introduced as a feature of the blanking punch in a 'pierce-and-blanking' set to obtain stock alignment from the pierced hole. Fig. 15.30 shows a pilot that is simply a reduced diameter at the end of the punch. Fig. 15.31 shows a pilot that is shaped to produce a more effective location, and is detachable so that it can be removed from the punch to allow for the re-grinding of the bottom of the punch.

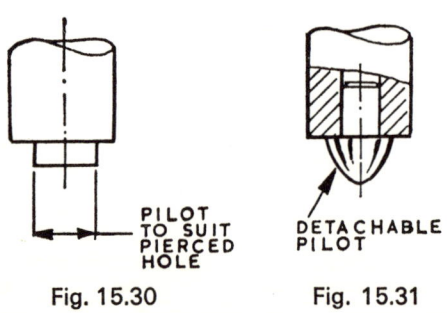

Fig. 15.30   Fig. 15.31

Punch Pilots

15.5.9. Feed methods. Stock in strip form can be hand-fed or be machine-fed. When hand-feed is used it is usual to introduce stock stops to position the stock before each stroke. For rapid large-quantity presswork a mechanical feed system is usually used; the most common method used is the roll-feed. In this method the machine has a pair of rolls at each end of the table that are operated by the punch slide mechanism (fig. 15.32

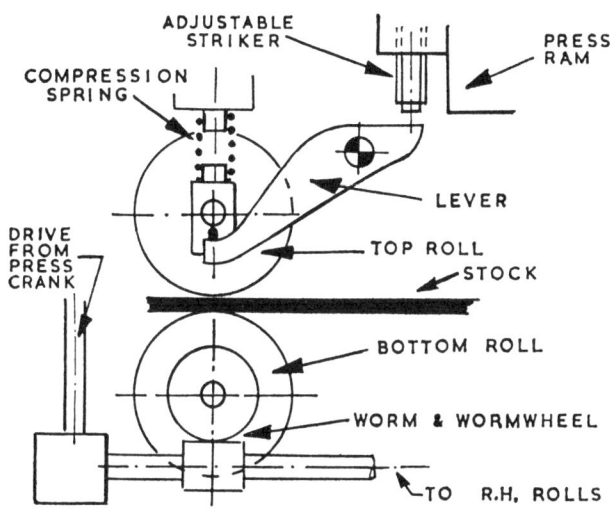

Fig. 15.32 Mechanical Feed System

illustrates roll-feed). The stock can be in coil form, and if required the scrap can be coiled at the other end, or alternatively cut up into short lengths for disposal. At the moment of punching, the top roll is lifted so that the stock is not constrained.

15.5.10. Stock stops. When hand-feed is used it is necessary to have some method of positioning the stock (from the previously punched hole) before each stroke. Fig. 15.33 shows a simple stop; when this stop is used the stock must be lifted over the stop, and then the next hole located by pushing the stock against the stop. The stop and stock can be sighted by introducing a 'window' in the stripper plate as shown in fig. 15.16 on page 219. This stop is simple, but demands skill on the part of the operator.

Figs. 15.34 and 15.35 show two spring-loaded stops; the stock is fed against the stop which retracts by wedge-action to allow the stock to pass over it, and is free to return when the punched hole is over it; the stock must be pulled back to locate it against the stop. Fig. 15.36 shows a pawl stop; the pawl is lifted by wedge-action as the stock is fed, and it springs back into position in the next hole; the stock is pulled back slightly to ensure location.

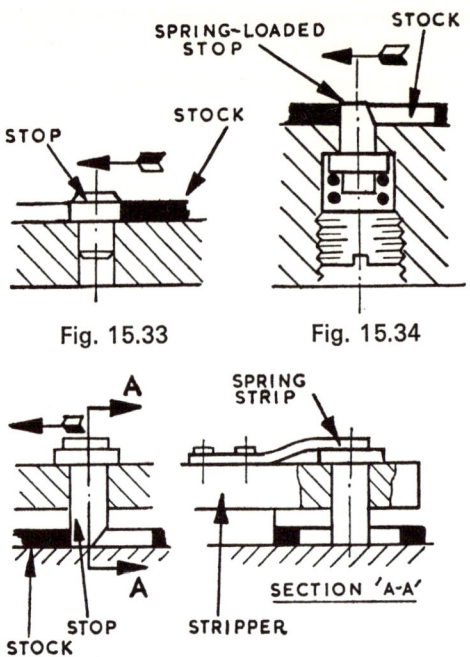

Fig. 15.33   Fig. 15.34

Fig. 15.35

Figs. 15.33/15.34/15.35 Stock Stops

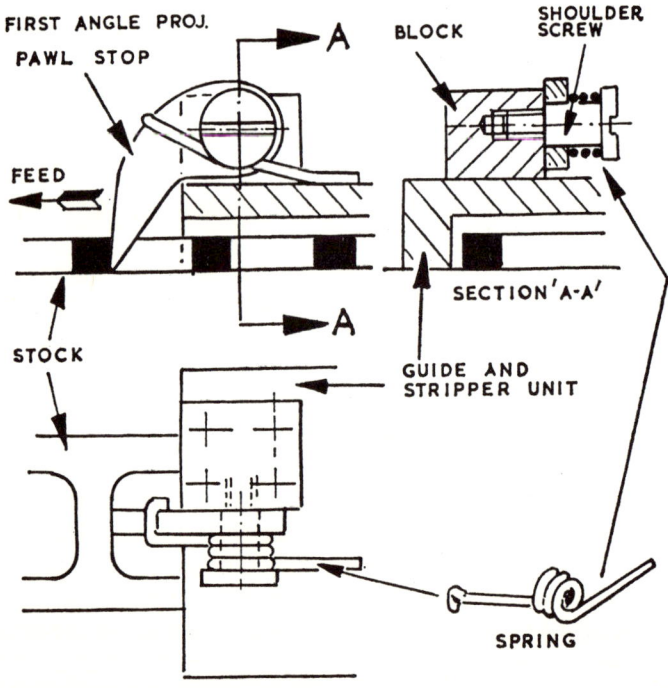

Fig. 15.36 Pawl-Type Stop

PRESSWORK 229

Fig. 15.37 shows a more elaborate system that positions the stock until the punch starts to cut and allows it to be free during the cutting. The location lever is a loose fit on the fulcrum pin and a very free fit in the slot in the stripper plate. The illustration shows the arrangement at the time of locating the stock before cutting; the location end of the location lever is

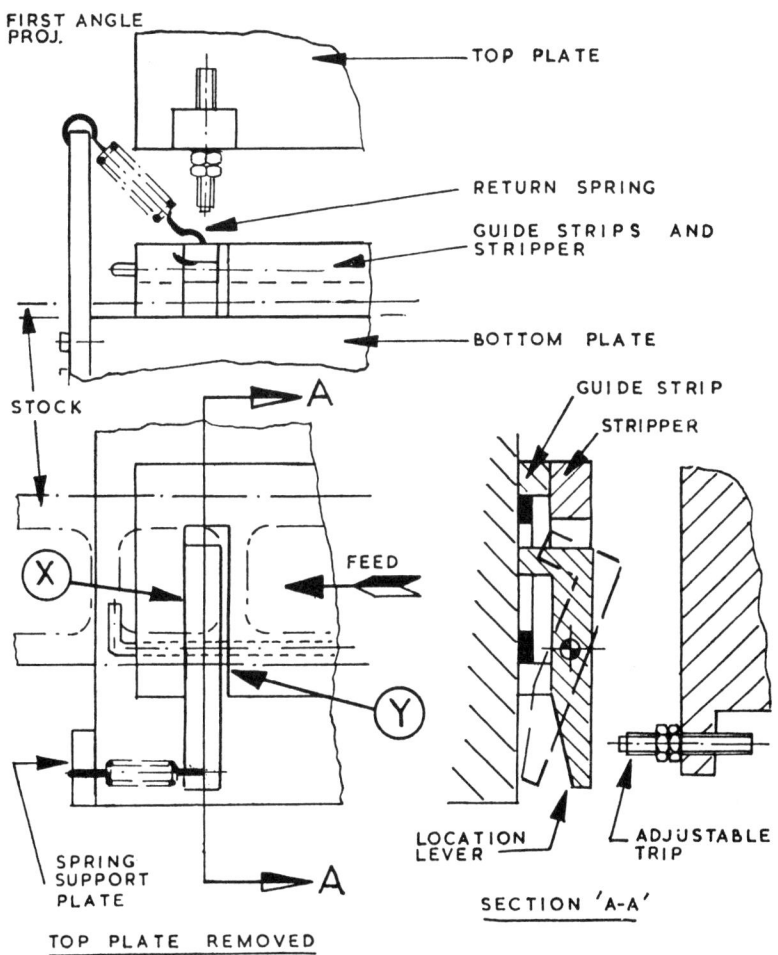

Fig. 15.37 Trigger-Type Stop

forced against the bottom plate by the return spring, the stock is punched against the location lever by the operator and, in turn, the location lever is pushed against face X of the slot in the stripper plate. At the moment of cutting, the trip attached to the top plate pushes the end of the location lever so that it swings in the vertical plane and no longer controls the position of the stock. At the same time, the lever is pulled by the return spring so that its location end is forced against face Y of the stripper plate; when the top plate returns, the return spring causes the location lever to rotate in the vertical plane so that the location end sits on the top of the stock and allows it to pass when the next feed movement takes place. The location end of the lever drops into position in the next hole in the stock during the feed movement ready to locate it for the next stroke.

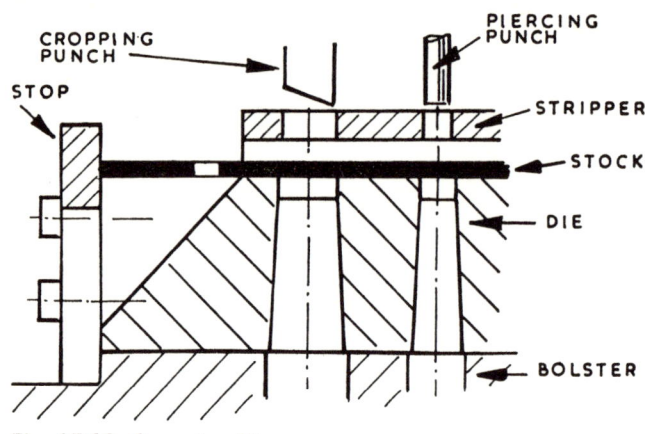

Fig. 15.38 Cropping Stop

Fig. 15.38 illustrates a stop used in conjunction with a piercing and cropping die-set. After piercing, the stock is advanced to the stop and cropped to length at the next stroke. The cropped material falls into the slot in the die, to escape through a hole in the stop plate.

15.5.11. Piercing and blanking pressures

The pressure required to shear the stock material is given by:

$$P_s = \text{Length of edge to be sheared} \times t \times f_s$$

where
- P_s = Pressure to shear the material
- t = Stock thickness
- f_s = Shear strength of the stock material

Material	Shear strength $tonf/in^2$
Mild steel	22
Medium-carbon steel	35
High-carbon steel	42
Brass (soft)	15
Brass (hard)	20
Copper (soft)	12
Copper (hard)	18
Aluminium (soft)	8
Aluminium (hard)	15
Aluminium alloy	20
Cupro-nickel	20

Table 15.3. Shear Strength of some Typical Materials

For example:

$$P_s \text{ for circular blank} = \pi . d . t . f_s$$

(where blank diameter is d)

$$P_s \text{ for square blank} = 4 . a . t . f_s$$

(where side of square is a)

$$P_s \text{ for rectangular blank} = 2(a + b)t . f_s$$

(where a and b are lengths of sides of rectangular blank)

15.5.12. Application of shear. It has been found that the punch need only penetrate partly into the stock to produce a complete shearing of the material. The percentage of stock thickness to cause complete shearing depends upon the thickness of the stock material as shown in Tables 15.4 (a) and (b).

Material thickness (in)	1	$\frac{3}{4}$	$\frac{5}{8}$	$\frac{1}{2}$	$\frac{3}{8}$	$\frac{5}{16}$	$\frac{1}{4}$	$\frac{3}{16}$	$\frac{1}{8}$	$\frac{3}{32}$	$\frac{1}{16}$	below $\frac{1}{16}$
Percentage penetration to shear	25	31	34	37	44	47	50	56	62	67	75	80

Table 15.4 (a). Percentage Penetration to Shear

Material thickness (mm)	25	20	15	10	8	6	5	3	2	below 2
Percentage penetration to shear	25	30	34	41	45	51	54	64	74	79

Table 15.4 (b). Percentage Penetration to Shear

If the punch is shaped as shown in fig. 15.39 the pressure to shear the material will be as already calculated, although the punch will not need to completely penetrate the stock to shear it (the pressure required to shear the material under this condition will be referred to as P_{max}). If the end of the punch is shaped as shown in fig. 15.40 the pressure will gradually build up to a maximum that is less than that when the punch does not have

Fig. 15.39 Flat-Ended Punch

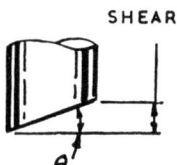

Fig. 15.40 Punch with **Shear**

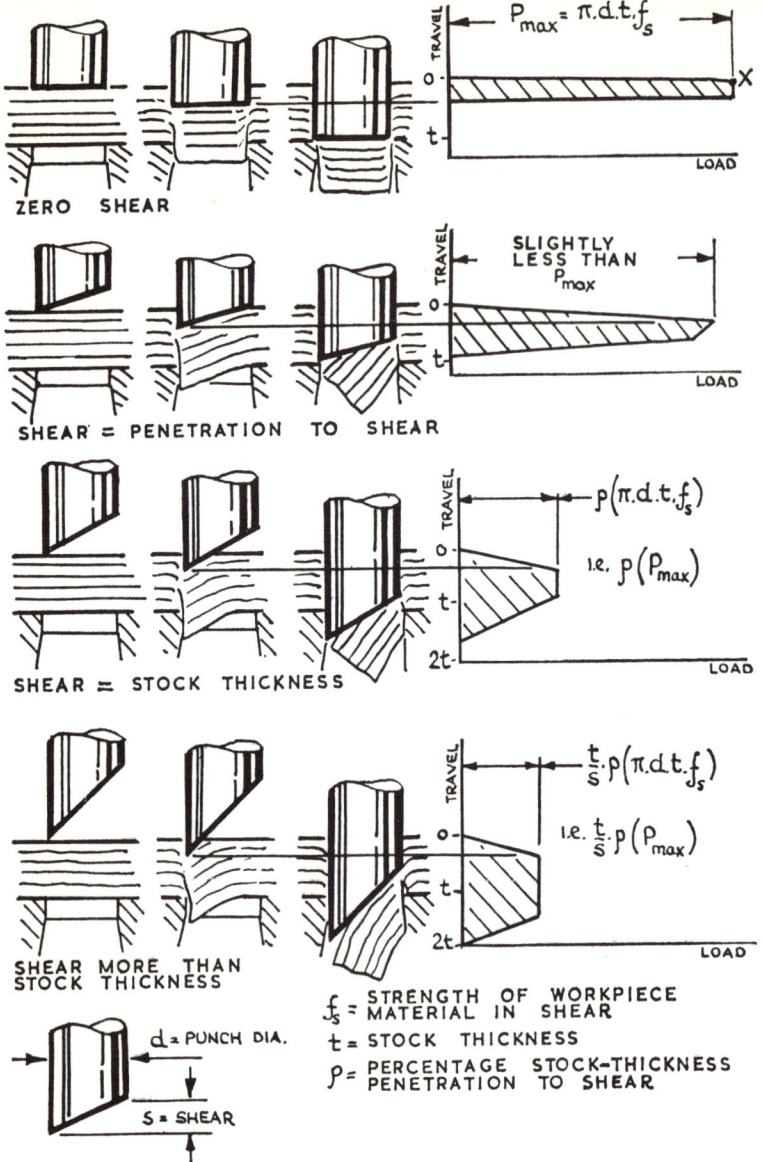

Fig. 15.41 Single Shear on Circular Punch

the shaped end; after reaching the maximum pressure, the pressure will gradually fall. The shaped end is known as **shear**. Fig. 15.41 above shows the effect of variation of **shear** upon the pressure required to shear the material.

When the **shear** is equal to the penetration to cause the material to shear, the pressure will gradually increase to reach a maximum value and then fall off. This maximum value is slightly less than P_{max}.

When the **shear** is equal to the stock thickness the pressure to cause the material to shear is given by:

$$P_s = p(P_{max})$$

where P_s = pressure to cause material to shear;
p = percentage penetration to cause the material to shear;
P_{max} = pressure required when **shear** is zero.

When the **shear** is more than the stock thickness, the pressure to cause the metal to shear is further reduced because only part of the workpiece will be under shear at any time. Fig. 15.42 shows how the area under shear is reduced as the **shear** is increased—see *a*, *b*, *c* and *d* on fig. 15.42.

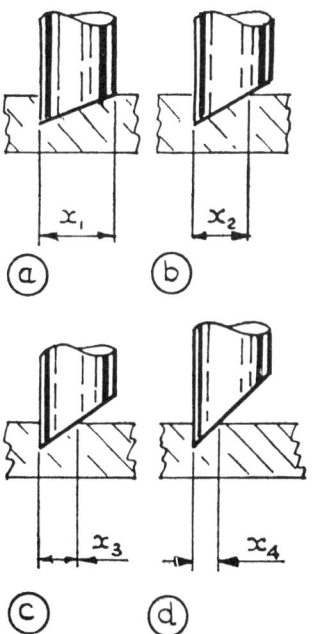

Fig. 15.42 Connection between Punch Shape and Area Being Sheared

The pressure required will then be

$$P_s = \frac{t}{s} \cdot p \cdot P_{max}$$

where t = stock thickness;
s = **shear**;
p = percentage penetration to shear the material.

The **shear** cannot be made too large because it will weaken the punch; it is also limited by the punch travel that can be accommodated because travel increases with amount of **shear** on the punch.

Shear can be applied either to the punch or to the die. When it is applied

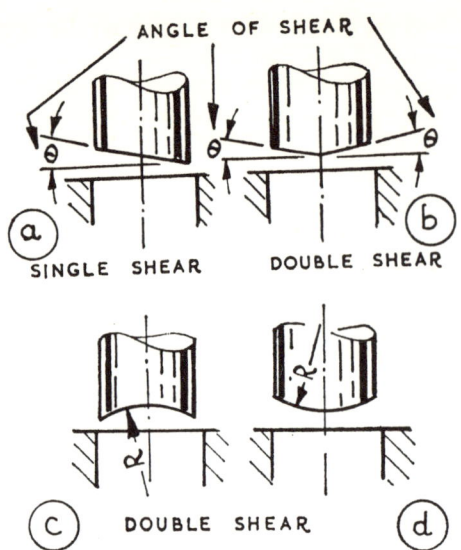

Fig. 15.43 Shear on Punch

to the punch the punching will be bent, and when applied to the die the remaining stock will be bent; therefore it is applied to the punch when piercing, and to the die when blanking.

The examples given so far are of single **shear**; single **shear** causes a side thrust to be set up and so it is better to apply double shear so that the thrusts will be neutralised.

Figs. 15.43 and 15.44 illustrate single **shear** and double **shear** applied to the punch and to the die.

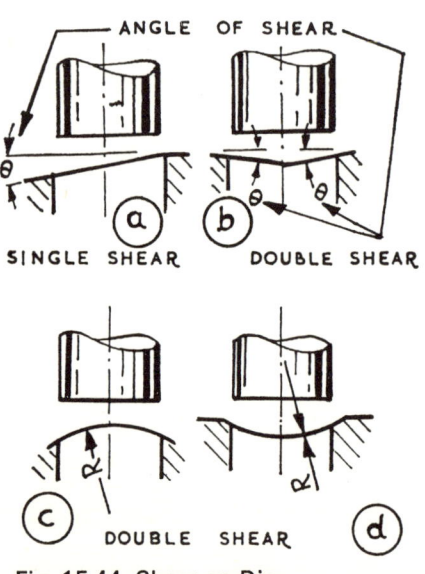

Fig. 15.44 Shear on Die

PRESSWORK 235

Example: What will be the blanking pressure for the die-set shown in fig. 15.16 (page 219) when the blank diameter is ¾ in and the material is 0·060-in-thick mild-steel strip? Consider the pressure when the **shear** applied to the punch is: (*a*) zero; (*b*) 0·060 in; (*c*) 0·100 in.

Solution: refer to Tables 15.3 and 15.4; $f_s = 22$ tonf/in² and $p = 75\%$.

Case (*a*) $P_{\max} = \pi \cdot d \cdot t \cdot f_s$
$= \pi \times 0\cdot75 \times 0\cdot060 \times 22$ tonf
$= 3\cdot11$ tonf

Case (*b*) $P_s = 0\cdot070 \times 3\cdot11 = 2\cdot34$ tonf

Case (*c*) $P_s = \dfrac{0\cdot060}{0\cdot100} \times 0\cdot75 \times 3\cdot11 = 1\cdot40$ tonf

15.5.13. Blanking layout. The layout of the blanks on the strip must take into account material economy, grain flow of the strip, the tool layout and the width of strip that is available.

Grain flow. If the blanked material is to be bent, the blanks must be placed so that the bending will not cause fracture. Fig. 15.45 illustrates the connection between the grain fibre that is produced during the rolling of the strip and the blank layout for best strength. If the blank is to be bent in more than one direction, the most severe bending should be along the grain.

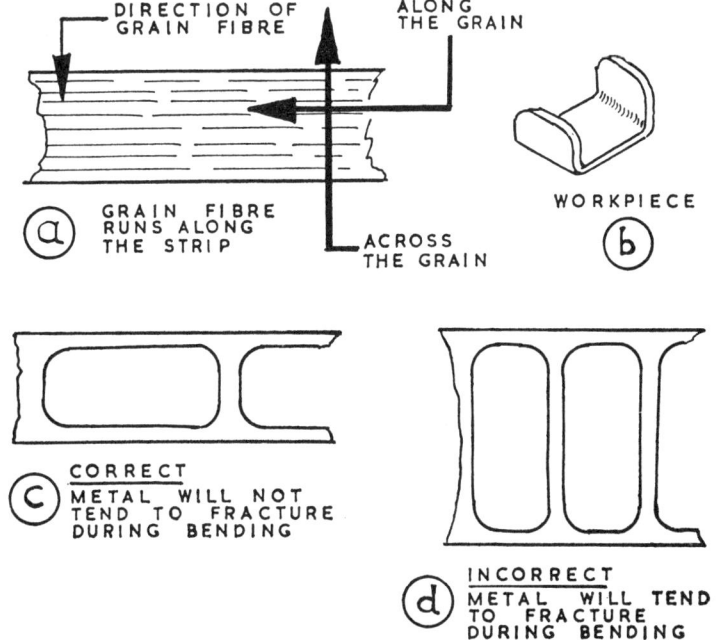

Fig. 15.45 Consideration of Grain Fibre when Producing Blanking Layout

Width of strip. The width of strip that is required depends upon the accuracy of the strip, the material and the thickness of the strip. The strip width is obtained by adding $2Y$ to the blank size (see fig. 15.46): the size

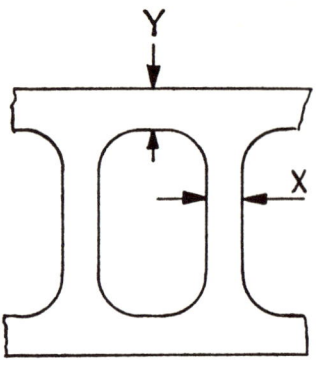

Fig. 15.46 Strip Layout

so obtained is 'rounded up' to obtain a standard stock width. Table 15.5 (*a*) gives typical values of Y for metal up to $\frac{1}{16}$ in thick, and Table 15.5 (*b*) gives typical values of Y for metal up to 1·5 mm.

Blank size	*Y in*	
	Aluminium up to 0·020 in thick	*Aluminium over 0·020 in thick, brass, copper and steel*
Up to 1½ in	$\frac{3}{16}$	$\frac{1}{8}$
1½–3 in	$\frac{1}{4}$	$\frac{3}{16}$
3–5 in	$\frac{3}{8}$	$\frac{1}{4}$
5 in and over	$\frac{1}{2}$	$\frac{3}{8}$

Table 15.5 (*a*)

Blank size	*Y mm*	
	Aluminium up to 0·5 mm thick	*Aluminium over 0·5 mm thick, brass, copper and steel*
Up to 38 mm	5	3
38–76 mm	7	5
76–127 mm	10	7
127 mm and over	13	10

Table 15.5 (*b*)

PRESSWORK 237

Bridge between the blanks (distance X on fig. 15.46). The bridge is less than the side allowance (Y). Tables 15.6 (a) and (b) give typical values for X, but as a rough guide it can be taken as equal to the stock thickness.

Strip thickness (in)	0·10	0·20	0·30	0·40	0·50	0·60
Bridge (in)	0·20	0·25	0·30	0·35	0·40	0·45

Strip thickness (in)	0·70	0·80	0·90	1·00	1·20
Bridge (in)	0·50	0·55	0·60	0·65	0·75

Table 15.6 (*a*)

Strip thickness (mm)	2	5	8	10	13	15	18	20	23	24
Bridge (mm)	4	6	8	9	11	12	13	15	16	17

Table 15.6 (*b*)

Blank arrangement for economy (see fig. 15.47). When considering the arrangement of the blanks for economy, it is convenient to determine the

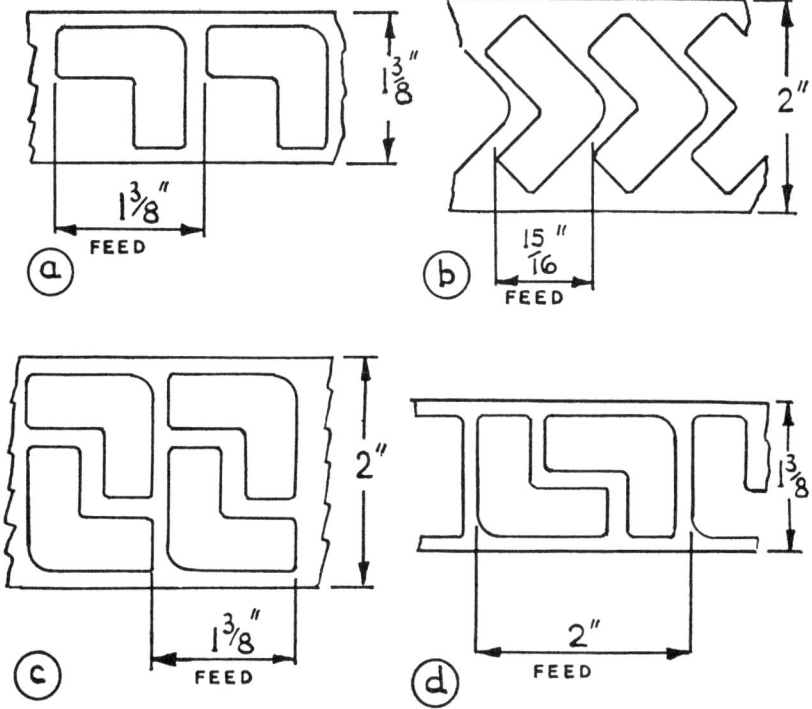

Fig. 15.47 Consideration of Economy when Producing Blanking Layout

amount of metal, including scrap required for each blanking. For example, fig. 15.47 (a) requires $1\frac{3}{8}$ in $\times$ $1\frac{3}{8}$ in of metal and the arrangement shown in fig. 15.47 (b) requires $\frac{5}{16}$ in $\times$ 2 in of metal for each blank. Fig. 15.47 (c) and (d) is for multiple blanking; each arrangement requires $1\frac{3}{8}$ in $\times$ 2 in of metal for two blanks; the choice between these two arrangements will be governed by the difference between the feeds, the grain fibre required and the strip width that is available. The choice between single and multiple blanking may be governed by the capacity of the press that is available.

15.6. Bending and Forming

Usually the material will be pierced and/or blanked or cropped before bending and forming is done; it will be located on the forming block with the aid of guide plates or guide strips, and formed between punch and form block.

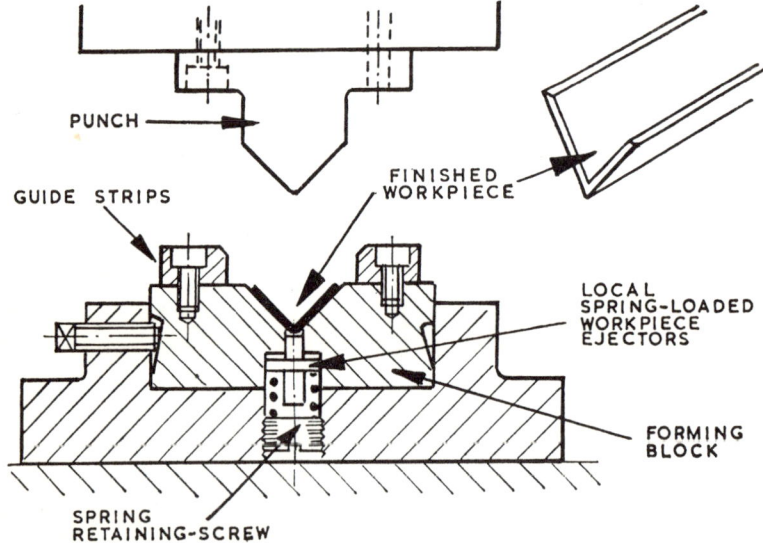

Fig. 15.48 Press-Tool Set for Vee Bending

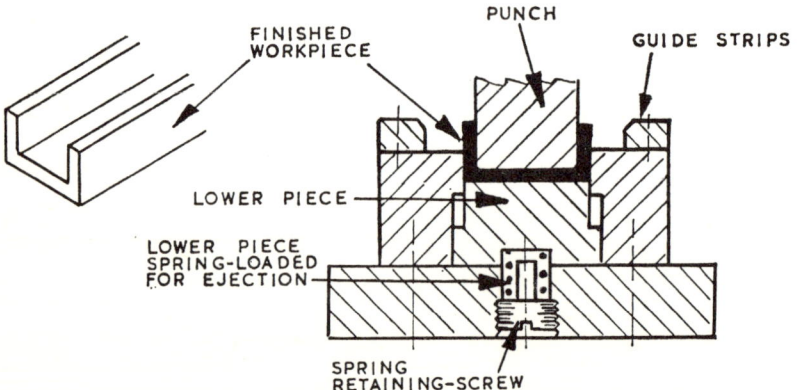

Fig. 15.49 Press-Tool Set for Bending a Channel Section

PRESSWORK 239

Fig. 15.48 shows the basic features of a press-tool set for 'vee bending':

Bolster—to hold the forming block.
Guide strips (or guide plate)—to locate the workpiece.
Ejectors—to lift the finished workpiece from the forming block.

Fig. 15.49 shows a press-tool set for bending a channel section. In this example the lower piece is spring-loaded and serves to eject the finished workpiece from the forming block.

15.6.1. Ejectors. Ejectors are necessary when bending and cupping because the workpiece is not knocked through the die. Figs. 15.48 and 15.49 show

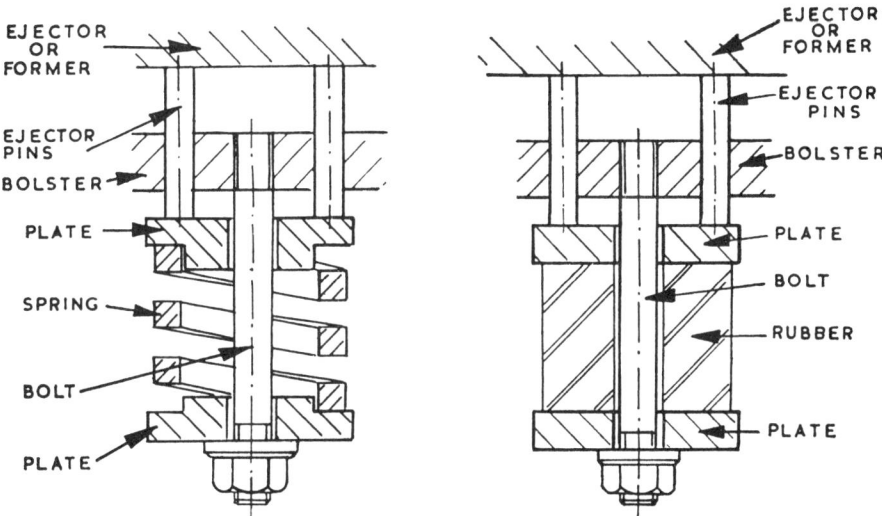

Fig. 15.50 Spring-Loaded Ejector   Fig. 15.51 Rubber Cushion Ejector

two simple spring-loaded ejectors; a more powerful spring-loaded ejector system is shown in fig. 15.50, and a rubber-cushion version of this method is shown in fig. 15.51. As stated on page 217, the ejector may be operated by the machine itself.

15.6.2. Punch and former material. Manganese steel or a 3·5% tungsten steel suitably heat-treated.

15.6.3. Material size. The size of material to be given a simple bend, e.g. vee or channel, can be determined by basing the length on the dimensions at 'mean thickness', but that for a more complicated double-curvature part may need to be determined by experiment.

15.6.4. Springback allowance. Material may need to be bent more than that shown on the workpiece drawing to allow for springback. Soft materials do not tend to suffer springback, but hard material suffers quite a large amount of springback. For example, a 90° bend may need to be given an 87° bend by the tools. The amount required may need to be determined by experiment.

15.6.5. Bending forces. The force to produce a simple bend can be determined by considering the length along which the bend takes place—the force being taken to be half that required to cut that length.

15.7. Drawing

Material may be blanked, located in a drawing die and then pushed through the die to form a cup or cup-like shape. The operation to produce a flanged cup is often called 'raising'.

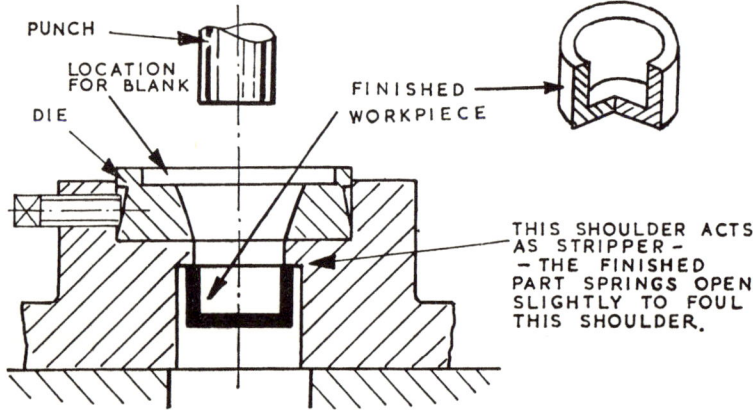

Fig. 15.52 Press-Tool Set for Drawing a Cup

Fig. 15.52 shows the basic features of a die-set to produce a cup:

Punch and drawing die—these control the metal. In this example the blank is located in the die.
No stripper plate is incorporated, but the finished cup is stripped from the punch during the return stroke because the finished part will spring open very slightly to foul the shoulder on the bolster.

15.7.1. Punch material. Manganese steel or high-carbon and high-chromium steel suitably heat-treated.

15.7.2. Die material. High-carbon steel, high-carbon and high-chromium steel, or 3·5% tungsten steel suitably heat-treated. When abrasion is expected a cobalt-base alloy or a cemented carbide is used.

15.7.3. Die shapes. Fig. 15.53 illustrates some typical drawing die shapes. Fig. 15.53 (*a*) shows the die shape used when the workpiece cannot be knocked through by the punch, and it incorporates a clearance at the bottom.

Fig. 15.53 (*b*) shows the die used when the workpiece is knocked through the die, and is stripped from the punch on the return stroke; it is shaped to produce a gradual reduction in diameter as the material is pushed through the die. Fig. 15.53 (*c*) and (*d*) illustrates dies that incorporate a location for the blank. Fig. 15.53 (*e*) shows how dies of the type shown in fig. 15.53 (*a*)

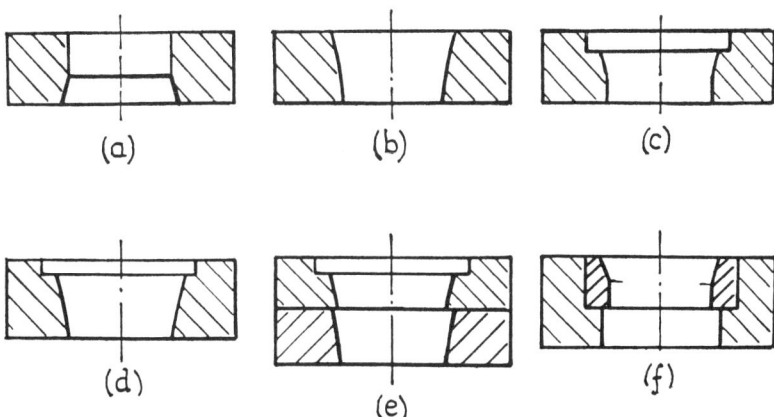

Fig. 15.53 Types of Drawing Die

and (*d*) can be combined to produce a gradual change of shape to allow a greater deformation. Fig. 15.53 (*f*) shows a cobalt-base or cemented carbide die held in a holder; this arrangement is necessary because of the brittleness of these hard materials. The dies shown in fig. 15.53 are shown simplified—they can be secured to the bolster in a similar way to the blanking dies (see fig. 15.25 on page 224).

15.7.4. Combination tools. Blanking and cupping can be done in one operation on a single-acting press using a combination tool. Use of a combination tool reduces handling time, ensures accuracy by gripping the blank immediately upon cutting it and improves the safety because the operator does not need to position the blank manually on the face of the die.

Fig. 15.54 on page 242 shows a typical combination tool, the operation of which is as follows: the material to be drawn is in strip form, and it is fed and located as for blanking. The blanking punch descends and shears the material in conjunction with the blanking die; as the punch descends further the blank is pressed against the blank holder, at the same time cupping the material between its own bore and the stationary cutting punch. As the punch returns, the spring-loaded blank holder pushes the cup with it; the stock is stripped from the outside of the blanking punch by the stripper plate, and the cup is ejected from the inside of the blanking punch by the spring-loaded ejector.

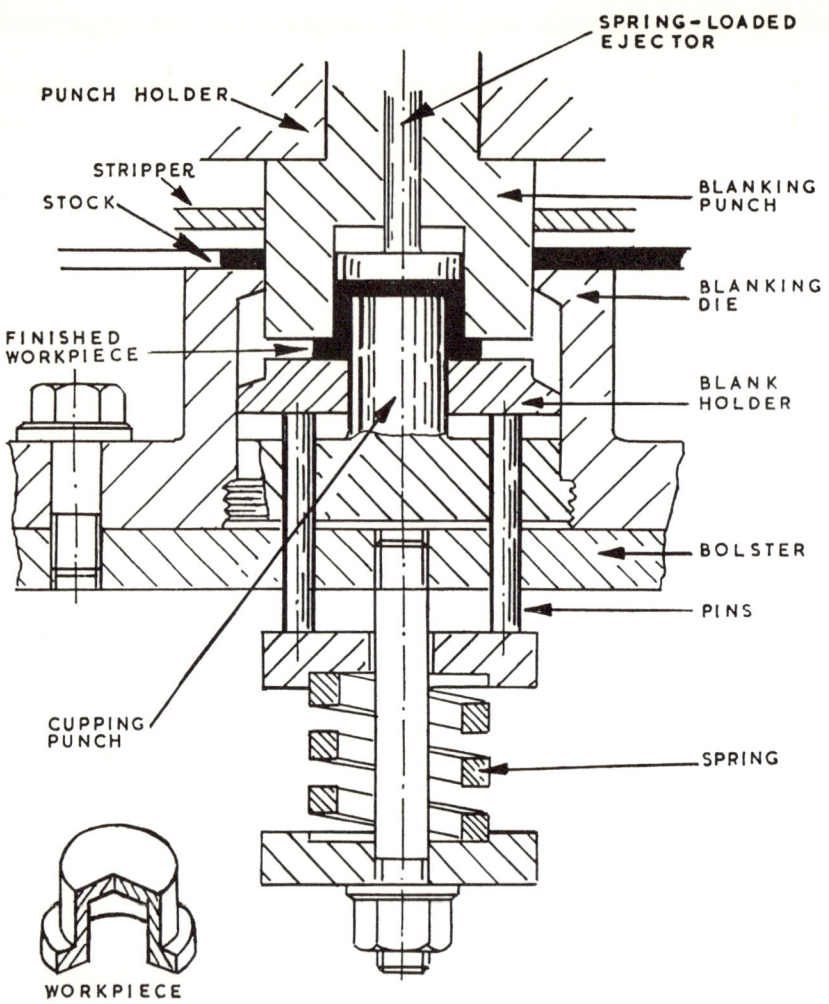

Fig. 15.54 Combination-Tool Set

15.7.5. Blank size

Blank size for a thin metal cup can be calculated by equating the surface area of the blank with that of the cup:

Area of the blank = Area of the cup produced

$$\frac{\pi D^2}{4} = \text{Area of base} + \text{area of side}$$

$$= \frac{\pi d_i^2}{4} + \pi d_i h$$

Multiplying by $\frac{4}{\pi}$:

$$D^2 = d_i^2 + 4d_i h$$

$$D = \sqrt{d_i^2 + 4d_i h}$$

Fig. 15.55 Simple Cup

This does not take into account the radius at the bottom of the cup; when the radius is less than $\frac{h}{4}$ it can be allowed for by using a modification of the foregoing expression:

$$D = \sqrt{d_i^2 + 4d_i(h - r)}$$

The blank size for a thin-metal flanged cup can be calculated by using a similar method:

Area of the blank

$\qquad$ = Area of the cup produced

$\qquad$ = Area of the base + area of the flange + area of the side

$$\frac{\pi D^2}{4} = \frac{\pi d_i^2}{4} + \left(\frac{\pi d^2}{4} - \frac{\pi d_i^2}{4}\right) + \pi d_i h$$

$$= \pi d^2 + \pi d_i h$$

$$D^2 = d^2 + 4d_i h$$

$$D = \sqrt{d^2 + 4d_i h}$$

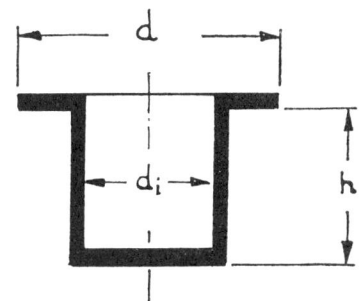

Fig. 15.56 Flanged Cup

Allowing for a radius of less than $\frac{h}{4}$, this expression becomes:

$$D = \sqrt{d^2 + 4d_i(h - r)}$$

When the cup has a hemispherical base, the blank size can be calculated by introducing the surface area of the inside of the hemisphere, i.e. $\frac{\pi d_i^2}{2}$, as the area of the base in the expressions developed above.

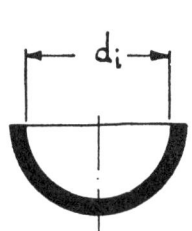

Fig. 15.57 Hemispherical Cup

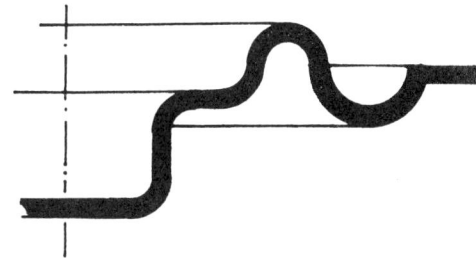

Fig. 15.58 Section through a Complicated Pressing

Awkward shapes. Fig. 15.58 represents the section through a complicated cupped part. The blank size for some pressings of this type can be calculated by an extension of the methods described above; but more complicated parts are best dealt with by applying the First Theorem of Pappus, which states that 'The Area of a Solid of Revolution is Equal to the Length of the Generating Surface, Multiplied by the Length of the Path of the Curve.'

The form is first drawn full size, and a length of soft wire is bent to the shape of its half-section. The wire is now balanced on a knife edge to find

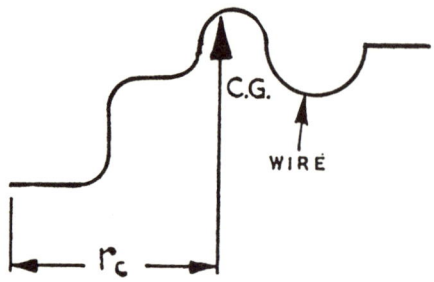

Fig. 15.59 Finding the C.G. of a Wire Model of Complicated Pressing

its C of G; the position of which is noted (dimension r_c on fig. 15.59). The wire is now straightened and its length l noted.

The surface area of the pressing $= 2\pi r_c \times l$

Equating this with the area of the blank:

$$\frac{\pi D^2}{4} = 2\pi r_c \times l$$
$$\therefore D^2 = 8r_c \times l$$
$$D = \sqrt{8r_c l}$$

In practice this method is approximate and the pressing may require cropping.

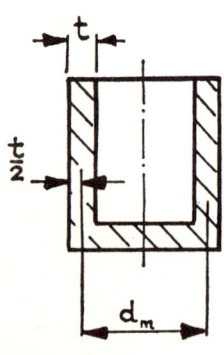

Fig. 15.60 Thick-Metal Cup

Thick-metal cupped parts. When the pressing is of thick-metal it may be necessary to consider the 'mean thickness' sizes; in fig. 15.60 the mean diameter is indicated (d_m). Alternatively the volume of the pressing can be equated with that of the required blank. The calculated values for blank sizes tend to be high because the metal will be stretched and thinned in the region of the corners.

Rectangular box with radii at its corners. If the pressing cannot be cropped at a later operation it is necessary to take into consideration the corner shape to prevent the formation of an incorrect shape as shown in fig. 15.61. Radius r_b can be determined by treating the corner as part of a cylinder:

PRESSWORK

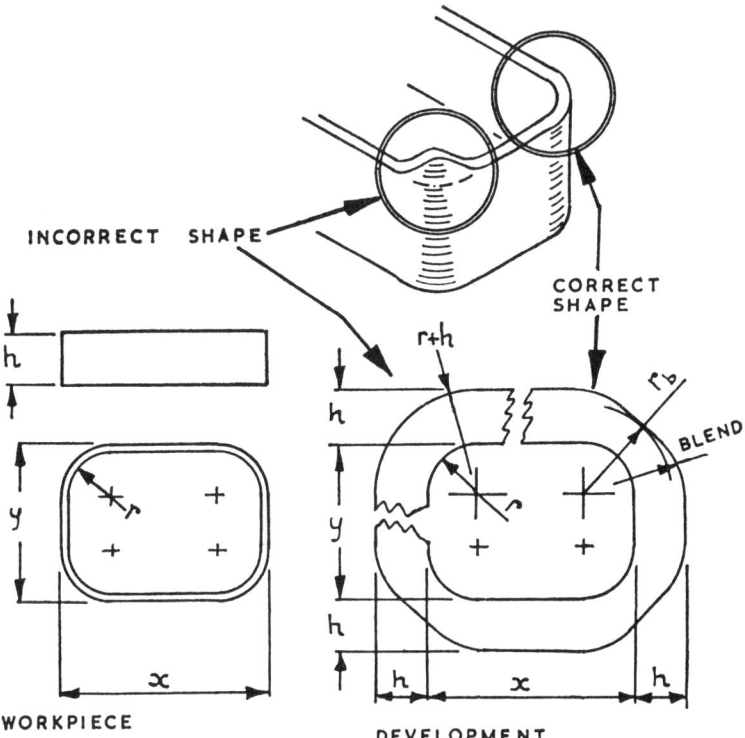

Fig. 15.61 Consideration of the Development of a Rectangular Box with Radii at its Corners

From which:

$$2r_b = \sqrt{(2r)^2 + 4(2r)h}$$

$$r_b = \frac{\sqrt{(2r)^2 + 8rh}}{2}$$

15.7.6. Pressure required for drawing. In order to select the press to be used for drawing it is usual to take the maximum pressure required as the same as that required to **blank** the bottom of the drawn part. The pressure required to draw the cup will be much less than this amount, particularly if the cup is shallow.

15.7.7. Cupping on a double-acting press. Fig. 15.62 shows a typical arrangement in which the blank is located on the face of the die. The material is controlled by the pressure pad mounted on the outer slide, and the forming punch, mounted on the inner slide, forms the cup. The finished pressing is ejected by the ejector which forms part of the press.

Fig. 15.63 shows a blanking and cupping die-set for use on a double-acting press. The stock is fed by hand and blanked by the punch on the

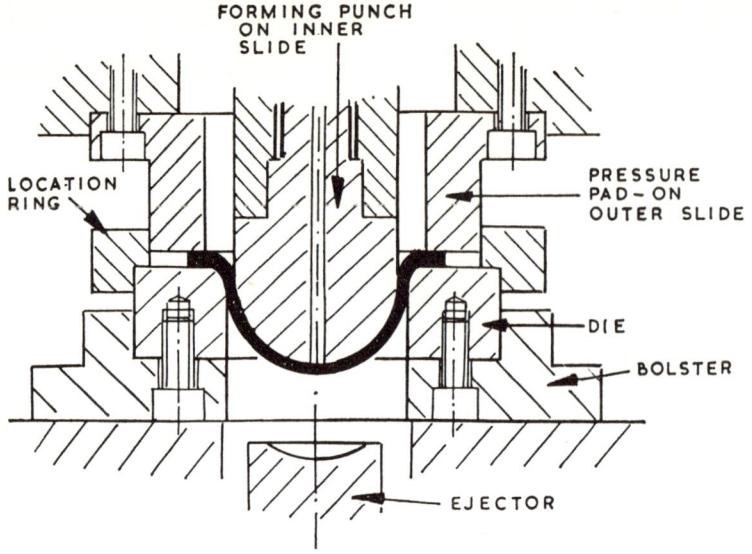

Fig. 15.62 Forming on a Double-Acting Press

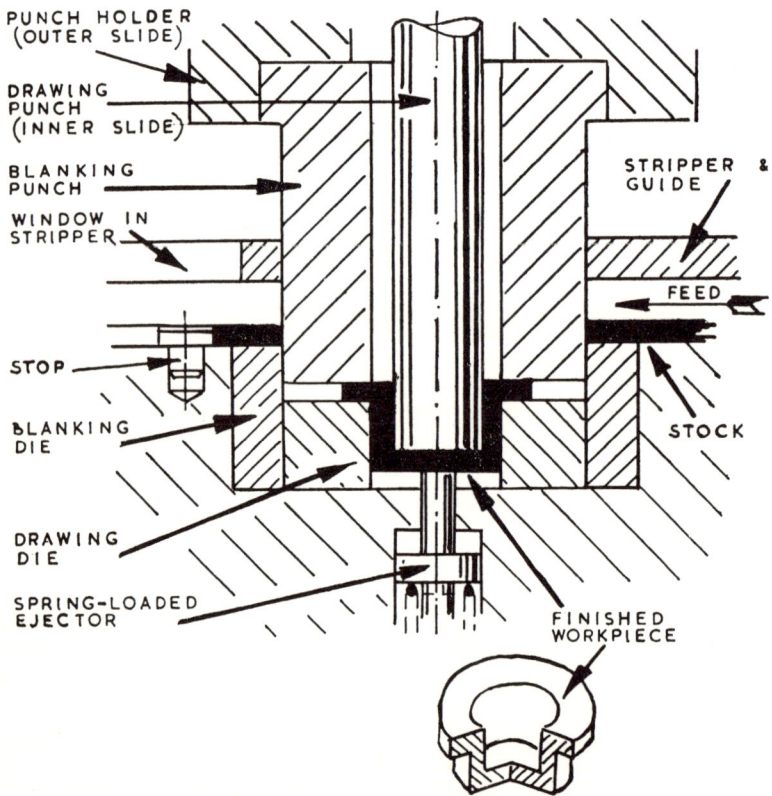

Fig. 15.63 Blanking and Cupping on a Double-Acting Press

outer slide. The blanking punch controls the metal whilst it is cupped by the punch that is mounted on the inner slide. The cup is ejected by the machine ejector, and the stock removed from the blanking punch by the stripper during the return stroke of the punch.

15.7.8. Slide tools. Fig. 15.64 shows an arrangement in which the blanked material is bent into a channel-like shape and then turned over by two slide tools operated by the outer slide of the press. During the return stroke the slide tools are returned by spring-action. The completed pressing is slid from the former by the operator.

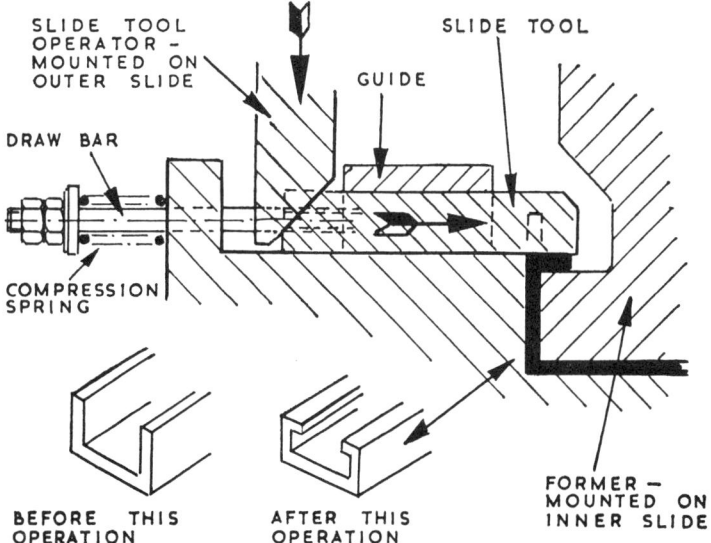

Fig. 15.64 Application of a Slide Tool (Only the Slide-Tool Movement Is Shown)

Chapter 16

ORGANISATION AND MANAGEMENT

In this chapter the general organisation of an engineering concern is described, and this is followed by a detailed study of the Planning and Jig Design departments.

16.1. The Organisation of an Engineering Concern

Fig. 16.1 on page 249 shows the main divisions of responsibility of the main divisions and departments of an engineering concern. The titles of the departmental heads have not been included because they vary from company to company; similarly, in a small company the duties of two or more departments may be merged. The main engineering divisions are those of **Product Design** and **Production**; these divisions are given specialised assistance by the **Research** and **Experimental** departments, and by the **Service Department**. The 'running' of the concern is done by the commercial departments and the various ancillary departments.

16.1.1. The design scheme for a product is prepared by the **Design Department**; this department is advised by the specialised departments, such as the **Stress Department**, and the various **Research** departments. Valuable information is also obtained from the **Sales and Service Department**, where the records of similar products and customer reactions are maintained.

The design scheme is used by the Detail Department to prepare the working drawings, and at this stage the **Standards Department** supply details of national and company standards. A prototype is usually produced by the **Experimental Department**; major modifications of design are at this stage dealt with by the Design Department, but minor modifications at later stages are usually dealt with by the **Modifications Department**, where salvage schemes are usually prepared.

The instructions to produce or modify are issued to the **Production Division** by the **Issue Department**.

16.1.2. The instructions to produce are passed to the Production Division by the Issue Department, and the policy regarding the volume of production and the tooling budget is settled by the **Board of Directors** acting upon the advice of the financial experts.

It must be emphasised that the principal section of the Production

ORGANISATION AND MANAGEMENT

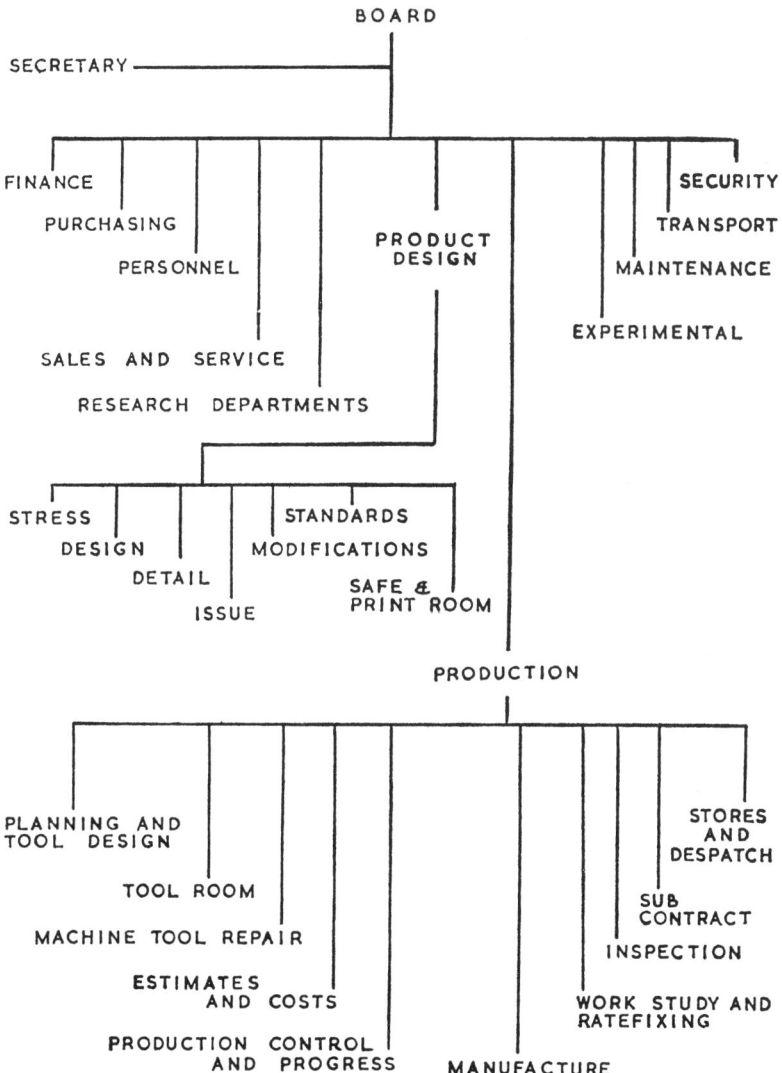

Fig. 16.1 Organisation of an Engineering Concern

Division is that of **manufacture**, and that all the other production departments are present to serve manufacture (see again fig. 16.1). The production planning and the equipment designs are produced by the **Planning** and **Jig Design** departments, and the equipment is produced by the **Tool Room**, or by a sub-contract firm; the **Production Control** and **Progress** departments control the [programming of the work and ensure the continued flow of work around the manufacturing departments. It is the duty of the [**Inspection Department** to maintain the required quality of the product.

The **Work Study** and **Ratefixing** departments ensure that the rate of production and the cost is as planned, and the **Estimates** and **Cost** departments deal with the financial aspects of the production. A number of important ancillary departments (see fig. 16.1) contribute to the smooth running of the Production Division.

As in the case of the Design departments, the efficiency of the Production Division depends upon the continued liaison between the many sections.

16.2. The Function of the Planning and Jig Design Departments

The function of the Planning and Jig Design departments is to plan the operation sequence, specify the raw material shape and size, and design the tooling equipment.

Fig. 16.2 shows how the instructions to the production departments are dealt with by the Production Control and the Planning departments so

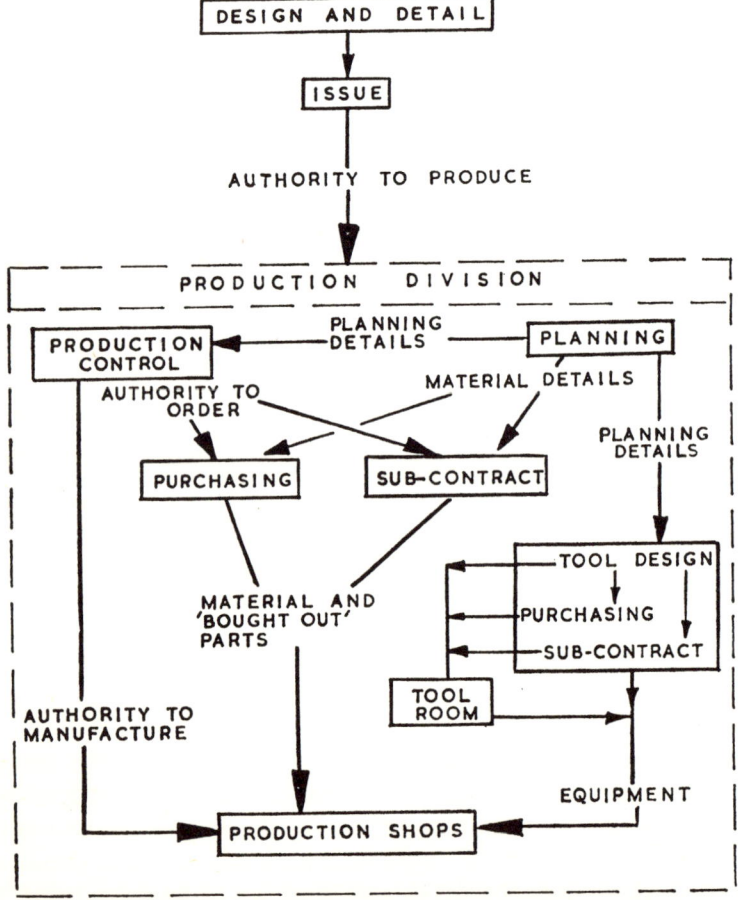

Fig. 16.2 The Issue of Equipment, Material and Authority to the Production Shops

ORGANISATION AND MANAGEMENT 251

that the authority to manufacture, the material and the tooling equipment will all arrive at the manufacturing sections at the required time.

Details of the machines to be used must be settled as soon as possible and passed to the Production Control Department so that the machine-tool and other requirements can be assessed; raw-material forms must be decided as soon as possible because there may be a delay in obtaining castings or forgings. Fig. 16.5 on page 254 shows how the advanced planning section examines these requirements (*Note:* advanced planning will be considered later in this chapter).

16.3 The Organisation of the Planning and Jig Design Departments

Figs. 16.3 and 16.4 show two basic methods of organising the functions of planning and tool design; the principal difference between these two

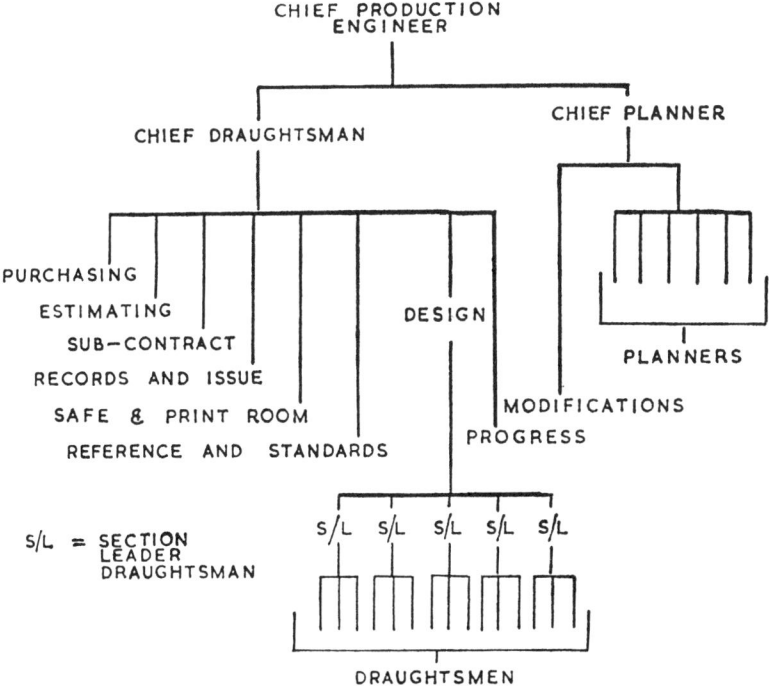

Fig. 16.3 Organisation of the Planning and Tool-Design Department

methods of organisation is that the system shown in fig. 16.3 divides the work between two closely-linked departments, and that shown in fig. 16.4 covers the work by a single department. In the system shown in fig. 16.3 the work of planning is done by the planners, who prepare details of the operations and indicate the operations that are to be done using special equipment; the design section of the Jig Design Office then prepare all the designs. In the system shown in fig. 16.4 the office is divided up into

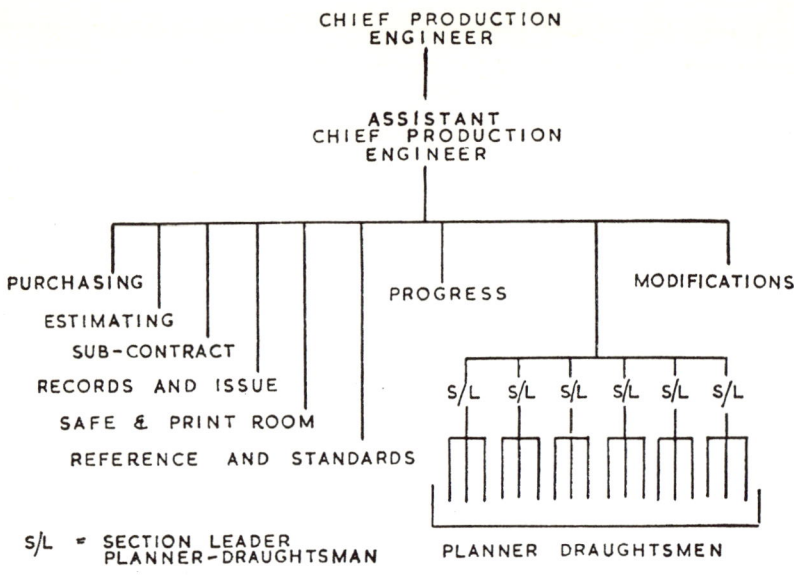

Fig. 16.4 Organisation of the Planning and Tool-Design Department

sections that specialise in particular types of component; all the work from planning to completion of the equipment design is done on one section.

16.3.1. The advantages associated with having the planners separated from the design sections are that they can concentrate upon one function (that of planning), and are more easily able to spend time on the shop floor, and with the work study and similar departments; the disadvantage is that the process is not established in conjunction with the equipment design, but it can be argued that a close liaison between planners and designers will produce the required results. The advantages associated with having the work on similar components done on one section are that all the information is in one place, planning and tool design can be done together, and that experience can be more easily drawn upon; the disadvantages being that there is a tendency to produce 'key men' with personal information, and that by having the same people involved every time, bad practice may be continued as well as good practice.

From these comments it will be obvious that the choice of method is very largely a matter of personal opinion, but care must be taken to prevent a change of organisation each time there is a change of departmental head.

16.3.2. It will be seen that the ancillary sections of the Planning and Jig Design departments are the same whatever the division between planning and jig design. Most of these sections are mentioned later in this chapter, but their functions are listed here for convenience.

ORGANISATION AND MANAGEMENT 253

The **purchasing** section, **estimating** section and **sub-contract** section perform the same functions as the equivalent departments perform in the main organisation, but they are of a more specialised nature. The **records and issue** section is responsible for the maintenance of tool record cards and of the issue and reissue of the equipment drawings to the tool room and sub-contract firms, the **reference and standards** section runs a form of library where records of national standards, standard equipment, etc., are kept and the **print room** produces prints directly from the drawings, for issue to the tool room, etc. In addition to these, the department has its own **progress** section, and a **modifications** section to advise regarding the action following modification to the products.

16.4. The Planning of Product Manufacture

When a design project is started a 'deadline' is fixed for the delivery of the first batch of products. In the case of an aeroplane power plant, this deadline will be fixed by the latest date that the airframe manufacturer can accept the first power plant set so that the maiden flight of the aircraft will be on schedule. In the case of a motor car or similar product, the 'deadline' will be fixed by the annual trade show.

16.4.1. The work of 'pre-planning' is usually done by a Project Engineer who is attached to the Production Division. By studying the project drawings, and from experience, the Project Engineer can assess the time required for planning, material delivery, tool design and manufacture, product machining, fitting and testing.

16.4.2. Starting from the deadline, and using his time estimates, the date at which the design must be finalised can be established. The chart shown on page 254 (fig. 16.5) is typical of this analysis, and shows how many of the pre-delivery functions can overlap in time. It will be appreciated that an indication of the raw material and equipment for the major components must be given at an early stage because of the long time-cycle involved; very often a preliminary drawing is passed to production, and certain sizes finalised at a later stage. The more modern system of 'network analysis' is a development of this work, and use may be made of this technique if necessary.

A continuous examination of the progress is made and the plans amended if necessary. It is the duty of the Project Engineer to co-ordinate the activities of the many departments involved and to keep them up to date with the arrangements.

16.4.3. At the appropriate time the schedule is 'sealed' and no further modifications are accepted without agreement. At this stage modifications become the responsibility of the Modifications Department, who classify them according to whether they affect the function of the product, simplify

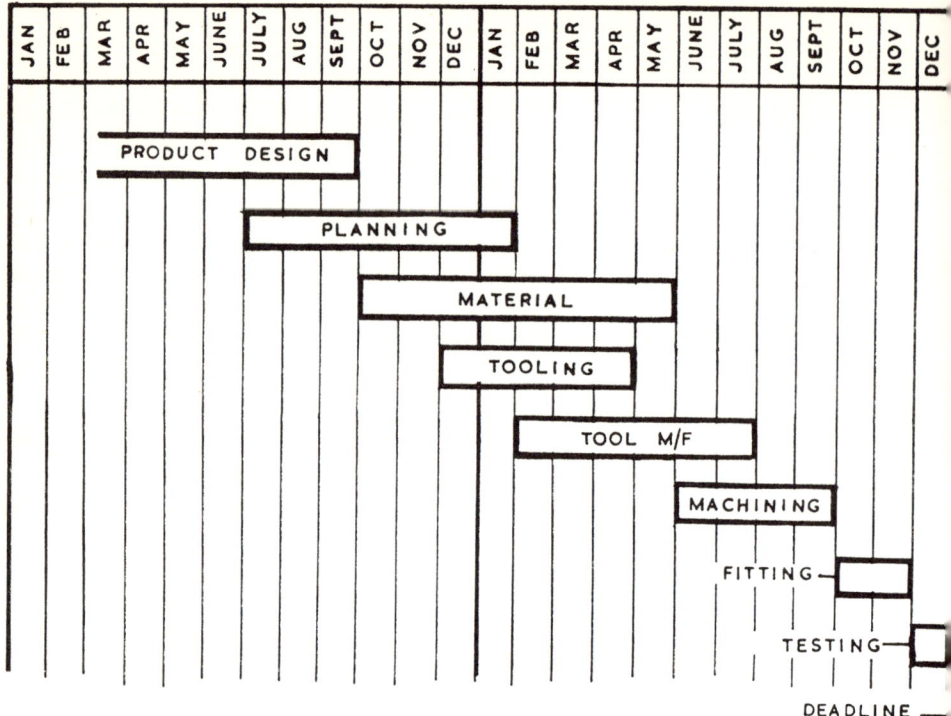

Fig. 16.5 Project Planning Chart

production, introduce 'selling points', etc. The stage in the production programme at which a modification is introduced is decided by the modifications committee.

16.4.4. When the manufacturing drawings are available, the Process Planning Department (or the section leader planner-draughtsmen) can commence the operation planning. At this stage the amount of metal to be 'left on' castings and forgings can be determined, and the position of 'holding lugs' can be decided upon; raw-material drawings are prepared by the Jig and Tool Drawing Office, or alternatively, a set of the working drawings (also called 'detail drawings') may be marked up to show where metal must be left on for machining. In the case of forgings, the raw-material form must be suitable to produce the desired grain fibre, and the Production Metallurgist usually gives approval for these drawings to be issued to the supplier of the raw material.

At this stage the B.O.F. ('bought outside finished') parts, such as nuts, bolts, springwashers and ball races, can be ordered by the Purchasing Department. When the machining processes have been completed, the Production Control Department can establish the plant loading; if the plant loading exceeds the plant availability a decision must be made either

ORGANISATION AND MANAGEMENT 255

to increase the machining capacity or to employ a sub-contract company to do some of the work; although the latter is more costly, it may prove to be a better proposition if the increased volume of work is only temporary. Sometimes a sub-contract company is engaged to do the rough machining only; in this case that company is supplied with a P.M. (part-machined) drawing so that their planning department can prepare processes, etc.

16.4.5. The machining process is initially in the form of a written instruction sheet or sheets, giving details of the operations and the machines to be used. If the workpiece is complicated and the operations are difficult to describe verbally, drawings are made showing the work done at each operation. If the volume of production is small these operation drawings may be all on the same sheet of drawing paper and presented in 'strip cartoon' fashion, but when a large volume of work is involved it is usual to produce a series of **operation sheets**, one for each operation. Each

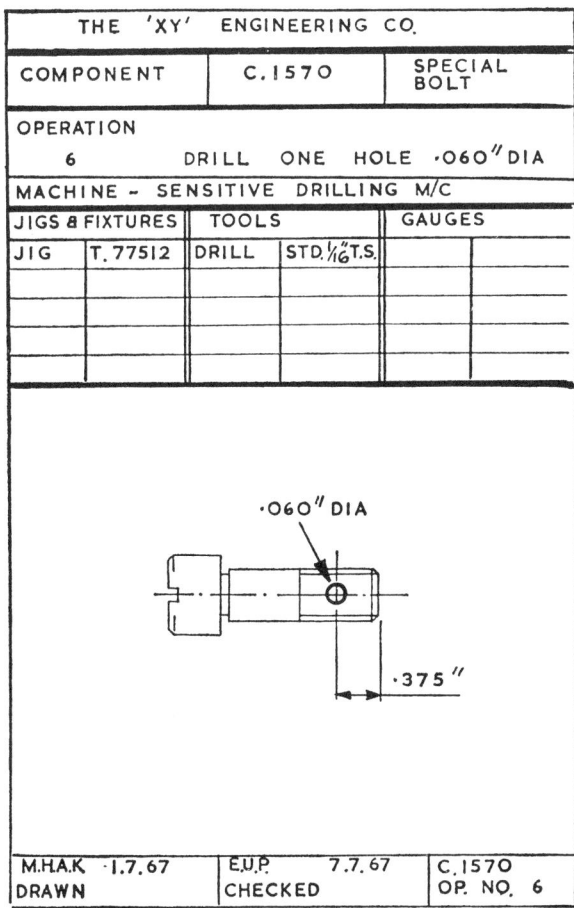

Fig. 16.6 Operation Instruction Sheet

operation sheet gives details of the work done at the operation and also details of all the equipment to be used for the operation; fig. 16.6 shows a typical operation sheet prepared for operation 6 of the process shown on page 53. A set of operation sheets in folder form is issued to the production shops supervision and similar departments, and a set is mounted on boards and issued to the machine operators with their tool kits for the operation that they are to perform. Intermediate inspection is done working to the operation sheets because the tolerance on certain features may be reduced when they are to be used for location or when accumulation of errors is likely to occur; the final inspection is done working to the component drawing.

16.5. The Planning of Tooling and Tool Records

When the operation sheets have been completed the tool designer draws up the tooling schedule for the component (it must be emphasised that the operation sequence is not finalised until the tool design has been completed

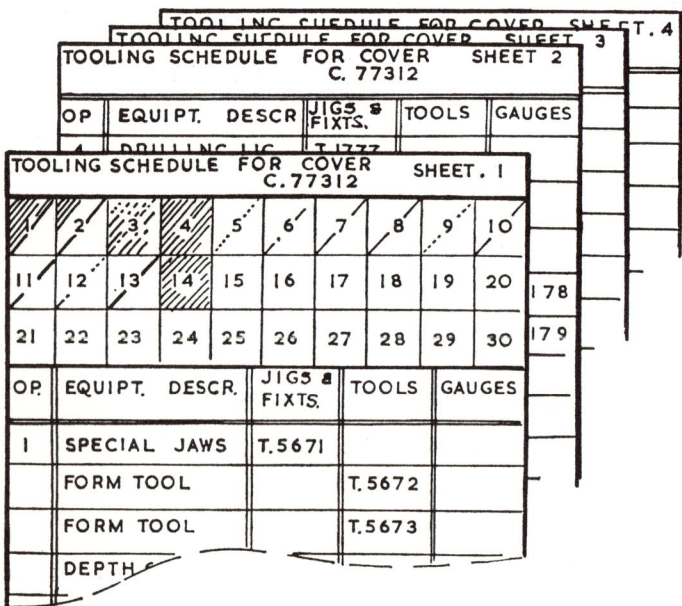

Fig. 16.7 Tooling Schedule

because it may be found that it is better to combine two operations and use one jig or fixture; or, on the other hand, a complicated operation may be performed better if split into two simpler operations). Fig. 16.7 shows sheets from a typical tooling schedule. The first sheet of the schedule is used to progress the tooling work, and includes a number of squares to represent the operations to be done upon the component. When the tooling work is started these squares have a diagonal line drawn across them

ORGANISATION AND MANAGEMENT 257

of a suitable colour to indicate the nature of each operation (i.e. machining, fitting, heat-treatment, inspection, etc.), and as the tooling work proceeds, the squares are gradually coloured in to indicate the progress upon the tooling up of each operation.

16.5.1. Title block and parts list. The equipment drawings are of standard sizes and all carry a title block at the bottom right-hand corner. The sheet containing the general arrangement also carries a parts list giving a list of parts, quantity of each required, material, treatment if appropriate and reference number if standard or a 'unit tooling' part. Fig. 16.8 shows a

7	STUD	2	M.S.
6	CLAMP	2	M.S.
5	LOCN. PIN	1	S.S. HARDN.
4	LOCN. PIN	1	S.S. HARDN.
3	SETTING BLOCK	1	C'HD. STEEL
2	TENON	2	M.S. C'HD & G.
1	BASE	1	C.I.
DET	DESCRIPTION	NO OFF	MATERIAL

TITLE **MILLING FIXTURE**

COMPT. BLOCK C.712

OP. 4 MILL SLOT

M.H.A.K. 1-7-67	E.U.P. 2-7-67	T. 17571
DRAWN	CHECKED	

Fig. 16.8 Equipment Drawing Title and Parts List

typical title block and a section of the parts list; it will be seen that the parts list is arranged so that the parts are listed upwards, starting from the major details; this method allows further details to be added without causing confusion. For convenience, drawings often carry a 'grid reference' so that reference can be made easily to modifications, etc.

16.5.2. Equipment numbering. The equipment numbering system varies from company to company; some companies use a single letter (often a 'T' reference) to indicate equipment, but some use a system that indicates the type of equipment (for example, 'T' for tools, 'F' for jigs and fixtures, and 'G' for gauges). The usual method of number allocation is for the records section to number the tool record cards, and for the tool designer to collect a batch of cards when compiling the tooling schedule for the component; the tool designer fills in the front of the card. The card is signed by the tool designer or draughtsman, and also by the checker.

16.5.3. Tool record cards. Fig. 16.9 shows the two sides of a typical tool record card. The front gives details of the equipment, component, operation, designer and checker, and the reverse side is used to record the

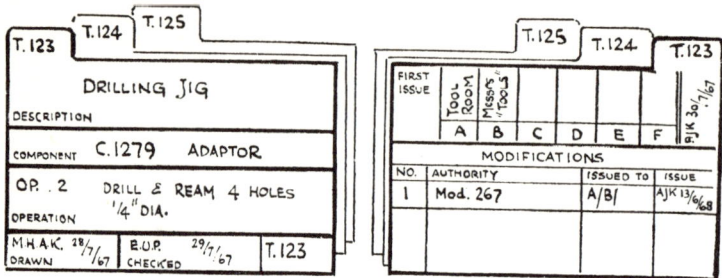

Fig. 16.9 Tool Record Card

details of issue and modifications. The record of drawing issue helps to prevent errors that can occur if the drawings are issued to a sub-contract tooling manufacturer, and the department forgets to issue amended drawings in the event of a modification to the component.

16.5.4. Component record cards. It is convenient to use a cross index system so that the equipment is also recorded against the component. Fig. 16.10 shows such a record card; modifications to the drawings are not recorded on this card, but changes of tool number must be recorded.

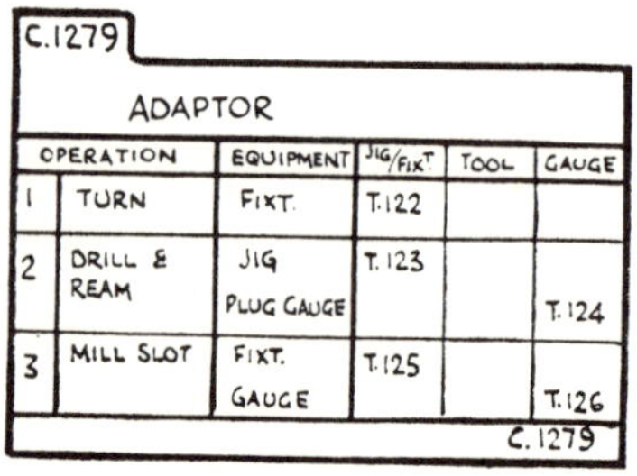

Fig. 16.10 Component Record Card

16.5.5. The ordering and manufacture of equipment. The ordering and manufacture of the tooling equipment is initiated by the section leader informing the clerical section of the availability of equipment drawings; it is the responsibility of the clerical section to place orders and instructions for the equipment to be manufactured and to issue the drawings. The placing of orders depends upon the Tool Room capacity at the time, and often a technical man acts as liaison, and advises the clerical section regarding the placing of the work.

ORGANISATION AND MANAGEMENT 259

The drawing parts list simplifies the work of ordering and estimating, and also helps the Purchasing Department to replenish the tool-room stocks of 'unit tooling details', tool tips, etc.

16.5.6. The issue of drawings. The records and issue section obtain prints of the equipment drawings, raw-material drawings, sub-contract drawings and operation sheets from the print room, and arrange for their issue. The drawings are then passed to the safe for filing; the safe personnel keep their own records of the drawings.

16.5.7. Progress. During the manufacture of the equipment the Planning and Jig Design Department progress the work, and advise the Project Engineer of any delays.

16.6. Modifications

Modifications to the product are classified and their introduction is programmed by the modifications committee. Certain modifications may be in response to requests by the Planning Department, and others may be urgent because of shortcomings of the previous design; these modifications will be introduced immediately. Other modifications are improvements that are not urgent, and arrangements must be made to withdraw tooling equipment at the programmed stage, and to modify the equipment rapidly so that there is little delay in the manufacture.

16.6.1. Modification notes. The authority to modify the product is passed to the Production Division in the form of a modification note, giving details of the modification and of action required upon the components already produced or advanced so far that the modification cannot be incorporated. The time of introduction of the modification is programmed by the Production Control Department, acting upon the decision of the modifications committee.

16.6.2. Modifications to processes and equipment. The modification notes are usually dealt with by the modifications section of the Planning Department. This section examines the modification notes and assesses the work involved in making the modification, so that the planning and tool design sections can be advised, and so that their representative can report to the modification committee regarding the time required to modify the tooling equipment.

The process and operation sheets, and, where appropriate, the raw-material and part-machined drawings, are amended and reissued in line with the modifications programme. The equipment drawings are also modified if required; fig. 16.11 shows a typical modification table on a drawing, and also indicates one method of indicating a modified dimension;

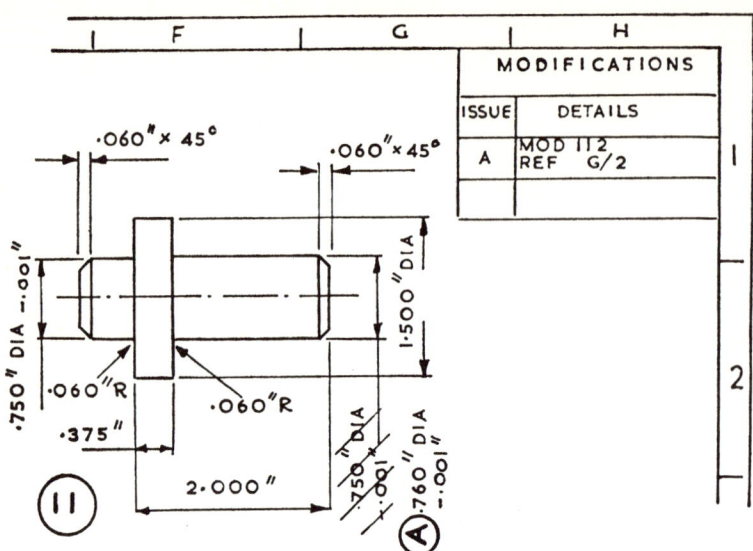

Fig. 16.11 Modification to a Drawing

in the method shown the original dimension is retained so that the exact modification can be seen.

Details of the modification are passed to the clerical section, where the tool record card is endorsed, the amended drawings issued, and the necessary instructions issued to the Tool Room or Sub-Contract Tooling Company.

16.7. Information Library and Records

The planner and tool designer relies upon a vast amount of information regarding machine tools, small tools, standards, etc., and very often acquires his personal library of information. Such a library will be more efficient in the hands of a reliable technical clerk who will be able to devote more time to maintaining the library, and even be able to give advice regarding the standards, etc., in his collection. The following are some of the many features that are included in an information library of this sort.

16.7.1. Capacity charts. A capacity chart is a diagrammatic representation of a machine tool, giving dimensions of the clearances, movements, etc.; this chart is most useful when producing a tool layout and when deciding how to locate and clamp the equipment on the machine. The alternative to using a capacity chart is to measure the actual machine tool in the machine shop. Fig. 16.12 shows a typical capacity chart for a capstan lathe, and fig. 16.13 shows details of the spindle and capstan of the same machine.

16.7.2. Manufacturers' catalogues and handbooks. Manufacturers' catalogues give valuable information regarding their range of small tools and

ORGANISATION AND MANAGEMENT

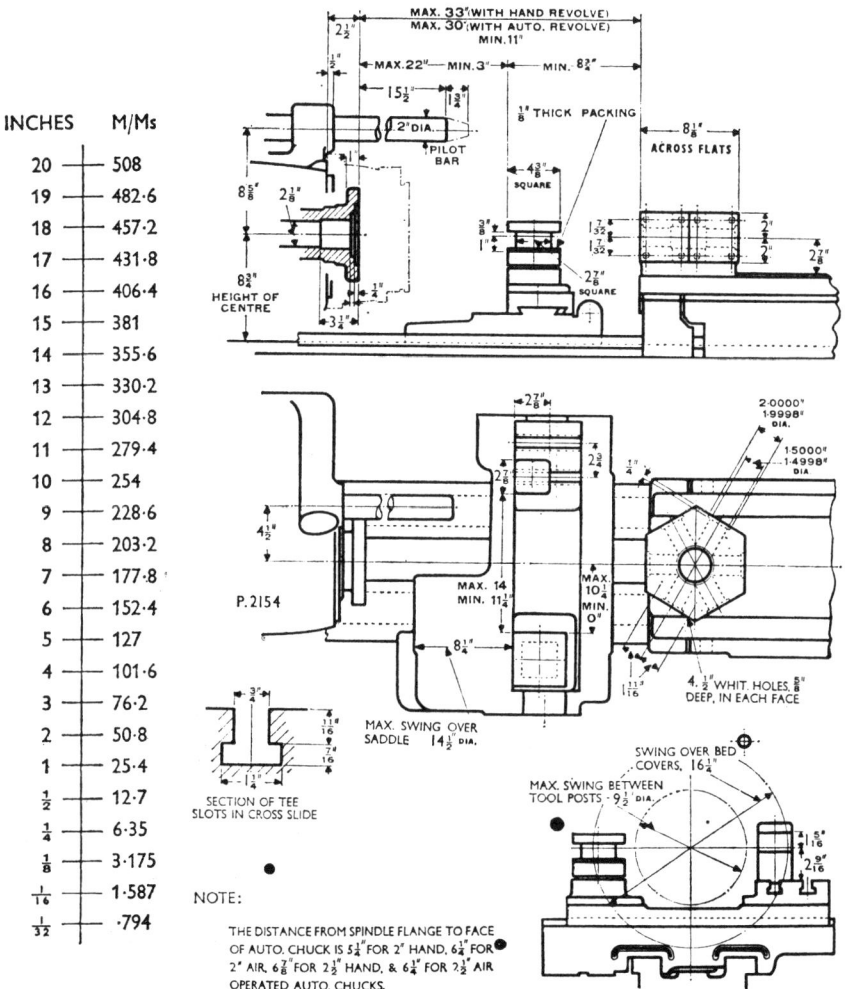

Fig. 16.12 Machine Capacity Chart (Courtesy H. W. Ward & Co. Ltd.)

similar products, and machine operators' handbooks give information regarding the speeds and feeds and other characteristics of their machines.

16.7.3. National standards. The national standards should be worked to at all times, and although each planner and tool designer should have his own copy of British Standards for Workshop Practice (or the company's abstracts from same), a copy of all relevant standards should be kept in the office library.

16.7.4. Company standards. A large engineering company usually produces its own standards (a Standards Department is usually part of the

DIMENSIONS OF SPINDLE

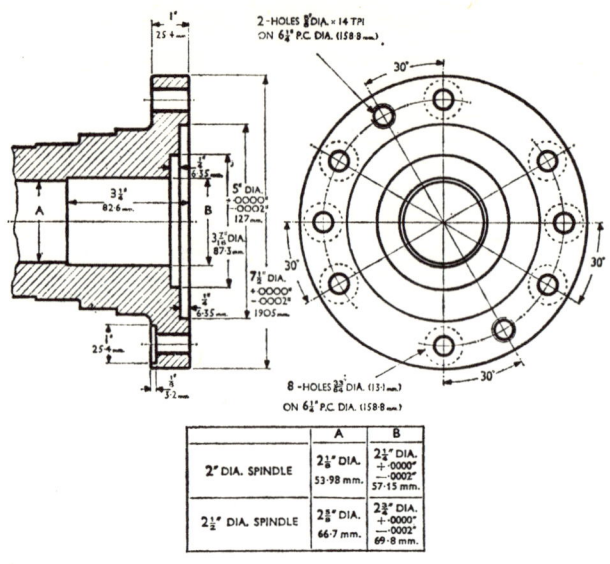

	A	B
2″ DIA. SPINDLE	2⅛″ DIA. 53·98 mm.	2¼″ DIA. +·0000″ −·0002″ 57·15 mm.
2½″ DIA. SPINDLE	2⅝″ DIA. 66·7 mm.	2¾″ DIA. +·0000″ −·0002″ 69·8 mm.

DIMENSIONS OF CAPSTAN

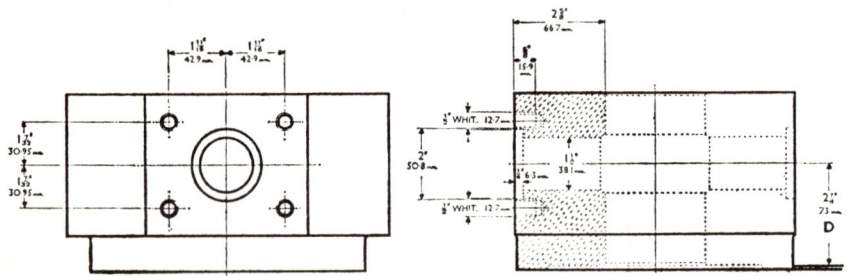

NOTE.—All tapped holes are Whitworth Standard.   D is the swing over the plate on the top of the Capstan Slide.

Fig. 16.13 Machine Details (Courtesy H. W. Ward & Co. Ltd.)

Product Design Division); some of these are based upon the national standards, but some of them originate in the company itself. A copy of all these standards will be kept in the library because many of them will be referred to on the working drawings. In addition to these product design standards, the Tool Design Office will gradually create its own standards when certain pieces of equipment are used extensively; when certain pieces of equipment become common, it is necessary to prevent wasted time by drawing duplication and equipment duplication, and so a register must be maintained. Fig. 16.14 shows part of such a register showing details of a drill stop nut (drill stop nuts are described on page 120); if a designer is unable to find a suitable drill stop nut in the system, he will draw a suitable

ORGANISATION AND MANAGEMENT 263

one, and fill in the principle features of the new one on the appropriate chart.

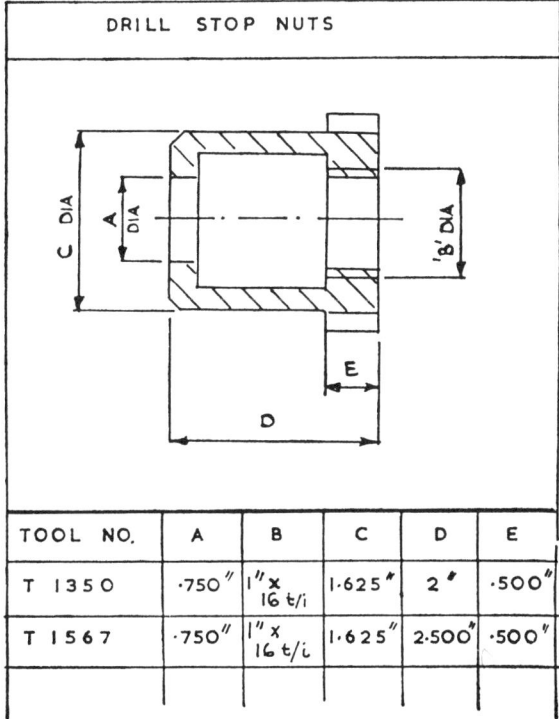

Fig. 16.14 Register of Tool-Design Office Standards

16.7.5. Stores holding. It is usual for a company to maintain a stores holding of standard items such as 'unit tooling details', carbide tips, billets, dowels, etc.; and if possible the designer should incorporate these items in his designs. A list of the stores holdings should therefore be kept on the library section.

16.7.6. It will be obvious that a well-run technical library incorporating the sort of information described here will provide a valuable service to the planners and designers, and will reduce unnecessary costs.

16.8. The Contract Tool Drawing Office

If the loading on the Tool Design Office becomes too great, additional staff will be required. When such loading is likely to be permanent because of an expansion of the company, it is economical to increase the size of the department and to employ more staff; but if the loading is only temporary, because of change of manufacturing programme, etc., the additional staff will only be required for a short time, and it will probably be very difficult to employ staff on a temporary basis. In these circumstances the services of a contract tool drawing office will be used.

16.8.1. The contract tool drawing office may either supply staff to work at the company's own drawing office or do the work on their own premises. It is easier to maintain technical liaison if the contract tool drawing office staff work at the company that requires the help, but differing conditions of employment can lead to problems in industrial relations. It must be realised that it is more expensive to employ the contract tool drawing office, but will be more effective if the loading is only for a short time, or if local housing and similar conditions make it difficult to employ additional staff.

16.8.2. The placing of contracts. It is necessary that the contract tool drawing office has a good reputation and is suitable for the class of work to be done; it is therefore a good idea to keep a record of suitable offices so that the type of assistance required can be obtained without loss of time. The method of payment may be a fixed charge for drawings of a certain size, or alternatively a charge negotiated for the particular component to be 'tooled-up'. It is usually best for the tooling section concerned with the component being tooled-up to deal directly with the contract tool drawing office, or through a liaison man, so that changes in process plans, modifications to the product or changes in the manufacturing programme can be effectively dealt with.

16.8.3. Company standards and the contract tool drawing office. The contract office must be informed of the drawing format used by the company, and also of the company's policy regarding tooling, the company's standards and the stores holding of unit tooling details, etc.

16.8.4. Modification action. Details of modification to the product being tooled must be passed to the contract office in the same way as these details are passed to the company's own tool designers; as already stated, this is done most effectively by the section for whom the work is being done. It may be necessary to reach agreement regarding charges for the additional design work involved when modifications are introduced.

16.9. The Tool Room

The Tool Room functions like a small production division, having its own planning, inspection and similar departments. The drawings are passed to the Planning Department, where the process instructions are produced; the instructions will be less detailed than those for the production shops, but it is necessary to indicate the main sequence so that plant balance can be maintained, and so that the correct heat-treatment and grinding allowances can be made.

The Tool Room is divided up into turning, milling, grinding and similar sections, and the work progressed from section to section. The machine tools used in the Tool Room are of a basic type, and include machines such as universal milling machines, universal grinding machines and jig boring machines; a small number of capstan lathes are included in case it is necessary to make a large number of small items. Fixture castings will need to be marked out, and so a marking-out section will be required; quite a lot of hand work will be required, and so a large fitting section is employed. A feature of the Tool Room organisation is the post of **Tool Room demonstrator.** A number of demonstrators are employed, and usually look after the tooling for a section of the products or for a specialised operation; a demonstrator usually supervises the fitting work involved in the manufacture of the equipment, and tries out the equipment in the production shop; later he acts as liaison between the shops and the tool designer to ensure that the equipment is maintained in good condition. A demonstrator usually gains valuable experience regarding the production of the products in his care, and the wise planning engineer and tool designer frequently consults the demonstrator.

Chapter 17

THE ECONOMICS OF TOOLING

17.1. Manufacturing Systems

Manufacturing systems can be classified as (*a*) job production, (*b*) batch production, (*c*) mass production and (*d*) flow production. These systems can be defined as follows:

(*a*) **Job production (or one-off production).** In this system one workpiece is produced, although it may be repeated later. This method of production is applied to the manufacture of jigs and fixtures, prototypes and to special jobs such as ship's hulls. In most cases high-grade labour is employed, with 'basic' equipment, and the minimum of process instructions; the production of a ship's hull is an exception, and in this case it is necessary to use more specialised equipment.

(*b*) **Batch production.** In this method of production a batch of workpieces are passed from operation to operation. Each machine tool is 'set up' for a batch of workpieces, and when they have been machined, it is re-set for a batch of different workpieces. The operator receives his machining instructions and the kit of equipment (comprising the jig or fixture, tools and gauges) for that operation, and either sets the machine himself or it is set up by the setter. Batch production is used in a concern where the range of work produced is large compared with the number of products. The size of the batch depends upon the delivery programme, the material availability, the modifications programme and upon the economics (see page 273). At any one time a number of batches of the same product may be at various stages in manufacture.

(*c*) **Mass production.** Mass production employs a number of machines, and each machine is engaged in the manufacture of one component; screws, pipe unions, etc., are produced in this way. In theory, a machine will spend its working life making one component, but in practice, the product will need to be changed from time to time to fit in with the orders, or because of the rate at which different parts of the same assembly can be produced. As an alternative to re-setting a complicated machine, it may be allowed to stand idle or be serviced when its product is not required for a time.

(*d*) **Flow production.** In this method a complete factory is laid out to manufacture one product, and so each machine or department is

THE ECONOMICS OF TOOLING

dependent upon the others. The machine tools are highly specialised, and a change of product may well demand a complete change of factory layout.

17.1.1. Tooling systems. Tooling systems can be classified as (*a*) basic, (*b*) skeleton and (*c*) complete; these can be defined as follows:

(*a*) **Basic tooling.** In this method use is made of equipment such as machine vices, chucks, faceplates, dividing heads and circular tables; but when awkward parts are to be machined, the machinist may make up his own holding device. Basic tooling is associated with job production.

(*b*) **Skeleton tooling.** Skeleton tooling implies that tooling is supplied only for the awkward operations. This tooling may consist of simple holding devices for some operations and more complicated equipment for other operations. It is less expensive to supply skeleton tooling than to supply complete tooling, but it imposes greater responsibility upon the machinists.

(*c*) **Complete tooling.** This implies that jigs, fixtures, tools and gauges are designed for the workpiece, and that when standard tools are to be used, arrangements are made for these to be available in the tool kit for each operation. Complete tooling is the most expensive method of tooling, but it enables less experienced operators to be employed.

Skeleton tooling and complete tooling are both used for batch production; skeleton tooling is used when the volume of work is small or when the work does not justify the cost of complete tooling; complete tooling is usually used for large-scale production, but may be used for small production when interchangeability or the required accuracy justifies its use. Mass production and flow production usually require complete tooling.

17.2. The Calculation of 'Break-Even' Quantity

When the choice of tooling expenditure is purely a matter of economics, it is usually necessary to determine whether the quantity to be produced justifies the cost of the special equipment. The minimum quantity to be produced to justify the tooling cost in this way is called the **break-even quantity**.

For example, if equipment costing £300 results in a saving of 5 shillings per component, it is necessary to produce $\frac{300 \times 20}{5}$ components so that the saving will pay for the special equipment; in this example the equipment will 'pay for itself' after 1 200 components are produced, and so 1 200 is termed the 'break-even' quantity.

The consideration of 'break-even' quantity can be illustrated graphically.

Fig. 17.1 shows the connection between the total quantity to be produced (x), and the total cost (y) for two tooling systems. Curve 1 is the curve for a system employing skeleton tooling, and curve 2 is that for a system employing complete tooling. Both curves are of the law $y = mx + c$; where m is the cost of machining the workpiece and c is the cost of tooling and setting the machine. Curve 1 starts low but rises sharply, and curve 2

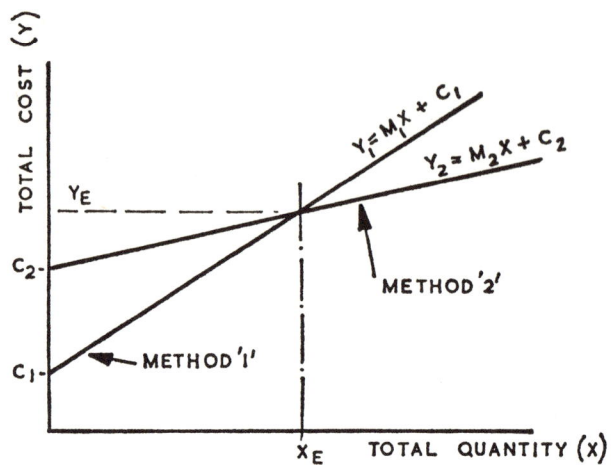

Fig. 17.1 Determination of Break-Even Quantity

starts high but rises less sharply; at point P the two curves cross, and so at higher values of x, the system 2 is economical over system 1. Let the values of x and y at point P be y_e and x_e, and equate as follows:

$$m_1 x_e + c_1 = m_2 x_e + c_2 = y_e$$

$$x_e = \frac{c_2 - c_1}{m_2 - m_1}$$

where x_e is the 'break-even' quantity, and is equal to the increase in initial cost divided by the saving per component.

17.2.1. The following example shows that when the fixtures being compared hold different numbers of workpieces, it is best to consider the cost to produce one workpiece by each method, and then to compare these costs.

Example: The flats on the component shown in fig. 17.2 are to be straddle milled using one of the following methods:

(*a*) Using a fixture which will hold 1 component.
(*b*) Using a fixture in which 10 components are string milled as illustrated by the line diagram fig. 17.2 (*a*).

THE ECONOMICS OF TOOLING 269

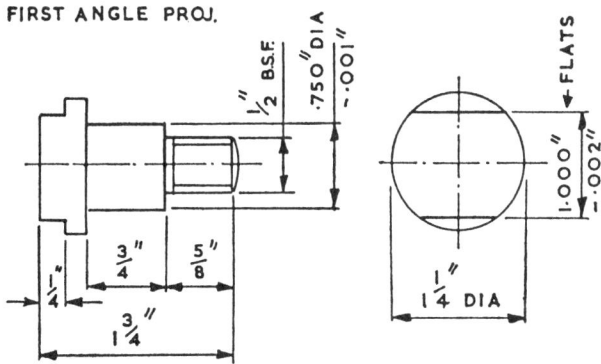

Fig. 17.2 Plug

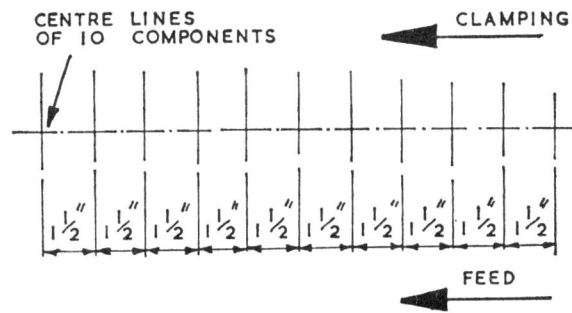

Fig. 17.2 (a) Layout Showing 10 Components in Position for String Milling by Method 'B'

Determine the minimum number of components that would be economically produced using method (b) instead of method (a).

Data: Machine spindle speeds in rev/min: 23, 31, 41, 54, 95, 126, 167, 222, 293, 388, 514, 680, 903, 1 200.
Machine table feed in ft/min: $\frac{3}{4}$, 1, $1\frac{3}{8}$, 2, $2\frac{7}{8}$, $3\frac{1}{8}$, $5\frac{3}{4}$, $8\frac{1}{8}$, 11, 16, 22, 30.
Cutter to be used: 3 in diameter $\times$ 10 teeth, side and face.
Feed per tooth: 0·010 in.
Labour cost, 8 shillings per hour.
Overheads, 50%.

Method (a): Loading time, 3 seconds.
Clamping time, 4 seconds.
Unclamping time, 4 seconds.
Unloading time, 4 seconds.
Cost of fixture, £20.

Method (*b*): Loading time, 14 seconds (for 10 components).
Clamping time, 4 seconds (for 10 components).
Unclamping time, 4 seconds.
Unloading time, 20 seconds (for 10 components).
Cost of fixture, £34.

Setting time the same for both methods.
Cutter must run out completely under cutting feed.

Calculations

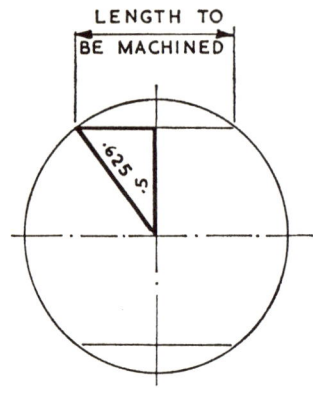

Fig. 17.2 (*b*)

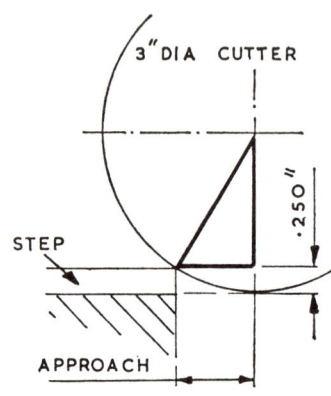

Fig. 17.2 (*c*)

From fig. 17.2 (*b*):

Length of surface to be cut per component $= 2\sqrt{0.625^2 - 0.5^2}$
$= 0.750$ in

From fig. 17.2 (*c*):

Approach $=$ over-run $= \sqrt{1.5^2 - 1.25^2} = 0.830$ in

Spindle speed $= \dfrac{12 \times S}{\pi \times d} = \dfrac{12 \times 80}{\pi \times 3} = 102$, say 95 rev/min (see speed range)

Table feed $= (0.010 \times 10 \times 95) = 9.5$, say $8\frac{1}{8}$ in/min (see feed range)

Consider method (*a*):

Travel $= 0.750 \times 2(0.830) = 2.41$ in

Cutting time $= \dfrac{2.41 \times 60}{8\frac{1}{8}} = 17.8$ s

Total time $= 17.8 + 3 + 4 + 4 + 4 = 32.8$ s (for 1 off)

Consider method (*b*):

Travel $= (2 \times 0.375) + (9 \times 1.5) + (2 \times 0.830)$
$= 15.91$ in for 10 off

THE ECONOMICS OF TOOLING

$$\text{Cutting time} = \frac{15\cdot 91 \times 60}{8\frac{1}{8}} = 117\cdot 5 \text{ s}$$

Total time = $117\cdot 5 + 14 + 4 + 4 + 20 = 159\cdot 5$ s (for 10 off)
= $15\cdot 95$ s (for 1 off)

Time saving when using method (b) = $32\cdot 8 - 15\cdot 95 = 16\cdot 8$ s (for 1 off)

$$\text{Cost saving when using method } (b) = \frac{16\cdot 8 \times 8(1\frac{1}{2})}{60 \times 60} = 0\cdot 056\ 1 \text{ shillings}$$

Number to pay for difference in equipment cost (i.e. £14)

$$= \frac{14 \times 20}{0\cdot 056\ 1} = 4\ 980$$

The break-even number is therefore 4 980 components.

17.3. Other Factors that Influence the Tooling Expenditure

It may not be necessary to justify the tooling expenditure on purely economic grounds: for example, when a high degree of interchangeability is demanded or when the use of special equipment produces a high finish. In these examples the cost of the equipment is considered against the prestige value. It may be good policy to produce one component of an assembly at a loss in order to improve the 'customer appeal' of the product as a whole.

17.4. Factors that Influence the Purchase of Special Equipment

Although the use of jigs, fixtures and special cutting tools and attachments will, in most cases, result in a reduction of machining costs, it may be necessary to consider the use of special equipment involving large capital expenditure. For example, the production of awkward shapes may demand the use of a broaching machine, or the introduction of a product that includes gears may demand the introduction of a gear section. Before deciding if the new work justifies the purchase of the special machines it is necessary to determine the use to which the equipment is to be put; if the use justifies the consideration of the new equipment, it is then necessary to conduct a more detailed examination on the following lines:

1. Consider the cost of the equipment and special attachments, and also the cost of its installation.
2. Since the money will probably need to be borrowed, it is necessary to determine if a loan can be obtained, and the interest to be charged.
3. Since special equipment is usually very accurate, it is necessary also to examine the depreciation allowances to be made.
4. At this point the initial financial considerations will be completed, and some idea of the equipment that can be purchased will be known. It is now necessary to consider the suitability of the existing building and the power that is available.

5. It will be necessary to renew parts from time to time, and so the spares availability must be considered; this is particularly important if the equipment under consideration is produced abroad.

6. If the special equipment requires trained operators, it is necessary to consider the availability of the personnel, and as an alternative, the training of personnel.

7. When several similar pieces of equipment are being compared it is also necessary to compare the costs as follows:

    (*a*) Direct and indirect labour costs.
    (*b*) Maintenance costs.
    (*c*) Insurance premiums.
    (*d*) Relative power consumption.
    (*e*) Costs of special tools.
    (*f*) Setting-up times.
    (*g*) Costs of spares.

It may be possible to undertake machining work for other concerns if the machine loading does not justify the purchase of the new equipment.

17.5. Consideration of Bought-Out and Sub-Contract Parts

It is usually cheaper to purchase small items, such as nuts and bolts and springs, from another company because such a company will be able to manufacture these in very large quantities if it supplies a number of concerns. Similarly, it is usually cheaper to purchase specialised parts, such as electrical and fuel accessories, from specialists rather than to set up design, development and manufacturing departments for these items.

In para. 17.4 the purchase of special manufacturing equipment was considered. If the proposed equipment will be used only occasionally it may be cheaper to have the special operations (such as broaching and gear-cutting) done by a sub-contract concern. The cost of sub-contract work will be higher than the cost of machining within the company, but it may well be cheaper when the cost of the equipment, its running cost, maintenance and depreciation is taken into account.

17.6. The Economical Design of Tooling Equipment

The cost of the tooling equipment can be minimised by careful design. The basic design should be as simple as possible and be robust so that it will operate without trouble; the component parts of the equipment should be of simple shape so that tool-room time is minimised, and a record of the tool-room holding of billets, etc., should be kept in the drawing office, so that the equipment can be 'designed around' these stock sizes and so minimise the amount of metal to be removed.

17.6.1. On page 190, para. 13.2, the various methods of construction were compared, and it is again emphasised that provided the equipment performs its function correctly, the cheapest method of construction must be

THE ECONOMICS OF TOOLING 273

used whenever possible. The use of British Standard and 'unit tooling parts' will produce a saving, particularly if the equipment is designed so that the parts of obsolete equipment can be used in new equipment. As explained in Chapter 16, the tool design office library will play an important part in this standardisation.

17.7. The Economical Use of the Equipment Designer's Time

The cost of the equipment designer's time is often overlooked when the economy is considered, but valuable time can be saved if unnecessary drawing work is eliminated. In most cases the drawing will only be used once, and it is unnecessary to spend time making an elaborate drawing on the general arrangement of nuts, springs and similar standard parts; these parts can be drawn using a template, or even simply indicated by centre-lines and detail number. It is often necessary to make a slightly more elaborate drawing so that clearances, etc., can be studied, and in this connection the tracing templates supplied by the 'unit tooling part' manufacturers will be useful; fig. 13.8 on page 194 illustrates such a template.

17.8. Batch Size

On page 266 it was stated that in batch production a machine is set up to machine a batch of components, and that when the batch is completed the

Batch size	Setting time for the batch in min	Machining time for the batch in min	Machine occupation time for the batch in min	Unit machine occupation time in min	Stock-holding in weeks
50	120	500	620	12·4	0·5
100	120	1 000	1 120	11·2	1·0
200	120	2 000	2,120	10·6	2·0
300	120	3 000	3 120	10·4	3·0
400	120	4 000	4 120	10·3	4·0
500	120	5 000	5 120	10·25	5·0
600	120	6 000	6 120	10·2	6·0
700	120	7 000	7 120	10·17	7·0
800	120	8 000	8 120	10·15	8·0
900	120	9 000	9 120	10·13	9·0
1 000	120	10 000	10 120	10·12	10·0

Table 17.1

machine is re-set and another batch is machined. The size of the batch must be studied because the **'unit machine occupation time'** (i.e. the setting time plus the machining time) is reduced as the batch size is increased.

Table 17.1 shows a set of calculations for an operation that takes 2 hours to set up and 10 minutes to machine each component; the parts being machined are used up at the rate of 100 per week.

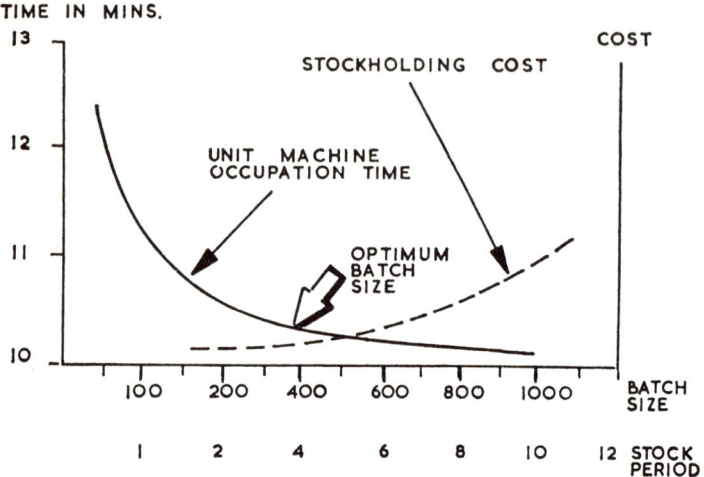

Fig. 17.3 Determination of Optimum Batch Size

The unit machine occupation time is plotted against the batch size (see fig. 17.3), and indicates that the saving in occupation time is large when the batch size is increased from a small batch to a slightly larger one, but that the saving is less marked when the batch size is increased from a large one to a larger one. At the same time the stockholding cost rises with the batch size because stockholding means that valuable space is taken up, capital is 'frozen', there is a possibility of deterioration, etc.

A possible optimum batch size is indicated on fig. 17.3, but the actual optimum size may vary from time to time to suit circumstances.

Exercises

Some typical operation layouts are shown on the following pages and on pages 53 and 55. It is suggested that some of these examples be used for projects, and the component be fully tooled-up, including raw-material drawings and operation drawings. As an alternative, some of the following exercises may be worked.

Exercises Set E.1 (see component E.1 and operation layout on page 276)

Design the following equipment:

(a) Design plug gauges and counterbore depth gauge (0·750 + 0·005 in) for operation No. 1.
(b) Expanding post for operation No. 2. For details of lathe spindle nose see fig. 16.13 on page 262.
(c) Drilling jig for operation No. 3.
(d) Receiver gauge for 2·375 in dia and holes.

Exercises Set E.2 (see component E.2 and operation layout on page 277)

Design the following equipment:

(a) Drilling jig and drill stop assembly and gauge for operation No. 5.
(b) Drilling jig for operation No. 6.
(c) Diameter gauge for 2·750 in dia − 0·001 in.

Exercises Set E.3 (see component E.3 and operation layout on page 278)

Design the following equipment:

(a) Drilling jig for operation No. 4.
(b) Milling fixture for operation No. 7. For details of machine table see fig. E.3 (a).

Exercises Set E.4 (see component E.4 and operation layout on page 279)

Design the following equipment:

(a) Milling fixture for operation No. 1. For details of machine table see fig. E.3 (a).
(b) Drilling jig for operation No. 4.
(c) Milling fixture for operation No. 5. For details of machine table see fig. E.3 (a).
(d) Base and sighting plate for operation No. 8.

Exercises Set E.5 (see component E.5 and operation layout on page 280)

Design the following equipment:

(a) Turning fixture for operation No. 1. For details of lathe spindle nose see fig. 16.13 on page 262.
(b) Drilling jig for operation No. 4.
(c) Drilling jig for operation No. 7.

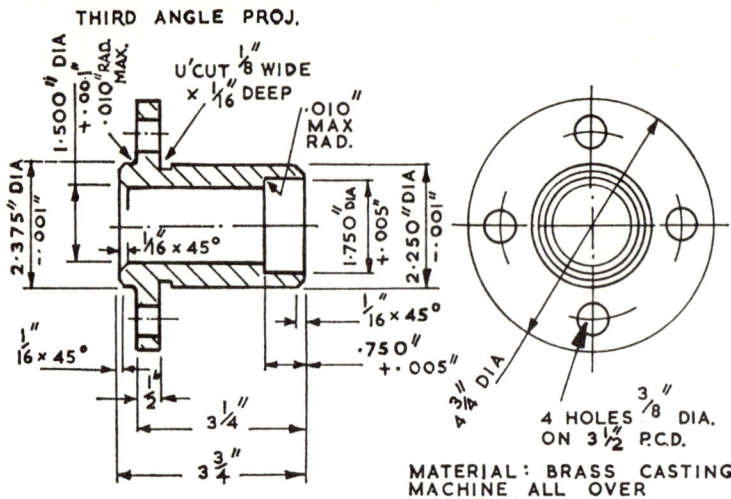

Fig. E.1 Bearing Housing

Operation number	Description	Machine
1	Hold in special jaws from long end; face end, turn outside diameter and face of flange, forming undercut. Open out bore and counterbore. Form chamfer	Turret lathe
2	Reverse; hold on expanding post from bore; face end, turn spigot, forming chamfer and small radius. Turn flange diameter	Turret lathe
3	Remove burrs	Burr bench
4	View	View bench
5	In jig, locating from bore and flange face. Drill four holes in flange	Drilling machine
6	Final view	View bench

Operation Layout for Bearing Housing—Fig. E.1

EXERCISES

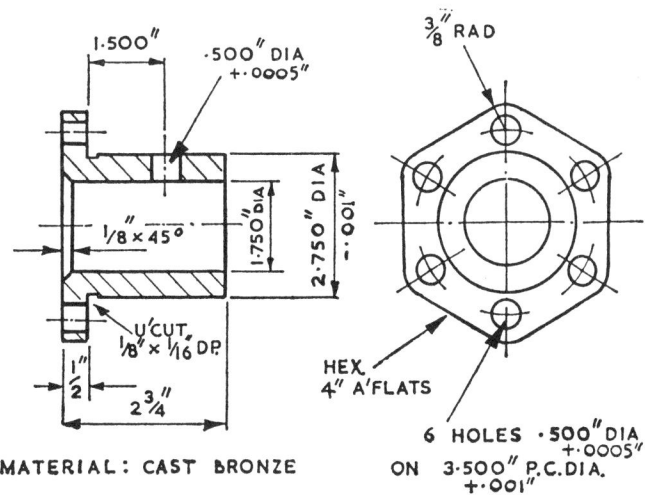

Fig. E.2 Bearing

Operation number	Description	Machine
1	In special jaws from outside diameter. Face flange. Open out bore to 1·750 + 0·005 in dia	Turret lathe
2	Reverse. Locate from bore on expanding post. Finish turn outside diameter, face flange to width forming undercut	Turret lathe
3	Remove burrs	Burr bench
4	View	View bench
5	Locate from outside diameter and flange profile in jig. Drill and ream six holes	Drilling machine
6	Locate from outside diameter and one hole in flange in jig. Drill and ream 0·500-in-dia hole in stem	Drilling machine
7	Final view	View bench

Operation Layout for Bearing—Fig. E.2

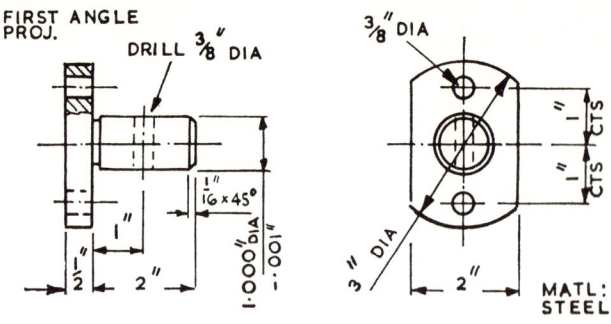

Fig. E.3 Special Pin

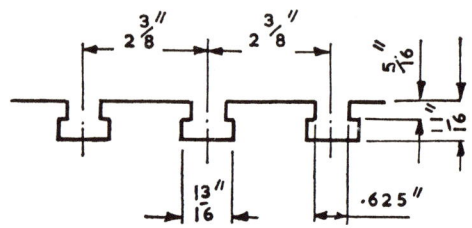

Fig. E.3 (*a*) Details of Milling Machine Table

Operation number	Description	Machine
1	Hold from head. Face end, turn stem diameter. Form chamfer. Face head, forming undercut	Turret lathe
2	Reverse. Face head	Turret lathe
3	View	View bench
4	Locate from stem, in jig. Drill and ream two holes ⅜ in diameter in head, and one hole in the stem	Sensitive drilling machine
5	Remove burrs	Burr bench
6	View	View bench
7	In fixture, locating from stem and one ⅜-in-dia hole, mill flats on head	Milling machine
8	Remove burrs	Burr bench
9	Final view	View bench

Operation Layout for Special Pin—Fig E.3

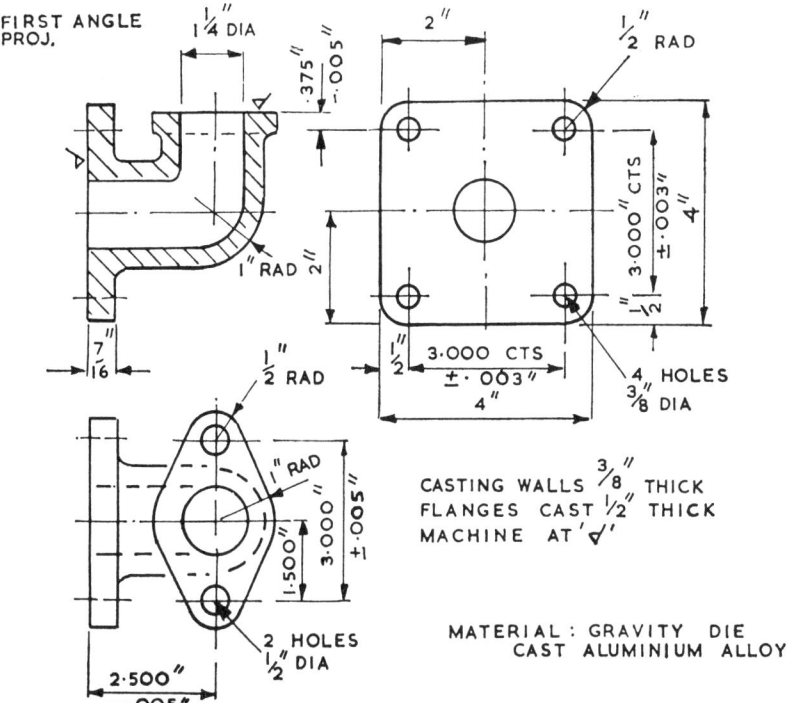

Fig. E.4 Elbow

Operation number	Description	Machine
1	In fixture. Mill face of square flange to thickness	Milling machine
2	Remove burrs	Burr bench
3	View	View bench
4	In jig, locating from square flange face and profile. Drill and ream four holes $\frac{3}{8}$ in dia	Drilling machine
5	In fixture, locating from square flange face and two holes. Mill face of elongated flange	Milling fixture
6	Remove burrs	Burr bench
7	View	View bench
8	Locate from face of square flange and two holes, and using a sighting drill plate. Drill two holes $\frac{1}{2}$ in dia in elongated flange	Drilling machine
9	Final view	View bench

Operation Layout for Elbow—Fig. E.4

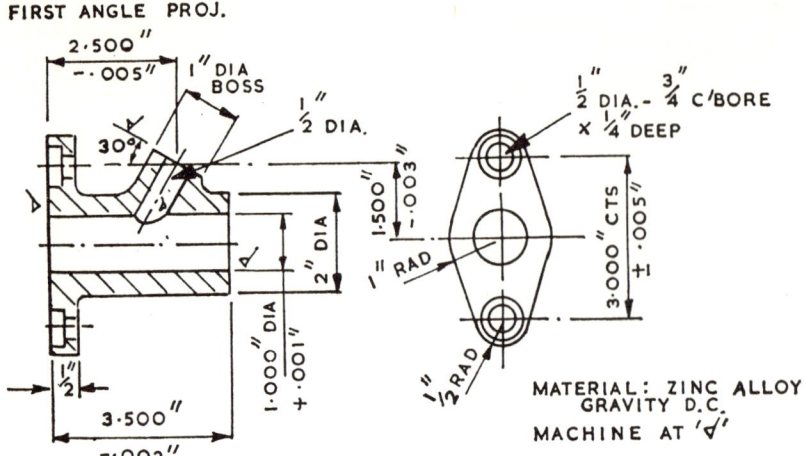

Fig. E.5 Connection

Operation number	Description	Machine
1	In special jaws, hold from body. Face flange to thickness and open out bore to size	Turret lathe
2	Reverse. Locate from flange face and bore. Face end to length	Turret lathe
3	View	View bench
4	Locate from bore, flange face and flange profile. Drill, ream and counterbore two holes in flange	Drilling machine
5	Remove burrs	Burr bench
6	View	View bench
7	Locate from bore, flange face and one ½-in-dia hole. Spotface 1-in-dia boss, and drill ½-in-dia hole	Drilling machine
8	Final view	View bench

Operation Layout for Connection—Fig. E.5

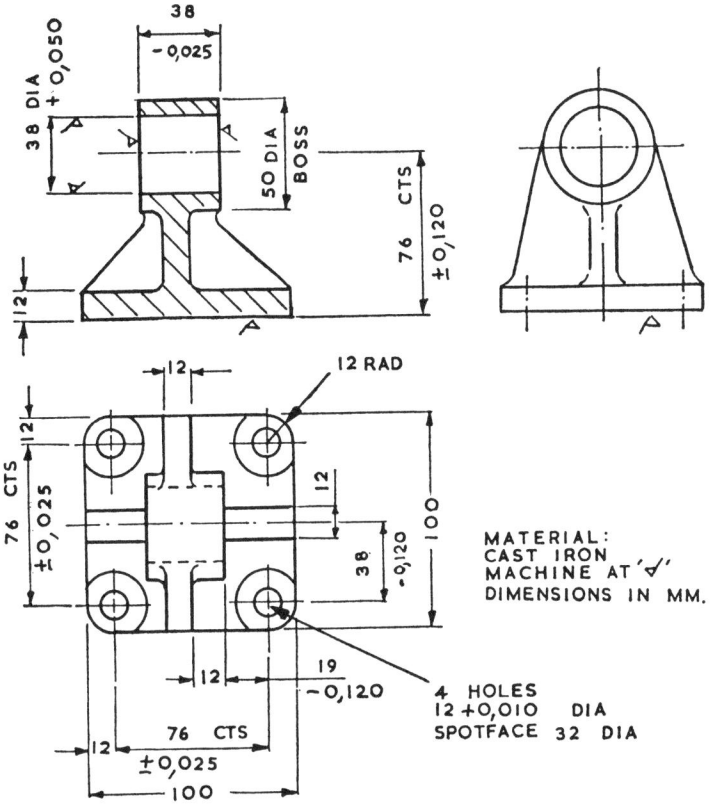

Fig. E.6 Bearing Bracket

Operation number	Description	Machine
1	In special vice jaws. Mill base to thickness	Horizontal milling machine
2	Remove burrs	Burr bench
3	View	View bench
4	Locate from base. Drill and ream four holes 12 mm dia and spotface 32 mm dia	Drilling machine
5	Remove burrs	Burr bench
6	View	View bench
7	In fixture locate from base and two 12-mm-dia holes. Face both sides of boss, and bore to size	Turret lathe
8	Final view	View bench

Operation Layout for Bearing Bracket—Fig. E.6

JIG AND TOOL DESIGN

Exercises Set E.6 (see component E.6 and operation layout on page 281)
Design the following equipment:
  (a) Drilling jig for operation No. 4.
  (b) Turning fixture and tool setting gauge for operation No. 7. For details of lathe spindle nose see fig. 16.13 on page 262.

Exercise E.7

Design an index drilling jig for drilling the four 12-mm-dia holes in component E.7 below. This is the final machining operation.

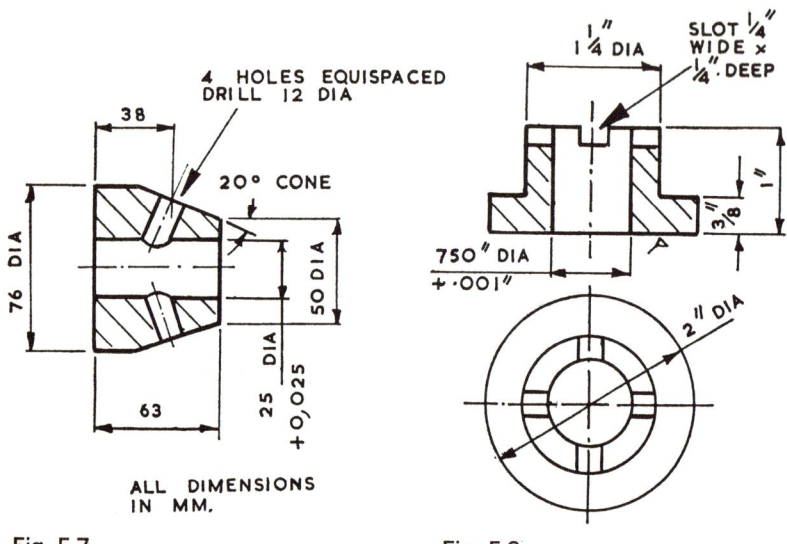

Fig. E.7   Fig. E.8

Exercise E.8

Design an index milling fixture for milling the ¼-in slots in component E.8 above. This is the final machining operation. For details of the milling machine table see fig. E.3 (a) on page 278.

Exercise E.9

Fig. E.9 on page 283 shows details of a vee groove to be turned using a form tool.
  (a) Draw and dimension a flat form tool to produce the groove; the tool is to have a 5° front clearance angle and a zero rake angle. Determine the shape of the form tool by calculation and also by construction.
  (b) Draw and dimension a flat form tool to produce the groove; this tool is to have a 5° front clearance angle and a 10° front rake angle. Determine the shape of the form tool by calculation and also by construction.

Exercise E.10

Fig. E.10 on page 283 shows details of a stepped groove to be turned using a circular form tool.

EXERCISES

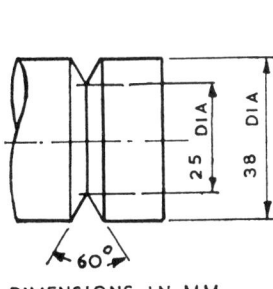

Fig. E.9

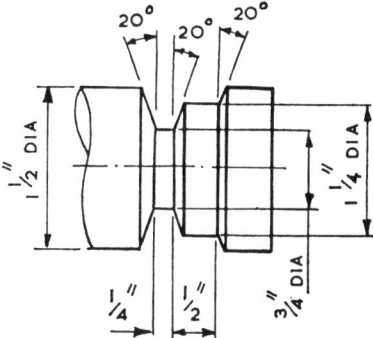

Fig. E.10

(a) Draw and dimension a circular form tool to produce this groove; this tool is to have a 5° front clearance angle and a zero rake angle. The shape of the form and the height of the tool centre above that of the workpiece is to be determined by calculation and also by construction.

(b) Draw and dimension a circular form tool to produce the groove; this form tool is to have a 6° front clearance angle and a 15° front rake angle. The shape of the form and the height of the tool centre above that of the workpiece is to be determined by calculation and also by construction.

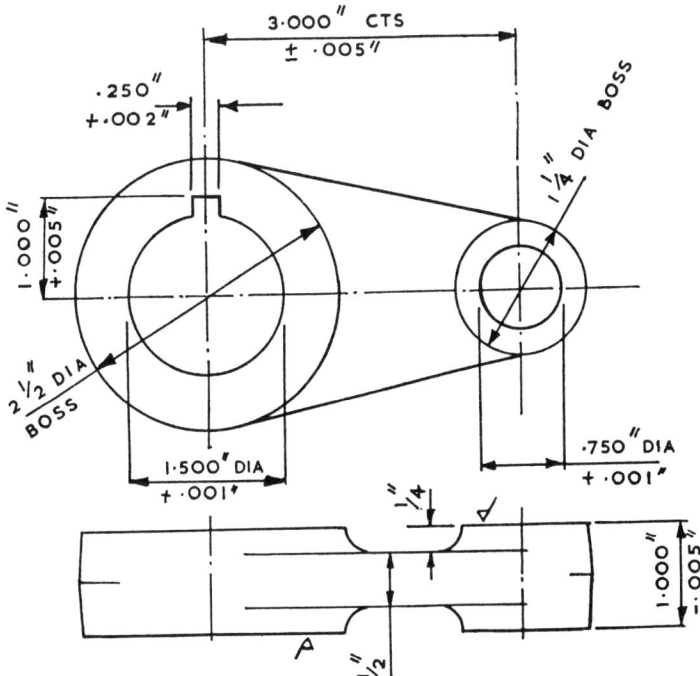

Fig. E.11 Crank

Exercise E.11

Design a broach, puller and broaching fixture to produce the 0·250-in keyway in the bore of component E.11 on page 283. The keyway is to be broached at the final machining operation.

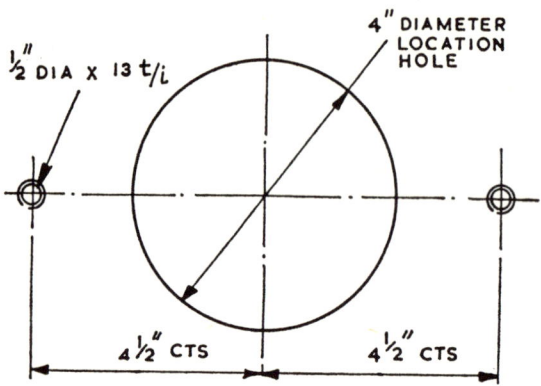

Fig. E.11 (*a*) Details of Broaching Machine Faceplate

The broaching machine has a capacity of 2½ tonf, a stroke of 30 in and the pull-head thread is 1¼ in × 10 t/in. Fig. E.11 (*a*) shows the dimensions of the broach faceplate upon which the fixture is to be mounted.

Exercise E.12

Design a press-tool set for blanking and piercing the bell crank shown in fig. E.12 below. This is made from 2-mm mild-steel strip.

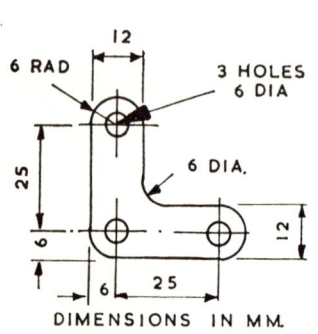

Fig. E.12 Bell Crank

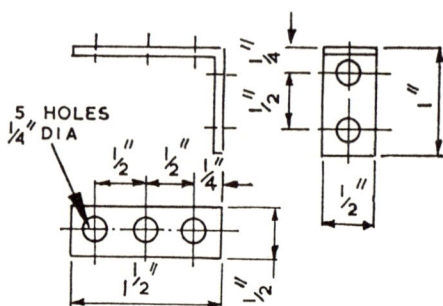

Fig. E.13 Bracket

Exercise E.13

Design the following press sets for the Bracket E.13 above. This Bracket is to be made from 16 S.W.G. mild-steel strip.

(*a*) Piercing and blanking press-tool set to produce the bracket before bending.
(*b*) A bending tool to bend the pierced and blanked bracket.

EXERCISES

Exercises Set E.14

Fig. 4.1 on page 53 shows a special bolt, and the operation layout to produce this bolt is also shown on page 53.

Design the following equipment for the special bolt:
  (a) Milling fixture for operation No. 4. For details of the milling machine table see fig. E.3 (a) on page 278.
  (b) Drilling jig for operation No. 6.

Exercises Set E.15

Fig. 4.2 on page 54 shows a fulcrum pin, and the operation layout for this component is shown on page 55.

Design the following equipment:
  (a) Turning fixture for operation No. 1. For details of the lathe spindle nose see fig. 16.13 on page 262.
  (b) Drilling jig, tools and drill stop nut for operation No. 3.
  (c) Turning fixture for operation No. 5. For details of the lathe spindle nose see fig. 16.13 on page 262.
  (d) Drilling jig for operation No. 7.
  (e) Grinding fixture for operation No. 8.
  (f) Grinding fixture for operation No. 9.

INDEX

Adjustable caliper gauge 205
Adjustable conical location 89
Adjustable location pad 83, 84
Adjustable location pin 85
Adjustable screw ring gauge 206
Air-operated clamp 110, 112
Allowance for springback (press-tools) 239
Allowance for wear (gauges) 199
Ancillary sections of planning and jig design department 252
Angular die clearance 224
Angular post jig 126
Annular-type gauge 203, 204
Applications of indexing 156
Application of shear to press-tools 231
Approach angle 11
Assembly fixtures 153
Automatic lathe 58

Back off (broach teeth) 169
Balancing of turning fixture 152
Balloon references 196
Ball-type indexing device 158
Basic tooling 267
Batch production 266
Batch size 273
Bending 214
Bending and forming 238
Bending forces 240
Bending—press tool set for 238
Blank size for drawing 242
Blanking 213
Blanking dies 223
Blanking layout 235
Blanking—press tool set for 219, 220
Blanking pressures 230
Board of directors 248
Bolts—hook type 103
Boring and boring tools 24; boring bars 25
'Bought out' parts 272
Box drilling jig 131, 133
'Break even' quantity 267
British Standards: Jig Bushes 135; Limits and Fits for Engineering 211; Milling Cutters 146; Plain Limit Gauges—Limits and Tolerances 211; Reamers, Countersinks and Counterbores 135; Screw Gauges—Limits and Tolerances 211; Specifications for Gauge Blanks 211; Twist Drills and Combined Drills and Countersinks 135
Broach: holder 175; material 167; pull-end 180; puller 175; teeth 167
Broach design: internal 165; spiral 181; surface 181, 182
Broaching: advantages and limitations of 163, 164; fixtures 185; machines 165; pull method 175; push method 174; spiral 181; surface 181
Built-up edge 8
Burnishers (broach) 171
Burr formation 121
Burr grooves applied to drilling jigs 121, 122
Button clamp 100

Calculation of 'Break even' quantity 267
Caliper gauge—adjustable type 205
Cam-actuated clamping 106
Cam profile for screw type automatic 63
Cams for Swiss automatic 70, 72, 73
Capacity charts 260, 261, 262
Capacity of press 217
Capstan lathe 56
Captive washer 101
Carbon tool-steels 15
Cast body 190
Casting 190
Catalogues 260
'Cee' washer 102
Cemented carbides 17
Ceramic cutting-tool material 18
Channel jig 123
Chip breakers 11, 13; applied to broach 170
Chip formation 7
Chucks—collet type 147
Circular form tool 44
Clamp: air operated 110, 112, button 100; edge 97; equalising 105; latch 98; pivoted 98; plate 94, 104; spider 97; swinging 100; two-point 97; two-way 98, 99
Clamping: applied to drilling jigs 113; bush 120; cam actuated 106, 107; differential 105, 106; direct 100, hydraulic 111; more than one workpiece 104; of workpiece when milling 141; pneumatic 111; principles 2; system requirements 94

INDEX

Clearance 3; between punch and die (presswork) 224
Cobalt-base alloys 16
Coining 214
Collet chucks 147; dead-length type 148; pull-in type 147; push-out type 147
Collet-type plug gauges 200
Combination blanking and cupping tools 241, 242
Company standards 261
Compensation for wear (indexing devices) 159
Complete tooling 267
Component record cards 258
Conical location 89
Consideration of jig and fixture construction 190
Construction method 190
Continuous chip formation 8
Contract tool drawing office 263
Contracts—placing of 264
Control of depth (drill jigs) 120
Controlling the cutters (drill jigs) 116
Copying 1
Costs department 250
Counterboring cutters 27
Countersinking cutters 27
Crimping 215
Cupping on a double-acting press 245
Cut-and-cupping press 217
Cut per tooth (broach) 170
Cutter arrangement: gang milling 136; pendulum milling 137; profile milling 139; straddle milling 136; string milling 136
Cutting conditions and tool shape 9
Cutting-tool materials 13
Cylinder—location from 86
Cylindrical grinding 152

'D' bit 22
Dead-length collet chuck 148
Depth gauge—stepped pin type 208
Design Department 248
Design: of limit gauges 199; principles 2; study 4, 187, 190; of broaches 165, 181
Design study 187, 190
Designing: for casting 190; for welding 192; to suit manufacture 196
Devices for indexing 158
Diamond cutting tools 18
Die: angular clearance 224; material 220, 240; sets 218; shapes 240
Dies—for thread cutting 34
Differential clamping 105
Direct clamping 100; from a post 101
Discontinuous chip formation 8
Double-acting cylinder for clamping 110
Double-acting press 216
Dovetail form tool 43
Down-cut milling 30

Drawing 214, 240; blank size 242; die material 240; die shapes 240; press-tool set 240; pressure required 245; punch material 240
Drawings—issue of 259
Drill bush: clamping type 120; extended 117; fixed renewable 118; headed 116; headless 116; locating type 119; shaped 117; slip renewable 117, 118
Drill nomenclature 21
Drill over-run 121
Drill stop assembly 120
Drilling jigs 113, 122; control of cutters 116; handling 114
Drilling jigs—types of: box 131, 133; channel 123; indexing 160; latch 130, 132; local 123, 124; nutcracker 129, 130; plate 123; post 123, 125, 126; pot 127; sandwich 128; solid 123, 124; table, 130, 131; trunnion 133, 134

Eccentric-operated hook bolt 103, 104
Economical design 272
Economical use of equipment designer's time 273
Economics of tooling 266
Edge clamp 97
Ejection (presswork) 239
Ejectors: jig and fixture 92; presswork 239
Embossing 214
Equalising clamps 105
Equipment numbering 257
Estimates department 250
Examples of presswork 212
Expanding posts 148, 149
Experimental department 248
Extended drill bush 117

Fabricating from several parts 192
Face angle (broach) 167
Facing cutters 32
Features of an indexing jig or fixture 157
Feed methods (presswork) 217, 227
Feet for drilling jigs 114
Fixed renewable drill bush 118
Fixture defined 2, 113
Fixtures: assembly 153; broaching 185; grinding 153; indexing 160; milling 141, 142, 143; turning 148; welding 153
Flat form tool 40
Floating cutter 27
Floating pads 101
Floating reamer 27
Flow production 266
Fly press 215
Foolproofing 81
Foot nut and bolt assembly 115
Foot-operated press 215
Forces during cutting 10
Forces during bending 240

Form correction: circular form tool 45, 46, 47; flat form tool 40, 41, 42; notes on method 48
Form-relieved cutter teeth 29
Form tools 40; dovetail type 43; flat type 40; circular type 44; skiving 48
Formation of burr 121
Forming 1
Forming applied to presswork 214, 238
Function of planning and jig design departments 250

Gang milling 136
Gap gauge 204
Gauge: length 207; plug 200; position 209, 210; receiver 210; ring 204; recess 207, 208; screw thread 205, 206; shaft 204; spherical-ended rod 202; step 208; thickness 207; trilock plug gauge 201
Gear cutting 35
Gear hobbing 37
Gear planing 36
Gear shaping 39
Generating 1
Grinding fixture for internal grinding 153
Grinding mandrel 152, 153
Grinding: workholding devices 152
Guides for presswork 225

Hand nut 97, 98
Handbooks 260
Handling clearances 122
Handling: drilling jig 114; of jigs and fixtures 3
Hand-operated press 215
Headed drill bush 116
Headless drill bushes 116
Heel pin for clamp 96
High-speed steels 15
Hobbing 37
Holder for broach 175
Holding in stores 263
Hole-finishing cutting tools 24
Hole-producing cutting tools 20
Hook angle (broach) 167
Hook bolts 103
Hydraulic clamping 111

Indexing: application 156; linear 156; rotational 156, 157
Indexing devices 158; ball type 158; lever type 158; plunger type 158; rack and pinion 159; spring-loaded plunger 159
Indexing jigs and fixtures 156, 160, 161
Information library and records 260
Inserted blades 16
Inspection department 249
Interlocking cutters 33
Internal broach 165, 166
Internal broaching fixture 185

Issue department 248
Issue of drawings 259
Internal grinding fixture 153

Jamming 86
Jaw chucks 148
Jig—defined 2, 113
Jig and fixture stability and rigidity 3
Jig design Department 249; organisation of 251
Jigs—typical examples 122
Job production 266

Latch clamp 98
Latch jig 130, 132
Layout for blanking 235
Length gauge 207
Lever systems 94
Lever-type indexing device 158
Limit gauges 197; examples of 200; materials for 199; limits 199; tolerances 199
Limits of size 197
Linear indexing 156
Liner bush, 117
Local drilling jig 123, 124
Locating bush 119
Location and clamping of milling fixtures 140
Location applied to drilling jigs 113
Location devices 83
Location: from a cylinder 86; from plane surface 83; from a profile 85; from two cylindrical holes 88
Location pins 85
Location post 86; retention of 87
Location pot 87
Location principles 2
Location system—choice of 77
Long location post 87

Management 248
Mandrel—grinding 152, 153
Manufacture of equipment 258
Manufacturers' catalogues and handbooks 260
Manufacturing systems 266
Mass production 266
Material: for broach 167; for limit gauges 199; for punches and dies 220; for punch and former 239
Material size for bending 239
'Matrix' thread gauge 205
Maximum metal limit 197
Method and equipment 49
Method of construction 190
Milling cutters 27
Milling fixture 136; general features 142; location and clamping 140; clamping of workpiece 141; tool setting 141
Milling fixture types: indexing 160, 161; line 143, 145; simple 143, 144

INDEX

Milling methods 136: gang 136; pendulum 137, 138; profile 139; rotary table 138; straddle 136, string 137
Minimum metal limit 197
Modification action and contract drawing office 264
Modification notes 259
Modifications 259
Modifications department 248
Modifications to processes and equipment 259
Multi-spindle automatics 74

National standards 261
Negative rake angle 7
Number of teeth (broach) 171
Numbering of equipment 257
Nutcracker drilling jig 129, 130
Nuts: hand 98; quick action 102

Oblique cutting 6
Open drilling jig 130, 131
Operation sheets 255
Ordering of equipment 258
Organisation and management 248
Organisation of planning and jig design departments 251
Orthogonal cutting 6

Pads—floating type 101
Parts list 257
Pendulum milling 137, 138
Piercing 212
Piercing-and-blanking 213; press tool set 220, 221; pressures 230
Piercing dies 223
Pilots for blanking punch 226
Pitch: internal broach teeth 167; surface broach teeth 184
Pivoted clamp 98
Placing of contracts 264
Plain gap gauge 204
Plane surface—location from 83
Planning department 249, 250; organisation of 251
Planning; method 50; of product manufacture 253; of tooling 256
Plastics materials 193
Plate clamps 95; with heel pin 96; securing two workpieces 104
Plate jig 123
Plug gauges 200
Plunger-type indexing device 158
Pneumatic clamping 109
Position gauge 209, 210
Position of a second locator 88
Post—expanding 148
Post jigs 123, 125, 126
Post—location 86
Pot jig 127
Pot—location from 87

Power press 216
Presses 215; capacity 217; cut-and-cupping 217; double-acting 216; fly 215; foot-operated 215; hand-operated 215; power 216; triple-acting 217; single-acting 216
Press-tool set: for bending 238; for drawing 240; piercing-and-blanking 220, 221; for simple blanking 219, 220; for Vee bending 238
Pressure: for drawing 245; for piercing and blanking 230
Pressure pads 226
Presswork 212: typical examples of 212
Principles of jig and fixture design 2
Print room 253
Process planning 49
Product design division 248
Product manufacture 253
Production—batch type 266
Production control 248
Production division 248
Production—flow type 266
Production—job type 266
Production—mass type 266
Profile—location from 85
Profile milling 139
Progress department 249
Progress of tooling 259
Progressive plug gauge 200
Project engineer 253
Pull broaching 175
Pull-end—broach 180
Puller—broach 175
Pull-in collet chuck 147
Punch and former material (forming and bending) 239
Punch material: blanking and piercing 220; drawing 240
Punch pilots 226
Punch retention 222
Purchase of special equipment 271
Push broaching 174
Push-out collet chuck 147
Push-type ejector 92

Quick-action nuts 102

Rack and pinion indexing device 159
Rake and clearance angles 6
Ratefixing department 249
Raw material drawings 254
Reamers 27
Receiver gauge 210
Recess gauge 207, 208
Record cards: component 258; tool 258
Red-hardness 14
Redundant location 79, prevention of 88
Renewable bush—special slip type 119
Requirements of the clamping system 94

INDEX

Research department 248
Retention: of boring tools 25, 26; of inserted blades 16; of location post 87; of punch 222
Rigidity of jigs and fixtures 3
Ring gauge 204
Rotary-table milling 138
Rotational indexing 156, 157

Sales Department 248
Sandwich drilling jig 128
Schedule 253
Screw plug gauge 206
Screw ring gauge—adjustable 206
Screw thread gauging 205
Section broaches 173
Securing surface broaches 184
Segmental cylindrical gauge 202, 203
Service department 248
Setting block for milling fixture 141
Setting on assembly 196
Shafts—gauges for 204
Shaped drill bush 117
Shear: applied to press tools 231; chip formation 7
Simple blanking set 219, 220
Simple drill stop 121
Simple plate clamp 95
Single-acting cylinder for clamping 109
Single-acting press 216
Single-point cutting-tool angles 11, 12
Sintered oxides 18
Six degrees of freedom 77
Six-point location principle 82
Size of batch 273
Skeleton tooling 267
Skiving tool 48
Slide tools for presswork 247
Slide vee location 90
Slip bush 117
Slip renewable drill bush 117, 118
Soft jaw 148
Solid drilling jig 123, 124
Spade drills 20
Special drill bushes 116
Special equipment 271
Special lead 87
Special slip renewable bush 119
Special vice jaws for milling 142
Spherical-ended rod gauge 202, 203
Spherical washers 96
Spider clamp 97
Spiral broaching 181
Spotfacing cutters 27
Springback allowance 239
Spring-loaded plunger 159
Spring-operated ejector 92
Stability of jigs and fixtures 3
Stagger—of broach teeth 168
Standard die sets 218
Standard parts 193
Standards—company 261

Standards Department 248
Standards—National 261
Step gauges 208
Stepped-pin depth gauge 208
Stepped taper plug gauge 208, 209
Stepped taper ring gauge 208, 209
Stock stops for presswork 227
Stores holding 263
Straddle milling 136
Stress department 248
String milling 137
Stripper and guides for presswork 225
Study of design 187, 190
Sub-contract parts 272
Surface broaching 181; surface broach design 182; securing broach to holder 184; fixture 186
Surface grinding 152
Swinging clamp 100
Swiss automatic 63
Systems of manufacture 266
Systems of tooling 267

Table drilling jig 130, 131
Taps 33
Taylor principle 197
Tear—chip formation 7
Teeth of typical broach 167
Tenon for milling fixture 140
Thickness gauge 207
Thread-producing tools 33
Throw-away tool tips 19
Title block 257
Toggle clamps 108, 109
Tolerance 197
Tolerances for limit gauges 199
Tool and cam layouts: for screw-type auto 59, for Swiss auto 65
Tool records 256, record cards 258
Tool room 249, 250, 258, 264; demonstrator 265
Tool setting: milling 141; turning 148, 151
Tool-steels 15
Tooling: basic type 267; complete 267; economics 266; planning of 256; skeleton type 167
Tooling equipment—economical design of 272
Tooling expenditure 267, 271
Tooling layout: for capstan lathe 57; for 5-spindle auto 75
Tooling systems 267
Tooth space (broach) 169
Tracing templates 192, 194
Trepanning 23
Trigger-type ejector 92
Trilock plug gauge 200, 201
Triple-acting press 217
Trunnion drilling jig 133, 134
Turning fixtures 148; faceplate type 148, 151; simple type 148, 150

INDEX

Turning—workholding devices for 147
Turnover drilling jig 130, 131
Turret lathes 56
Twist drill 21
Two-point clamp 97
Two-way clamp 98
Typical limit gauges 200

Unit tooling parts 192
Up-cut milling 30

Variation tolerated 197
Vee applications 91
Vee bending press set 238

Vee location 90
Vice jaws—for milling 142

Washer: captive 101; 'Cee' 102; spherical 96
Wear allowance applied to gauges 199
Wear of indexing device 159
Welding 192; fixtures for 153
Work study department 249
Workholding devices: for grinding 152; for turning 147
Working drawings of production equipment 196
Workpiece ejectors 92
Workpiece location for milling 141

CLASSIFIED INDEX

Assembly Fixtures
Design 153

Broaching
Broach design (internal broaches) 165; back off 169; burnishers 171; chip breakers 170; cut per tooth 170; face, or hook angle 167; material 167; number of teeth 171; power required 173; pull-end 180; section broaches 173; side clearance 169; stagger of teeth 168; tooth pitch 167; tooth relief 169; tooth space 169; tooth width 168
Broach design (surface broaches) 182; pitch of teeth 184; stagger of teeth 183
Broach holders and pullers; holders for surface broaches 185; holders and pullers for internal broaches 175; securing surface broaches to holder 184, 185
Broaching: advantages and limitations 163, 164; broaching fixtures 185; broaching machines 165; types of broaching 164

Clamping
Clamp design: air operated clamp 110, 112; button clamp 100; captive washer 101; 'Cee' washer 102; floating pads 101; heel pin 96; hook bolt 103; latch clamp 98; nuts 97, 98; pivoted clamp 98; plate clamp 95, 104; quarter-turn screw 99; quick-action nut 102; spherical washer 96; spider clamp 97; swinging clamp 100; toggle clamp 108, 109; two point clamp 97; two-way clamp 98, 99
Clamping: cam actuated 106, 107; clamping bush 120; clamping plate 103; differential clamping 105, 106; direct clamping 100; drill jigs 113; edge clamping 97; hydraulic clamping 111, lever systems 94; milling 141; more than one workpiece being clamped 104; pneumatic clamping 111; principles of clamping 2; requirements of clamping system 94

Cutting and Cutting Tools
Boring, boring tools and boring bars 24, 25
Broach—see Broaching section
Built-up edge 8
Burr formation 121
Chip breakers 11, 13, 170
Chip formation 7: continuous chip formation 7; continuous chip formation with built-up edge 8; discontinuous chip formation 8
Control of depth (drilling) 120; control of position (drilling) 116
Copying 1
Counterboring cutters 27
Countersinking cutters 27
Cutter arrangement for milling 136
Cutting action: oblique cutting and orthogonal cutting 6; tear 7; shear 7
Cutting tool angles 6; approach angle 11; cutting conditions and tool shape 9; negative rake 7; rake and clearance angles 6
Cutting-tool materials 13: carbon tool steels 15; cemented carbides 17; ceramic materials 18; cobalt-base alloys 16; diamond 18; red-hardness 14
'D' bit 22
Dies for thread cutting 34

CLASSIFIED INDEX

Drill nomenclature 21
Facing cutter 32
Floating cutter 27
Floating reamer 27
Forces during cutting 10
Form tool—see Form Tool section
Forming 1
Form-relieved cutter teeth 29
Gear cutting 35; hobbing 37; planing 36; shaping 39
Inserted blades and their retention 16
Interlocking cutters 33
Milling Cutters 27
Single-point tool nomenclature 12
Sintered oxides 18
Skiving tool 48
Spade drill 20
Spotfacing cutter 27
Taps 33
Tool setting: milling 141; turning 148, 151
Trepanning 23
Turning tools 9
Twist drill 21

Design and Detailing
Balloon references 196
Design: consideration of construction 190; of broaches (see Broach section); for casting 190; of clamps (see Clamping section); of cutting tools (see Cutting and Cutting Tools section); of drilling jigs (see Drilling and Drilling Jigs section); of fixtures (see appropriate section); of gauges (see Limit Gauges section); of locators (see Location and Locators section); of press tools (see Presswork section); for welding 192
Economical design 272
Fabricating from several parts 192
Parts list 257
Plastics materials 193
Principles of jig and fixture design 2
Standard parts 192
Title block 257
Tracing templates 194
Working drawings 196

Drilling and Drilling Jigs
Burrs and burr grooves 121, 122
Control: depth 120; position 116
Drill bushes: clamping 120; extended 117; fixed renewable 118; headed 116; headless 116; liner 117; locating type 119; shaped 117; slip renewable 117, 118; special slip bush 119
Drill nomenclature 21
Drill over-run 121
Drill stop assembly 120
Drilling jig—handling 114
Drilling jigs: box type 133; channel type 123; indexing 160; latch type 130, 132; local type 123, 124; nutcracker type 129, 130; plate type 123; post type 123, 125, 126; pot type 127; sandwich type 128; solid type 123, 124; table type (also called 'open type' and 'turnover type') 130, 131; trunnion type 133, 134
Feet for drilling jigs 114, 115

Location of workpiece 113
Reamers 27

Economics of Tooling
Batch size 273
Bought out parts 272
Break-even quantity 267
Economical design 272
Economical use of designer's time 273
Production methods: batch 266; flow 266; job 266; mass 266
Purchase of special equipment 271
Sub-contract parts 272
Tooling expenditure 271
Tooling systems: basic, complete, and skeleton 267

Form Tools
Circular form tools 40: form correction 45, 46, 47, 48; holder 44
Dovetail form tool 43; holder 43
Flat form tool 40; form correction 40, 41, 42
Form tool calculations—notes on methods 48
Skiving tool 48

Gear Cutting
Gear production 35
Gear production methods: hobbing 37; planing 36; shaping 38, 39

Grinding
Grinding fixture 153, 154
Grinding mandrel 153
Surface grinding 152

Indexing Jigs and Fixtures
Applications of indexing 156
Basic features of indexing jigs and fixtures 157
Indexing devices: ball type 158; lever type 158; plunger type 158; rack and pinion type 159; spring-loaded plunger 159
Indexing devices: compensation for wear 159
Indexing drilling jigs 160
Indexing milling fixture 160, 161

Limit Gauges
Air venting 203
Allowance for wear 199
Centres 203
Gauge design 199
Gauge types: length 207, plug 200, position 209, 210, receiver 210, recess 207, 208; recess diameter 208; ring 204; screw thread 205, 206; shaft 204; spherical-ended rod 202, steps 208, thickness 207, trilock plug gauge 200, 201
Limits 197, 199
Materials 199
Maximum metal limit 197
Minimum metal limit 197
Taylor principle 197
Tolerances 197, 199
Wear allowance 199

CLASSIFIED INDEX

Location and Locators
Extent of location 77
Foolproofing 81
Location from a cylinder 86; conical locators 89; jamming of workpiece 86; location post 86; location pot 87
Location from a plane surface 83; location pads 83; prevention of mal-location 84
Location from a profile 85; pins 85; sighting plate 85
Redundant location 79: prevention of 88
Second locator position 88
Six degrees of freedom 77
Six-point location principle 82
Vee location 90; application of 91; sliding vee location 90
Workpiece ejectors 92

Milling and Milling Fixtures
General features of a milling fixture 142
Location and clamping of milling fixture 140
Milling cutters 27
Milling fixtures 143; indexing 160, 161, 162; line 145; simple 144
Milling methods: down-cut milling 30; gang milling 136; line milling 136; pendulum milling 137; profile milling 139; rotary table milling 138; straddle milling 136; string milling 136, 137, 145; up-cut milling 30
Tee bolts 140
Tenons 140
Tool setting 141
Vice jaws—special 142
Workpiece: clamping 141; location 141

Organisation and Management
Capacity charts 260
Catalogues 260
Company standards 261
Component record cards 258
Contract tool drawing office 263
Equipment numbering 257
Function of planning and jig design departments 250
Information library and records 260
Issue of drawings 259
Issue of equipment, material and manufacturing authority 250
Manufacture of equipment 258
Modifications, modification notes and action 259
National Standards 261
Operation instruction sheet 255
Ordering of equipment 258
Organisation of an engineering concern 248
Organisation of planning and jig design departments 251
Parts list 257
Planning and jig design departments—function of 250
Planning of product manufacture 253
Planning of tooling 256
Progress 259
Records: of component 258; of tools 256
Stores holding 263
Title block 257
Tool records 256

Tool record cards 258
Tool room 264; demonstrator 265
Tool schedule 256
Schedule 253: 'sealing' of 253

Presswork
Blanking and piercing 220; application of shear 231; clearances 224; dies 223; feed methods 227; guides 225; layout 235; pilots 226; press tool sets 219, 220; pressure 230; pressure pads 226; stock stops 227, stripper 225
Die sets—standard 218
Drawing 240; blank size 242; combination tools 241; cupping 245; die material 240; die shapes 240; pressure 245; punch material 240
Piercing-and-blanking 220, 221
Presses 215; capacity 217; cut-and-cupping 217; double-acting 216; fly 215; foot-operated 215; hand-operated 215; power 216; single-acting 216; triple-acting 217
Presswork operations 212; bending 214; blanking 213; coining 214; crimping 215; drawing 214; embossing 214; piercing 212; piercing-and-blanking 213
Slide tools 247

Process Planning
Choice of method and equipment 49
Planning method 50
Specimen operation layouts 52

Tooling and Cam Layouts
Automatic machines 58, 63, 74
Capstan and turret lathes 56; tooling layouts 57
Multi-spindle automatic machine 74
Screw-type automatic machine 58; cams 63; tool and cam design worksheet 61; tool and cam layout 59
Swiss-type or sliding head automatic machine 63; cams 70, 72, 73

Turning and Turning Fixtures
Collet chucks 147; dead length 148; pull-in 147; push-out 147
Expanding posts 148
Jaw chucks 148
Soft jaw 148
Tool setting 148, 151
Turning fixtures 148: balancing of 152

Welding Fixtures
Design of welding fixtures 153

TJ
1187
.K4
1968

ASHEVILLE-BUNCOMBE TECHNICAL COLLEGE

3 3312 00002 2657

69-1081

Kempster, M. H. A.
Principles of jig and tool design

DATE DUE
$11 93-

Asheville-Buncombe Technical Institute
LIBRARY
340 Victoria Road
Asheville, North Carolina 28801

ASHEVILLE-BUNCOMBE TECHNICAL INSTITUTE

NORTH CAROLINA
STATE BOARD OF EDUCATION
DEPT. OF COMMUNITY COLLEGES
LIBRARIES

DISCARDED

JUN 2 4 2025